Computer-Aided Drug Design

Computer-Aided Drug Design

Principles, Techniques, and Applications

Edited by

Ajmer Singh Grewal
Professor
Guru Gobind Singh College of Pharmacy
Yamuna Nagar, Haryana, India

Viney Lather
Professor & Head
Centre for Pharmaceutical Chemistry and Analysis
Amity Institute of Pharmacy, Amity University
Noida, Uttar Pradesh, India

Geeta Deswal
Professor
Guru Gobind Singh College of Pharmacy
Yamuna Nagar, Haryana, India

Kumar Guarve
Professor & Principal
Guru Gobind Singh College of Pharmacy
Yamuna Nagar, Haryana, India

WILEY

This edition first published 2026

Registered Office(s)
John Wiley & Sons, Inc., 111 River Street, Hoboken, NJ 07030, USA
John Wiley & Sons Ltd., New Era House, 8 Oldlands Way, Bognor Regis, West Sussex PO22 9NQ, UK
Wiley-VCH GmbH, Boschstr. 12, 69469 Weinheim, Germany
John Wiley & Sons Singapore Pte. Ltd., 134 Jurong Gateway Road, #04-307H, Singapore 600134

For details of our global editorial offices, customer services, and more information about Wiley products visit us at www.wiley.com.

The manufacturer's authorized representative according to the EU General Product Safety Regulation is Wiley-VCH GmbH, Boschstr. 12, 69469 Weinheim, Germany, e-mail: Product_Safety@wiley.com.

Wiley also publishes its books in a variety of electronic formats and by print-on-demand. Some content that appears in standard print versions of this book may not be available in other formats.

Library of Congress Cataloging-in-Publication Data has been applied for:

Hardback: 9781394371082
ePdf ISBN: 9781394371105
ePub ISBN: 9781394371099
oBook ISBN: 9781394371112

Cover Design: Wiley
Cover Images: TechnicColor/Getty Images, cosmin4000/Getty Images

Set in 9.5/12.5pt STIXTwoText by Lumina Datamatics

Table of Contents

About the Editors

Dr. Ajmer Singh Grewal is a Professor (Pharmaceutical Chemistry) at Guru Gobind Singh College of Pharmacy, Yamuna Nagar, Haryana (India), with over 14 years of experience in pharmaceutical education and research. Dr. Grewal holds a PhD in Pharmaceutical Sciences from I. K. Gujral Punjab Technical University, India, with expertise in drug design, medicinal chemistry, and computational modeling. With over 120 publications in esteemed journals, a Google Scholar h-index of 26, and more than 4,000 citations, Dr. Grewal is emerging as an expert in his field. Dr. Grewal has authored 30 book chapters, and edited 20 books. His research focuses on structure-based drug design for anticancer, anti-diabetic, anti-inflammatory, and Alzheimer's therapies, with extensive use of tools like Glide, AutoDock Vina, and Discovery Studio. His expertise in computer-aided drug design makes him a valuable contributor to advances in this field.

Dr. Viney Lather is a distinguished Professor and Center Head at the Centre for Pharmaceutical Chemistry and Analysis, Amity Institute of Pharmacy, Amity University, Noida (India). With 23 years of extensive national and international experience in pharmaceutical sciences, he has made significant contributions to drug discovery and development. From 2007 to 2010, Dr. Lather worked as a Research Scientist at Universidade da Madeira, Funchal, Portugal. His research expertise spans medicinal chemistry, molecular modeling, AI-driven drug discovery, chemoinformatics, and bioinformatics. He has particular proficiency in developing nanoplatforms for drug delivery and employing in silico tools for structure-based drug design. With over 160 publications, a Google Scholar h-index of 32, and more than 5,000 citations, Dr. Lather is a recognized leader in his field. He has secured research grants from prominent agencies such as AICTE (New Delhi) and SERB (New Delhi) and received accolades, including the Best Scientist Award from Jubilant Biosys Ltd. Dr. Lather has delivered over 15 invited talks.

Dr. Geeta Deswal is a Professor Head of the Department (Pharmacognosy) at Guru Gobind Singh College of Pharmacy, Yamuna Nagar, Haryana (India), since August 2022. With over 14 years of teaching experience, Dr. Deswal has mentored and guided numerous UG and PG students. She has supervised five MPharm research projects and is currently guiding three PhD students in pharmaceutical sciences. Her research interests include bioactivity-guided extraction and isolation of phytoconstituents, plant tissue culture, molecular modeling, and natural product-based drug discovery. She has authored over 50 articles in reputed journals, 20 book chapters, and edited 15 books.

Dr. Kumar Guarve is a seasoned academician and Principal-cum-Professor at Guru Gobind Singh College of Pharmacy, Yamuna Nagar, Haryana (India), with over 20 years of extensive experience in pharmaceutical education and institutional leadership. Under his leadership, faculty members have achieved numerous Scopus-indexed publications and patented innovative products. Dr. Guarve has also contributed to securing copyrights for educational processes and curriculum frameworks, further solidifying his commitment to advancing the academic and research profile of the institutions he serves. Dr. Guarve has published over 90 articles in national and international journals of repute, contributed 20 book chapters and edited five books.

List of Contributors

Abhinav
Guru Gobind Singh College
of Pharmacy
Yamuna Nagar
Haryana
India

Anilkumar Adimulapu
Department of Pharmaceutics, School
of Pharmacy
The Assam Kaziranga University
Jorhat
Assam
India

Saurabh Agrawal
ABESIT College of Pharmacy
Ghaziabad
Uttar Pradesh
India

Sejal Agrawal
Department of Pharmaceutical
Sciences and Natural Products, School
of Health Sciences
Central University of Punjab
Bathinda
Punjab
India

Anand Anbarasu
Medical and Biological Computing
Laboratory, School of Biosciences and
Technology (SBST)
Vellore Institute of Technology (VIT)
Vellore
Tamil Nadu
India

Firoz Anwar
Department of Biochemistry, Faculty
of Sciences
King Abdulaziz University
Jeddah
Saudi Arabia

Yogeshwar Bachhav
AiCuris Anti-infective Cures AG
Wuppertal
Germany

Soumya Basu
Department of International Health,
Bloomberg School of Public Health
Johns Hopkins University
Baltimore, MD
USA

Shaik Asha Begum
KVSR Siddhartha College
of Pharmaceutical Sciences
Vijayawada
Andhra Pradesh
India

Ajay Bilandi
Bhai Gurdas College of Pharmacy
Sangrur
Punjab
India

Ajay Bommareddy
Schmidt School of Medicine
Florida Atlantic University
Boca Raton, FL
USA

Arsh Chanana
NIMS Institute of Pharmacy
NIMS University
Jaipur
Rajasthan
India

Abhishek Chandola
School of Pharmaceutical Sciences
Swami Rama Himalayan University
Dehradun
Uttarakhand
India

Vandana Chaudhary
Antaria Pty. Ltd.
Brisbane, QLD
Australia

Shikha Baghel Chauhan
Amity Institute of Pharmacy
Amity University
Noida
Uttar Pradesh
India

Deepika Choudhary
Department of Pharmaceutical
Sciences and Natural Products, School
of Health Sciences
Central University of Punjab
Bathinda
Punjab
India

Kagan Tolga Cinisli
Atatürk University Eastern Anatolia
High Technology Application and
Research Center Directorate
Erzurum
Türkiye

Drashti Dave
Department of Pharmaceutics, Sigma
Institute of Pharmacy
Sigma University
Vadodara
Gujarat
India

Reetika Debroy
Department of Biotechnology
Jaypee Institute of Information
Technology
Noida
Uttar Pradesh
India

Suddhasattya Dey
Department of Pharmacy
IQ City Institute of Pharmaceutical
Sciences
Durgapur
West Bengal
India

Rohit Dutt
Gandhi Memorial National College
Ambala Cantt
Haryana
India

Teena Garg
Amity Institute of Pharmacy
Amity University
Noida
Uttar Pradesh
India

A. Sree Gayatri
Faculty of Pharmacy
Aditya University
Surampalem, Andhra Pradesh
India

Soumyadip Ghosh
Medical and Biological Computing Laboratory, School of Biosciences and Technology (SBST)
Vellore Institute of Technology (VIT)
Vellore
Tamil Nadu
India

Chirag Goda
Department of Pharmaceutical Sciences
Ibn Sina National College for Medical Studies
Jeddah
Saudi Arabia

Ajmer Singh Grewal
Guru Gobind Singh College of Pharmacy
Yamuna Nagar
Haryana
India

Kumar Guarve
Guru Gobind Singh College of Pharmacy
Yamuna Nagar
Haryana
India

Pooja Gupta
NIMS Institute of Pharmacy
NIMS University
Jaipur
Rajasthan
India

Ahmet Hacımüftüoğlu
Department of Medical Pharmacology, Faculty of Medicine
Atatürk University
Erzurum
Türkiye

Rahul Jawarkar
Department of Medicinal Chemistry
Dr. Rajendra Gode Institute of Pharmacy
Amravati
Maharashtra
India

Ambika Nand Jha
School of Pharmacy
Sharda University
Greater Noida
Uttar Pradesh
India

Rohit Kamboj
Guru Gobind Singh College of Pharmacy
Yamuna Nagar
Haryana
India

Shikha Kamboj
Guru Gobind Singh College of Pharmacy
Yamuna Nagar
Haryana
India

Sweta Kamboj
Guru Gobind Singh College of Pharmacy
Yamuna Nagar
Haryana
India

Prabhjot Kaur
Guru Gobind Singh College of Pharmacy
Yamuna Nagar
Haryana
India

Titirsha Kayal
Department of Microbiology
Indira Gandhi Government Medical College and Hospital (IGGMCH)
Nagpur
Maharashtra
India

Vijay Kotra
Department of Pharmaceutical Chemistry, Faculty of Pharmacy
Quest International University
Ipoh
Malaysia

Pankaj Kumar
Anschutz Medical Campus
University of Colorado
Aurora, CO
USA

Rajender Kumar
Department of Chemistry, School of Engineering Sciences in Chemistry, Biotechnology and Health
KTH Royal Institute of Technology
AlbaNova University Center
Stockholm
Sweden

Purna Nagasree Kurre
Department of Pharmaceutical Chemistry
Avanthi Institute of Pharmaceutical Sciences
Vizianagaram
Andhra Pradesh
India

Ashishkumar Kyada
Department of Pharmaceutical Sciences, Faculty of Health Sciences
Marwadi University
Rajkot
Gujarat
India

Arindam Maity
Department of Pharmaceutical Technology
JIS University
Kolkata
West Bengal
India

Debashis Metya
Department of Pharmacy
Bengal College of Pharmaceutical Technology
Dubrajpur
West Bengal
India

Ravinesh Mishra
School of Pharmacy and Emerging Sciences
Baddi University of Emerging Sciences and Technology
Baddi
Himachal Pradesh
India

Arijit Mondal
Department of Pharmaceutical Technology
JIS University
Kolkata
West Bengal
India

Jyoti Monga
Ch. Devi Lal College of Pharmacy
Jagadhri
Haryana
India

Minky Mukhija
Ch. Devi Lal College of Pharmacy
Jagadhri
Haryana
India

Taufik Mulla
Faculty of Pharmacy, Institute of Pharmaceutical Sciences
Parul University
Vadodara
Gujarat
India

Murali Krishna Kumar Muthyala
AU College of Pharmaceutical Sciences
Andhra University
Visakhapatnam
Andhra Pradesh
India

Bhagwat Nagargoje
School of Health Sciences and Technology
Dr. Vishwanath Karad MIT World Peace University
Pune
Maharashtra
India

Ujjwal Nautiyal
School of Pharmaceutical Sciences
Swami Rama Himalayan University
Dehradun
Uttarakhand
India

Rahul Pandey
School of Pharmaceutical Sciences
Swami Rama Himalayan University
Dehradun
Uttarakhand
India

Ramesh Parmar
Department of Pharmaceutical Sciences, Faculty of Health Sciences
Marwadi University
Rajkot
Gujarat
India

Mallikarjuna Rao Pichika
Faculty of Medical Sciences
Newcastle University
Newcastle upon Tyne
UK

Newcastle University Medicine Malaysia
Iskandar Puteri
Malaysia

Satish Polshettiwar
School of Health Sciences and Technology
Dr. Vishwanath Karad MIT World Peace University
Pune
Maharashtra
India

Dishank Purandare
School of Health Sciences and Technology
Dr. Vishwanath Karad MIT World Peace University
Pune
Maharashtra
India

Sk. Abdul Rahaman
School of Medical and Allied Sciences
Galgotias University
Greater Noida
Uttar Pradesh
India

Sudha Ramaiah
Medical and Biological Computing Laboratory, School of Biosciences and Technology (SBST)
Vellore Institute of Technology (VIT)
Vellore
Tamil Nadu
India

Department of Bio Sciences
School of Bio Sciences and Technology
Vellore Institute of Technology (VIT)
Vellore
Tamil Nadu
India

Mimoza Basholli Salihu
Faculty of Medicine
University of Prishtina
Prishtina
Kosova

Azizeh Shadidizaji
Vaccine Development Application and Research Center
Ataturk University
Erzurum
Türkiye

Coordination Office of Computer-Based Modeling and Simulation of Biological Systems
Ataturk University
Erzurum
Türkiye

Anu Sharma
Metro College of Health Sciences & Research
Greater Noida
Uttar Pradesh
India

Bhartendu Sharma
School of Pharmacy and Emerging Sciences
Baddi University of Emerging Sciences and Technology
Baddi
Himachal Pradesh
India

Priya Sharma
School of Pharmacy and Emerging Sciences
Baddi University of Emerging Sciences and Technology
Baddi
Himachal Pradesh
India

Amrendra Pratap Singh
Morehouse School of Medicines
Atlanta, GA
USA

Indu Singh
Amity Institute of Pharmacy
Amity University
Noida
Uttar Pradesh
India

Priya Singh
Department of Pharmaceutical Sciences and Natural Products, School of Health Sciences
Central University of Punjab
Bathinda
Punjab
India

Ravindra Pal Singh
NIMS Institute of Pharmacy
NIMS University
Jaipur
Rajasthan
India

Anshuman Sinha
Department of Pediatrics
Ann & Robert H. Lurie Children's Hospital
Chicago, IL
USA

Feinberg School of Medicine
Northwestern University
Chicago, IL
USA

Vetriselvan Subramaniyan
Department of Medical Sciences School of Medical and Life Sciences
Sunway University
Petaling Jaya
Malaysia

Vijaykumar Sutariya
Department of Pharmaceutical Sciences, USF Health Taneja College of Pharmacy
University of South Florida
Tampa, FL
USA

Nasir Vadia
Department of Pharmaceutical Sciences, Faculty of Health Sciences
Marwadi University
Rajkot
Gujarat
India

B. N. B. Vaidehi
Faculty of Pharmacy
Aditya University
Surampalem
Andhra Pradesh
India

Krishan Kumar Verma
Metro College of Health Sciences & Research
Greater Noida
Uttar Pradesh
India

Mohamad Warda
Department of Physiology, Faculty of Veterinary Medicine
Atatürk University
Erzurum
Türkiye

Department of Biochemistry, Faculty of Veterinary Medicine
Cairo University
Giza
Egypt

Deepak Kumar Yadav
Metro College of Health Sciences & Research
Greater Noida
Uttar Pradesh
India

Sabina Yasmin
Department of Pharmaceutical Chemistry, College of Pharmacy
King Khalid University
Abha
Saudi Arabia

Preface

The discovery and development of new therapeutic agents is a complex, resource-intensive, and multidisciplinary endeavor. Over the past few decades, the integration of computational approaches into pharmaceutical research has fundamentally transformed drug discovery paradigms. Computer-aided drug design (CADD) has emerged as an indispensable component of modern medicinal chemistry, enabling the rational identification, optimization, and validation of drug candidates through the integration of computational chemistry, molecular modeling, bioinformatics, and experimental pharmacology. By streamlining these processes, CADD has significantly reduced development costs and timelines while improving efficiency and decision-making in drug discovery.

The present edited volume, *Computer-Aided Drug Design: Principles, Techniques, and Applications*, is conceived with the objective of providing a comprehensive and up-to-date account of computational methodologies that underpin contemporary drug discovery. The book is designed to serve as a valuable resource for academicians, researchers, and professionals working in pharmaceutical sciences, medicinal chemistry, biotechnology, bioinformatics, and allied disciplines.

The book begins with an introduction to the fundamental concepts and evolution of CADD, setting the stage for subsequent chapters that systematically explore key computational techniques. Core topics such as QSAR modeling, molecular mechanics, quantum mechanics, energy minimization, and molecular docking are addressed in detail, providing a strong theoretical framework for understanding drug–target interactions. These foundational methodologies are complemented by chapters focusing on pharmacophore mapping, ligand-based and structure-based drug design, and computational analysis of drug–receptor binding. The volume also includes dedicated discussions on in silico pharmacology, ADMET prediction, virtual screening, and de novo drug design strategies. Special emphasis is placed on homology modeling, AI-based protein structure prediction, and three-dimensional protein modeling, reflecting recent advances that have significantly expanded the scope and accuracy of computational drug discovery.

A distinctive feature of this book is its focus on emerging and forward-looking trends. Several chapters address the integration of artificial intelligence and machine learning into CADD workflows, highlighting how these technologies enhance target identification, lead optimization, and decision-making in drug development.

The volume further explores the convergence of computational chemistry, QSAR, molecular modeling, and AI-driven approaches, illustrating their collective impact on modern pharmaceutical research. Case studies and practical guides included toward the latter part of the book demonstrate successful applications of CADD, offering real-world context and reinforcing the translational relevance of computational methods.

This book is intentionally structured to support both teaching and self-learning. The inclusion of workflows, conceptual illustrations, and application-oriented discussions makes it suitable for postgraduate students and early-career researchers, while the depth of analysis and interdisciplinary coverage ensures relevance for experienced scientists and industry professionals. By bringing together contributions from proficient researchers and academicians across institutions and countries, the volume maintains academic rigor while addressing practical challenges faced in drug discovery.

In an era where artificial intelligence, personalized medicine, and advanced computational tools are redefining pharmaceutical innovation, this book seeks to fill an important gap in the existing literature. By integrating classical concepts with contemporary and emerging methods, this book aspires to serve as a foundational and forward-looking reference for those striving to innovate and advance the science of drug discovery.

Editors
Ajmer Singh Grewal, PhD
Viney Lather, PhD
Geeta Deswal, PhD
Kumar Guarve, PhD

1

Introduction to Computer-aided Drug Design

*Deepika Choudhary[1], Priya Singh[1], Sejal Agrawal[1], Jyoti Monga[2], Minky Mukhija[2], and Rajender Kumar[3]**

[1]*Department of Pharmaceutical Sciences and Natural Products, School of Health Sciences, Central University of Punjab, Bathinda, Punjab, India*
[2]*Ch. Devi Lal College of Pharmacy, Jagadhri, Haryana, India*
[3]*Department of Chemistry, School of Engineering Sciences in Chemistry, Biotechnology and Health, KTH Royal Institute of Technology, AlbaNova University Center, Stockholm, Sweden*

*Corresponding author: rajbioinformaticshau@gmail.com

Introduction

The approach to novel drug discovery evolved dramatically over the years. Earlier scientists relied on trial and error. In search for better therapeutic agents, they synthesized and evaluated thousands of natural or synthetic compounds in labs that might work in a more efficient way. Although this approach resulted in many important discoveries, it was generally slow, expensive, and unpredictable [1]. As science and technology advanced, so did the tools available for drug discovery. The most significant thing was the use of computers, which is actually the pioneer to Computer-aided Drug Design (CADD). CADD involves the use of computer programs to assist with the development and optimization of novel drugs [2]. CADD enables scientists to forecast how drugs may act in the body before the actual lab work, including synthesis and testing process, ever starts. Drug action can thus be efficiently and precisely assessed [3, 4]. This new method has made searching for new drugs much more efficient. As diseases evolve and germs develop resistance to existing drugs, the approach to design new drugs needs to evolve. CADD plays an important role in this changing scenario by providing the ability to rapidly redesign, modify, and optimize novel therapeutic compounds [5]. Consequently, the advancement of CADD not only signifies technological innovation but also addresses the persistent biological challenges faced in contemporary medicine.

Evolution of CADD

The CADD has advanced significantly in the 21st century, and several advanced techniques are now available to support the drug development process. To appreciate the reason why CADD is so effective, its evolution over time from its formative years to the advanced systems needed to be understood [6]. Therefore, we have to travel all the way past to the early 1960s, when CADD was first originally established.

The 1960s CADD: this decade is devoted to the creation of contemporary quantitative structure–activity relationship (QSAR). In 1964, Toshio Fujita and Corwin Hansch established the link between biological activity and molecular structure, laying the groundwork for QSAR. The Free–Wilson technique, created in 1964 by James W. Wilson and Spencer Free, was another innovation in this decade. It established the framework for the contemporary two-dimensional QSAR approach, which is still in use and significant globally [7, 8]. Another important turning point in this history is the founding of the Cambridge Crystallographic Data Centre by Dr. Olga Kennard in 1965, which collects and electronically encodes small molecule chemical, bibliographic, and crystal structural information. Eventually, Chemical Structure Database was created [9].

The 1970s CADD: the Protein Data Bank (PDB) was established in 1971 under the inspired leadership of Walter Hamilton, and this decade began with the founding of Brookhaven National Laboratory. PDB, which began with just seven structures, now has over 2,000,000 protein structures that are necessary for several CADD techniques, such as molecular docking and molecular dynamics (MD). Additionally, QSAR has been extended to environmental and toxicity research during this decade [10].

The 1980s CADD: this era can be seen as the decade in which structure-based drug design (SBDD) evolved. The European Molecular Biology Laboratory (EMBL) Nucleotide Sequence Database, which stores and provides access to DNA and RNA sequences, was founded in Germany in 1980 [11]. Along with that, the idea of molecular mechanics and several force fields, including assisted model building with energy refinement (AMBER), chemistry at Harvard macromolecular mechanics (CHARMM), and molecular orbital package (MOPAC), has emerged during these 10 years. The creation of other well-known molecular docking programs was also observed in the last years of the decade, such as AutoDock [12]. During this decade, significant progress was also made in the areas of X-ray crystallography and protein structure determination using nuclear magnetic resonance (NMR) spectroscopy. In 1988, Cramer et al. presented comparative molecular field analysis (CoMFA) and the first linear 3D-QSAR study [13]. The current CADD that we use today was made possible by all of these developments.

The 1990s CADD: in 1990, Henderson and his colleagues used cryoelectronic microscopy to unveil Bacteriorhodopsin's first high-resolution protein structure [14]. In 1994, Klebe et al. introduced a different version of CoMFA called the comparative molecular similarity indices analysis (CoMSIA) [15]. Human immunodeficiency virus (HIV) protease inhibitors, such as ritonavir, indinavir, and saquinavir,

Table 1.1 List of databases created during the first 10 years of the 21st century.

Database	Year	Purpose of the database
ZINC	2004	Commercially accessible chemicals in a free database for virtual screening
PubChem	2004	Small compound databases containing biological activity information
DrugBank	2006	Database that lists medications' structures as well as their pharmacological and medicinal characteristics
ChemSpider	2007	Quick text and structural searches are possible with this free chemical structure database
ChEMBL	2008	A library of carefully selected bioactive compounds having characteristics similar to those of drugs
SPIDER	2010	A database of documented adverse medication reactions for marketed pharmaceuticals

were also developed utilizing CADD methods and were introduced during this decade. These inhibitors are thought to represent a major breakthrough in HIV therapy.

The 2000s CADD: CADD has advanced significantly in all of its methodologies since the turn of the 21st century. The creation of many databases, which were crucial to CADD, was the most notable development in the first 10 years of the 21st century. Table 1.1 lists the major databases created in this era. Another notable development of this decade is the solid basis; these databases have provided for virtual screening techniques that are ligand-based and structure-based [16].

The 2010s CADD: by winning the 2013 Nobel Prize in Chemistry, this decade makes a name for itself in the CADD area. It was given to Martin Karplus, Michael Levitt, and Arieh Warshel for their work in computational chemistry [17]. The field of MD in SBDD has undergone a revolution because of their efforts. The development of high-speed computer systems has caused attention to move toward AI-based drug discovery, which has the potential to be exceedingly effective and save a significant amount of time and money.

Molecular Modeling

The prediction of molecular behavior and interactions has the potential to revolutionize scientific research. These types of predictions can give an idea of the effects of a new drug candidate or to design new, more potential molecules to combat the problem of drug resistance and others, prior to conducting laboratory synthesis and animal experiments. It utilizes advanced computational techniques to create and manipulate three-dimensional (3D) models of molecules, providing valuable insights into their physical and chemical properties. The prediction of molecular behavior and biological activity is based on the experimental evidences along with the theoretical chemistry and thus can help in screening of the best hits. This area of research is frequently called quantum chemistry, computational chemistry, or molecular graphics [18]. The methods we utilize today are founded on the groundbreaking contributions of researchers like Hückel and Mulliken, who formulated theories concerning molecular orbitals. Furthermore, Westheimer, Wiberg, and Boyd significantly contributed to the progress of classical mechanical modeling programs [19].

What Is Molecular Modeling?

Modeling serves as a crucial tool in chemistry. It greatly improves our comprehension of chemical principles. Molecular modeling contributes to both the practice and education of chemistry by offering sophisticated tools for investigation, interpretation, explanation, and the discovery of new phenomena [20]. Similar to experimental chemistry, it is a discipline that demands skill and practical experience rather than mere theoretical understanding. Although current software simplifies the process of molecular modeling, the true difficulty lies in choosing the right model and correctly interpreting the outcomes. Molecular modeling tools signify an intriguing advancement in scientific methodologies, evolving from conventional physical models like Dreiding and CPK to advanced computational techniques. Recently, there has been a thrilling transition toward the use of online tools and interactive workbenches (Table 1.2), leveraging technologies like HTML and JavaScript to develop dynamic and user-friendly modeling platforms [21]. Molecular modeling provides us with better resources for investigating, analyzing, elucidating, and uncovering new phenomena; molecular modeling improves our capacity to perform and conduct chemistry.

Table 1.2 Software programs for molecular modeling.

Software program	Function
PyMOL	Structure manipulation, surface rendering, molecular visualization
ChimeraX	Docking result, structure viewing, trajectory analysis
AutoDock	Binding energy estimation, molecular docking
GROMACS	Energy minimization, MD simulations, trajectory analysis
AMBER	MD, force field development, free energy calculations
CHARMM	MD, structure refinement, docking, QM/MM
Avogadro	Molecule building, visualization, quantum chemistry interfaces
VMD	Large biomolecular system visualization, trajectory analysis
MOE	Molecular modeling, QSAR, docking, simulations, cheminformatics
Schrödinger Suite	Protein–ligand docking, MD, QM, pharmacophore modeling

Molecular Modeling Strategies

The following are the two primary approaches that are employed for the development of new pharmaceuticals at present.

Direct Drug Design

Using the direct approach, researchers employ X-ray crystallography to investigate the 3D configuration at a particular site of the receptor. Utilizing this method, their objective is to develop a lead molecule. This process begins with the established shape of the receptor site. The aim is to discover a molecule that meets the necessary geometric specifications and also has suitable chemical compatibility. After identifying promising candidates, scientists can utilize computational chemistry software to generate and disseminate molecular data. The docking strategy through energy minimization assesses the binding affinity [22].

Indirect Drug Design

A critical opportunity for a thorough comparison of the structural characteristics of known active and inactive compounds that exactly match a putative receptor site is provided by the indirect drug design technique. Designers must use information from other ligand molecules that bind to the site efficiently in order to move the design process ahead with confidence when the site's geometry is not easily accessible, which frequently happens [23].

Assessment of Molecular Characteristics

Molecular characteristics serve as crucial indicators for different chemical substances, such as drugs. These characteristics are distinctly classified into three types: physical, chemical, and biological. The primary computational techniques for precisely estimating molecular characteristics are discussed in the following subsections.

Empirical (Molecular Mechanics)

This outlines a method for calculating the potential energy of isolated or interacting molecules by examining their nuclear arrangements [24]. The molecular mechanical model treats molecules as collections of atoms connected by classical forces. Within this framework, atoms are viewed as classical particles influenced by a molecular mechanical potential or force field, which estimates potential energy based on a theoretical unstressed state. In this ideal scenario, optimal bond lengths, angles, and torsions are present, with nonbonded atoms positioned infinitely far apart from one another. Each atom has a mass based on its atomic weight, and every bond acts like a mechanical spring with a specific force constant. These bonds create forces that help maintain the ideal shapes of molecules, addressing problems related to stretching, bending, and twisting. Furthermore, nonbonded interactions are crucial, including charge interactions and van der Waals forces. Coulombic interactions occur due to the attraction or repulsion between charged atoms [25]. Molecular systems in reality are more complex than this simplified view, but the basic idea of molecular mechanics is the same. The molecular mechanics method says that the total potential energy of a molecule comes from the contributions of each atom and bond in the molecule. The molecular mechanical framework considers molecules as groups of atoms kept in position by classical forces. Atoms are regarded as classical particles influenced by a potential or force field that correlates with their spatial arrangement. This force field utilizes a set of straightforward functions linked to the atomic positions to enable potential energy assessments in comparison to a theoretical unstrained condition. The unstrained condition is characterized by all bond lengths, angles, and twists being at their minimal energy configurations, while nonbonded atoms are positioned very far apart. In basic terms, every atom possesses a mass that corresponds to its atomic weight, and each bond connecting atoms behaves like a mechanical spring with a specific force constant. Each bond experiences a force that maintains it in its ideal configuration, including one for stretching or compressing the bond, another for adjusting the bond angle, and a third for rotating the bond. Additionally, nonbonded interactions, such as van der Waals forces and charge interactions, hold significant importance. These interactions, referred to as coulombic interactions, result from the attraction or repulsion between fully or partially charged atoms,

$$E_{\text{total}} = \Sigma E_{\text{stretch}} + \Sigma E_{\text{bend}} + \Sigma E_{\text{torsion}} + \Sigma E_{\text{vdW}} + \Sigma E_{\text{Coulomb}}$$

where E = energy due to the distortion of a specific bond.

Molecular mechanics calculations are important for the analysis of various systems, including isolated molecules in the "gas phase," solutions, and macromolecules. By employing periodic boundary conditions with a unit cell, it becomes possible to accurately model crystalline materials and surfaces. The data produced from these calculations is essential for understanding and controlling important processes such as synthesis, chemical reactions, solubility, drug binding, crystal structure, and the behavior of bulk materials [26].

Molecular Dynamics

The motion of particles in a molecular system is calculated using the equations of classical mechanics in MD. This makes it possible to study various significant properties such as conformational and molecular stability, the likely impact of flexibility on reactions or properties, thermally averaged molecular structure, and any other property that depends on molecular motion, such as binding mechanisms, diffusion, permeability, and vibrational modes. In this method, the computer is employed to calculate Newton's equations of motion for one or more molecules as a function of time. The output is a sequence of calculated coordinates, which when graphed trace out the motion of atoms in the molecule. From an examination of a plot of energy against time, a family of low-energy orientations can be identified and examined in more detail. The most significant assumption is that the time interval Δt should be small enough to excite the motions of all the atoms, i.e., it should be able to calculate the motion of the fastest vibrating atoms. The fastest moving atoms in any molecule are the hydrogen atoms with a C—H stretching periodicity of approximately 10^{-14}seconds. In order to record the movement of these atoms precisely, roughly 10 samples must be collected for every vibration, which takes a time gap (Δt) of 10^{-15}seconds. However, the movement of the heavy atoms (C, N, O, etc.) takes 10^{-12}seconds. MD run will have to traverse at least a 10picoseconds (ps) time interval, and a 100ps interval would be even better [27–31].

Quantum Mechanics

A more demanding strategy in molecular modeling is in the area of quantum chemical calculations. Some molecular properties are founded on the spatial arrangement of their electrons and on the interaction of these electrons with the electrons and nuclei of other molecules. In order to take into account these electronic properties, we have to take into account the quantum chemical nature of matter [32]. A number of computational approaches, inevitably utilizing quantum chemistry, have been designed to account for the molecule's electrons and nuclei explicitly. These quantum chemical methods are founded upon the determination of solutions to the Schrödinger wave equation:

$$H\Psi = E\Psi$$

where H represents Hamiltonian operator, Ψ is wave function, and E is system energy.

The precise solution of this equation would provide a full description of a molecular system. The completeness and precision of the description derived from a solution of the Schrödinger equation are determined by three important factors:

- The wave function's characteristics.
- The Hamiltonian operator's completeness.
- The process that employed to find a solution.

It is computationally not possible to find a solution for this equation for large systems of interest, and therefore approximations must be used. A hierarchy of methods in computational chemistry has been established corresponding to the selection of particular levels of such approximations. One of the most frequent approximations used in solving the Schrödinger equation is known as Born–Oppenheimer approximation. This provides a way to separate information into calculation of electronic properties solely on electrons and others solely on nuclear coordinates. Several approximations are employed to compute the electronic properties solely on electrons (electronic Schrödinger equation) [33]. The most frequently employed technique utilized for solving the electronic Schrödinger equation is an iterative Hartree–Fock (HF) and self-consistent field (SCF) method. In this approximation, the movement of each electron is approximated as if it is occurring in the mean field of all other electrons. The electrons are treated separately, and an SCF procedure is utilized to describe the electron interaction. The solution for the Schrödinger equation has been identified via two qualitatively distinct methods.

Ab initio method: all electrons are treated explicitly in this method. Ab initio method is beneficial in that it does not involve any parametrization and hence can be applied to all sorts of systems. A general-purpose program for calculating ab initio molecular orbitals is called GAUSSIAN [34].

Semiempirical method: this approach only explicitly treats the valence electrons. Experimentally determined parameters are used to approximate some integrals while ignoring others. A hierarchy of semiempirical methods has been developed, which differ mainly in the type of integrals that are neglected and in how the remaining integrals are treated. The MOPAC program is in fact a combination of semiempirical molecular orbital techniques such as Modified Neglect of Diatomic Overlap (MNDO), Modified Intermediate Neglect of Diatomic Overlap (MINDO/3), Poisson Model 3 (PM3), and Austin Model 1 (AM1). These techniques yield data on nuclear and electronic distribution of molecules and compute many molecular properties like heat of formation, energy of activation, and electronic properties viz. charges, molecular electrostatic potential (MEP), dipole moment, density, highest occupied molecular orbital (HOMO), and lowest unoccupied molecular orbital (LUMO) energies [35].

Techniques of CADD

Several techniques are now accessible under CADD, and they may all be generally divided into two categories: SBDD and ligand-based drug design (LBDD).

Structure-based Drug Design

The computational approach in which new hits are designed on the basis of the available 3D structure of the biological target is known as SBDD. In the first cycle, there are three principal methods: X-ray crystallography, homology modeling, and NMR, which are used for cloning, purification, and structure determination of the target protein or nucleic acid. Compounds or parts of compounds from a database are inserted into a designated area of the structure using computer algorithms. These compounds are ranked and scored based on their steric and electrostatic interactions with the target site, and the best compounds are then evaluated through biochemical research. In the second cycle, when the target's structure is established in relation to a potential lead from the first cycle that exhibits at least micromolar inhibition in vitro, the depicted sites on the lead can be altered to further increase potency. Thereafter, other cycles can be included to carry out the synthesis of the lead, the optimization of the lead complex, the structural characterization of the new target, and the optimization of other lead compounds. After several rounds of the drug discovery process, later, the enhanced molecules show a considerable improvement in binding and, often, specificity for the target. There are many drugs discovered through SBDD; few examples are raltitrexed, amprenavir, norfloxacin, captopril, and isoniazid [36]. There are various techniques that are used in SBDD to find and improve therapeutic candidates, based on the 3D structure of a target protein. The following are some essential SBDD techniques:

- molecular docking
- de novo drug design

Molecular docking: it is an in silico approach of forecasting how interaction between drug candidate and structure of the target protein takes place (Figure 1.1). In order for the relative orientation of the ligand and protein to have the lowest free energy for the entire system, molecular docking is used to optimize their conformations [37], considering the components' degree of flexibility.

Molecular docking falls under three main categories:

1. **Rigid docking:** the ligand and receptor stay fixed.
2. **Flexible docking:** the ligand is flexible, but the receptor stays fixed.
3. **Full flexible docking:** when both the ligand and the receptor are flexible.

Flexible docking is the most popular docking approach in both industry and academia.

Algorithms and scoring functions used in molecular docking software: the software used for molecular docking employ the use of different algorithms and scoring functions as shown in Table 1.3:

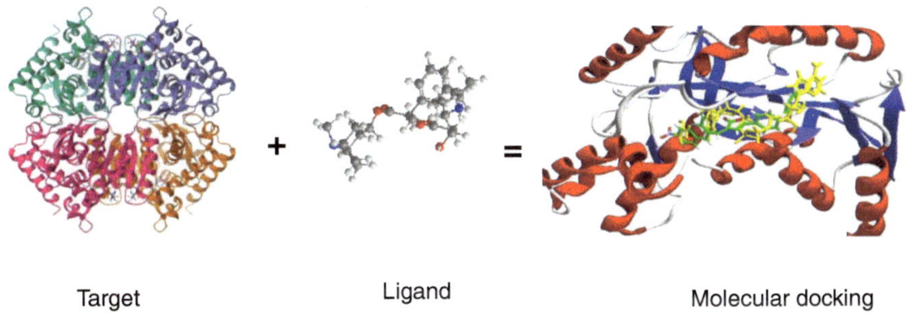

Figure 1.1 Binding of ligands with receptor in molecular docking.

Table 1.3 Algorithms and scoring functions used in different docking software.

Software	Algorithm	Scoring function
AutoDock	Genetic algorithm	Force field + empirical
DOCK	Incremental construction	Force field
FlexX	Incremental construction	Empirical
GOLD	Genetic algorithm	Empirical + knowledge-based
Glide	Simulated annealing/incremental construction	Empirical + knowledge-based

1. **Search algorithm:** a search algorithm, sometimes referred to as an optimization algorithm, is in charge of determining the optimal ligand and protein system conformations. The orientation and location of the ligand with respect to the protein are known as its conformation [38].
2. **Score function:** it is also known as an evaluation function. This function gives an indication of the strength of the interaction between a specific ligand and a given protein [39]. Four types of scoring functions exist: force field, empirical, knowledge-based, and machine learning.

As evaluation functions, energy force fields are frequently employed. These force fields compute the energy contribution using a variety of terms, including the known electrostatic forces between the ligand and protein atoms, forces resulting from ligand deformation, pure electron–shell repulsion between atoms, and the effect of the solvent in which the interaction occurs [40]. The overall methodology of docking is shown in Figure 1.2.

De novo drug design: de novo drug design is a computational method that uses the structure of a biological target, like a protein or enzyme, to construct new drug-like compounds from scratch. Rather than modifying existing compounds, this method is reliable on designing new molecules by exploring different targets and different libraries of ligands. By exploring a broader range of chemical possibilities, it can lead

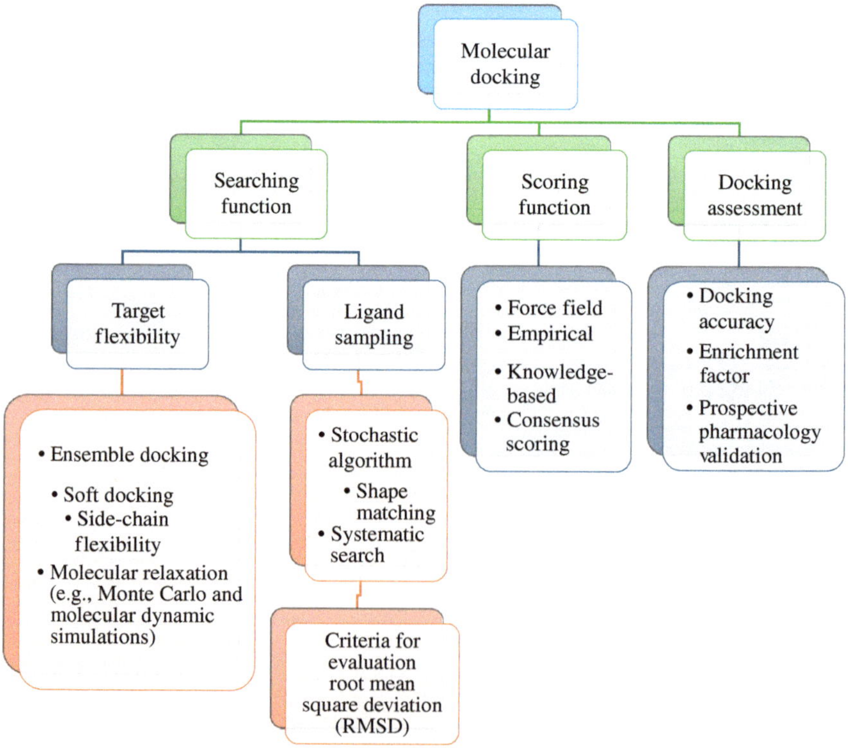

Figure 1.2 Molecular docking methodology.

to the discovery of novel, patentable drug candidates. This method is also referred to as fragment-based drug design [41, 42]. The two main approaches of de novo drug design are structure-based and ligand-based design. A 3D structure can be predicted by X-ray crystallography, NMR, or electron microscopy techniques. Also, homology modeling technique is employed to predict the protein structure when it is not known. Further, two sampling methods of candidate structure are the atom-based approach and the fragment-based approach. Atom-based sampling involves the random placement of an initial atom in the active site, which serves as a seed for the construction of the remaining molecule. A range of atoms and their hybridization states are investigated at each stage. As a result, the chemical space covered by this approach is vast, and the produced structures must be refined. This can typically be achieved by filtering the structures based on chemical accessibility. Atom-based sampling offers the advantage of exploring the chemical space more thoroughly, which results in a greater quantity and variety of structures. LEGEND is the algorithm used in this approach. The second approach for de novo drug design is fragment-based sampling, which is preferable than atom-based approach since the structures are created as fragment assemblies, which reduces the chemical search space, preserves good diversity, and produces candidate compounds with the best absorption, distribution, metabolism, excretion, and toxicity (ADMET) characteristics and chemical accessibility. This method requires a database that contains fragments and linkers, which are obtained either virtually or experimentally. A fragment is docked in the active site and is utilized as a seed to build the rest of the molecule. Algorithms used in this approach include CONCERTS, LUDI, PRO_LIGAND, SPROUT, etc. The several steps involved in de novo drug design are shown in Figure 1.3.

LBDD Techniques

When no structural information about the target is available, LBDD is an alternative approach. LBDD is based on the properties and structure of known ligands, which are tiny molecules that bind to a target protein or biomolecules [43]. The idea of

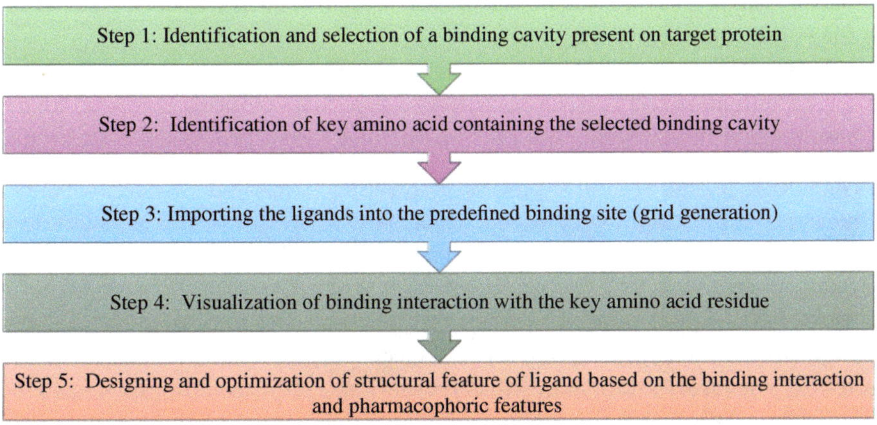

Figure 1.3 Steps of de novo drug design.

mechanism of action is required in SBDD; however, LBDD only needs structural information. LBDD is based on the concept that molecules with similar structures will most likely share comparable properties [43]. Retrieval and development of small molecule libraries are two phases in LBDD. Typically, chemical structures are created, processed, and utilized by using molecular graphs. A molecular graph is made up of nodes and edges, with atoms and bonds represented as nodes and edges, respectively. Linear notations and connection tables are two common methods to work with molecular graphs. A connection table contains sections with information on atom types, connection types, and coordinates. Examples of file formats that contain connection tables include MOL2, SDF, PDB, and others. An alphanumeric character combination is called a linear notation. Linear notations have two examples that include Wiswesser line notation and simplified molecular input line entry specification (SMILE). As compared with connection tables, linear notations are more compact, so they are used for storing and transmitting tiny molecules. A few examples of drugs discovered through the LBDD approach are ofloxacin, troglitazone, imatinib, sunitinib, sorafenib, gefitinib, and ziuleton [44]. The techniques that are mainly based on LBDD approach are molecular similarity-based search, QSAR, and pharmacophore mapping. These techniques are discussed below.

Molecular similarity-based search: it is based on the values of molecular descriptors to search small molecule libraries having desirable properties and features. It is a crucial part of various LBDD and SBDD techniques as well as a stand-alone method. Various software programs available for the techniques are ADAPT, ADMET, DRAGON, Beans, PreADMET, MOE, MOLGEN-QSPR, and CODESSA. It consists of three steps:

1. **Standard formatting:** initially, molecules are read and transformed into the SDF, PDB, MOL, MOL2, and other standard formats.
2. **Filtering:** when employing molecular filters in a similarity-based search, it is essential to eliminate such undesirable tiny molecules. Drug-likeliness is one of the most widely used criteria for filtering small molecule libraries [45], such as Lipinski rule of five, which is used to assess drug likelihood. Other than that, some more screening filters are extended drug-like filters [46], fragment-like filters [47], and Egan filters [46].
3. **Remove duplicates:** a molecule's different protonation states can produce different tautomers and therefore give rise to duplicate conformers. It is up to the user to determine whether or not these duplicates are significant for research.

QSAR: QSAR modeling is the process of creating prediction models based on mathematical expressions [47]. The concept of QSAR, also known as the quantitative structure–property relationship (QSPR), has been widely applied for correlating molecular information with biological activities as well as other physicochemical properties [48]. Typical molecular characteristics that take into consideration are topology, hydrophobicity, steric effects, and electrical properties that can be found theoretically through computational chemistry or experimentally through testing [49]. The method of developing prediction models of biological activities using structural and molecular data from a chemical library is known as QSAR modeling.

A crucial component of QSAR is to validate the newly constructed model. The collection of molecules used to construct a model in a QSAR research is referred to as the "training set," and the set of molecules used to make predictions is referred to as the "test set." To validate the QSAR model, two different techniques are available: internal validation and external validation. The most popular type of internal validation is leave-one-out, where one molecule is retained in the test set and the remaining molecules in the training set are used to estimate the coefficients of various descriptors in a QSAR model. Further, the test set molecule is used to determine its activity using the model constructed on the training set. This procedure is carried out multiple times until every molecule in the training set has functioned as a molecule in the test set. While external validation, a QSAR model is predicted using a test set that was never utilized to construct the model, as opposite to internal validation [50–53]. Tools used in OSAR are HQSAR, CoMFA, CoMSIA, Volsurf, HipHop, HypoGen, QSAR Toolbox, Derek, TOPKAT, and molecular field-based 3D QSAR.

Pharmacophore modeling: IUPAC defines a pharmacophore as, "The ensemble of steric and electronic features that is necessary to ensure the optimal supramolecular interactions with a specific biological target structure and to trigger (or to block) its biological response" [52]. A pharmacophore model can be made using either a structure-based method, which looks for possible points of interaction between the ligands and macromolecular targets, or a ligand-based method, which involves superposing a group of active molecules and identifying common chemical properties that are essential to their bioactivity. Pharmacophore techniques have been used extensively in virtual screening [54]. Tools used in pharmacophore modeling are LigandScout, Quasi, Unity, Phase, and MOE.

Applications of CADD

CADD covers a broad spectrum of applications that are essential to drug discovery and development process at every level, as discussed below.

Virtual Screening

Virtual screening is a computational method used for the screening of best hits through libraries of small molecules and find the structures that are most likely to bind to a drug target, usually an enzyme or protein receptor. Virtual screening can be done by structure-based virtual screening (SBVS) and ligand-based virtual screening (LBVS), as shown in Figure 1.4. SBVS's approach for drug development predicts and assesses possible ligand binding by using the 3D structure of a protein. The initial step in SBVS is an identification of the potential ligand-binding site on the target molecule [55], like a hydrophobic groove or pocket, or we can generate a structure-based pharmacophore model using the co-crystallized ligands present within the receptor protein. The resultant binding score, binding mode analysis, and comparative pharmacophoric features will help us to screen the best hits from the huge library of ligands. Another

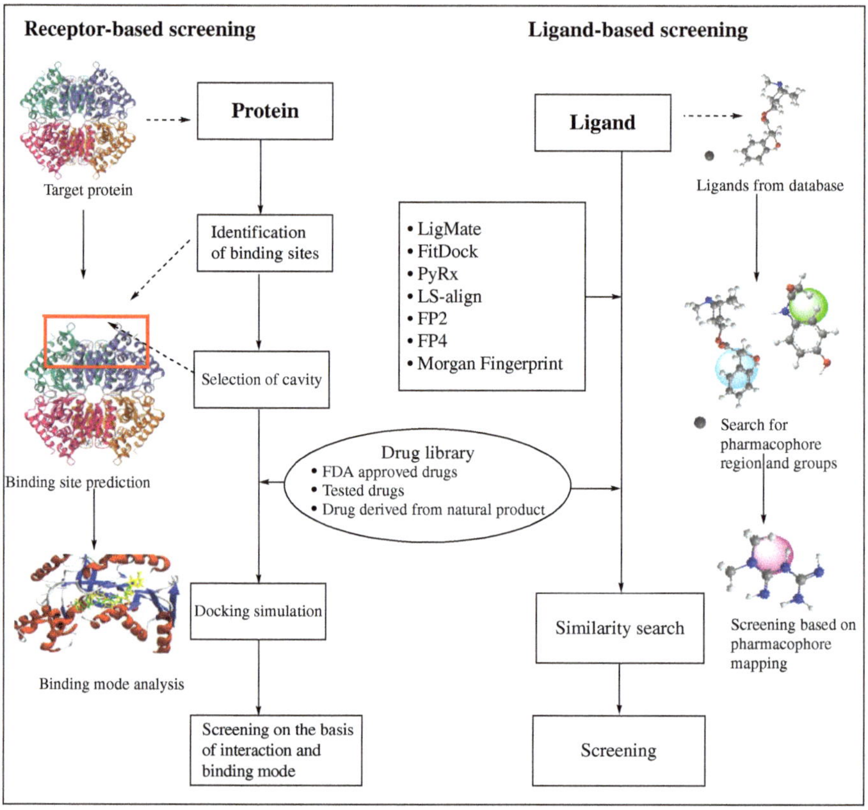

Figure 1.4 Methodology of virtual screening showing steps of SBVS and LBVS.

approach through which the virtual screening can be done is LBVS, which compares novel drug candidates to active compounds that are already known to exist. LBVS usually involves techniques like pharmacophore modeling, QSAR, or similarity searches. Similarity comparisons are carried out by taking into account a number of factors, including the structure and shape of each individual fragment or the molecule's electrostatic characteristics. The produced potent molecules are screened according to their similarity score, which is calculated by various algorithms or techniques. It will always be better to use a combination of both approaches in order to get more optimization and validated results.

Energy Minimization and Molecular Structure Drawing

Energy minimization and molecular structure drawing are the initial phases that play a vital role in understanding and optimizing molecular interaction in CADD. All CADD approaches, including 2D-QSAR, 3D-QSAR, molecular docking, pharmacophore-based virtual screening, MD, and others, need the drawing of molecular structures and the minimizing of their energy. The commonly used software for the energy minimization are Avogadro, ChemDoodle, Biovia Draw, ChemDraw, etc. [56].

Homology Modeling

The method of obtaining a 3D structure of the target protein receptor that is not included in the Research Collaboratory for Structural Bioinformatics' (RCSB) PDB is known as homology modeling. Building an atomic-resolution model of the target protein using its amino acid sequence and the experimental 3D structure of a similar homologous protein (the template) is known as homology modeling. The 3D structure produced by homology modeling can be used in MD or docking. This is based on the principle that similar proteins fold in the same manner; hence, there will be similarity in their 3D structure. Tools that can be used in homology modeling are MODELLER, LOMETS, Robetta, I-TASSER, and SWISS-MODEL [57].

Prediction of ADMET

The ADMET properties are useful in drug discovery to assess how a drug behaves in the body. Early prediction of these properties helps in saving time and resources by avoiding compounds with poor pharmacokinetic and toxicity profiles. A hit or lead cannot become a likely drug candidate just by having strong biological activity. The most crucial factors for any drug to effectively elicit the intended pharmacological response are ADMET characteristics. Consequently, the drug development process now relies heavily on in silico methods for forecasting the ADMET characteristics of every proposed molecule. Poor ADMET can sometimes result in failure at preclinical and clinical trial stages. Some of the software used for ADMET prediction are QikPro, SWISS-ADME, ADMET-predictor, Toxtree, pkCSM, ADMET Predictor, and TOPKAT [58, 59].

Role of CADD in Artificial Intelligence

Finding effective new medications is a challenging endeavor in drug development. AI-enabled technologies have evolved into multipurpose instruments that can be used widely in many phases of drug development, including identifying and validating drug targets, creating new drugs, repurposing existing drugs, increasing R&D efficiency, gathering and analyzing biomedical data, and streamlining the selection of patients for clinical trials [60]. For improving and expediting drug discovery, AI's CADD is essential. The computational tools that AI utilizes to forecast molecular interactions, improve drug design, and examine complicated biological data can be obtained by CADD. Promising leads are found more quickly because of AI approaches, especially machine learning, which enhance virtual screening by forecasting how well possible drug candidates would attach to their target proteins. Utilizing artificial intelligence, molecular docking and de novo design processes are being improved to generate novel drug-like molecules and predict the most beneficial drug–target interactions. AI also improves early-stage drug discovery decision-making by forecasting a compound's ADMET. The drug development process is made more accurate, efficient, and economical by fusing the predictive capabilities of AI with the structural and chemical data from CADD [61].

Conclusion

Drug discovery has been revolutionized by CADD, which provides a variety of benefits that greatly improve the efficacy and efficiency of identifying, optimizing, and creating possible therapeutic drug candidates. The capacity of in silico tools to screen large chemical library results in cost reduction, requirement for intensive lab testing that leads to accelerating the drug development process. Molecular docking, virtual screening, and pharmacophore modeling are examples of CADD techniques that allow for the early improvement of drug candidates, enhancing their specificity, efficacy, and binding affinity. Another important benefit is its ability to predict 3D protein structures using techniques like homology modeling, especially when experimental data is unavailable. This facilitates the identification of significant binding sites and offers insights into the interactions between proteins and ligands, both of which are critical for effective drug design. Additionally, CADD helps to reduce side effects and encourages the development of safer and more effective therapies by predicting potential off-target interactions and toxicological concerns. Instead of many advantages, there are areas where further development is required, such as enhancing prediction accuracy through the use of machine learning algorithms. Additionally, as dynamic interactions and protein flexibility are crucial for realistic drug binding, more studies are still needed in these areas; furthermore, attention should also be paid to mimic the complex biological environment, multi-target, and protein–protein interactions. This all will define a representation of real-world situations.

References

1 Jorgensen WL. The many roles of computation in drug discovery. *Science* 2004;303(5665):1813–8.
2 Lionta E, Spyrou G, Vassilatis DK, et al. Structure-based virtual screening for drug discovery: principles, applications and recent advances. *Curr. Top. Med. Chem.* 2014;14(16):1923–38.
3 Kitchen DB, Decornez H, Furr JR, et al. Docking and scoring in virtual screening for drug discovery: methods and applications. *Nat. Rev. Drug Discov.* 2004;3(11):935–49.
4 Meng XY, Zhang HX, Mezei M, et al. Molecular docking: a powerful approach for structure-based drug discovery. *Curr. Comput. Aided Drug Des.* 2011;7(2):146–57.
5 Pavan M, Bassani D, Bolcato G, et al. Computational strategies to identify new drug candidates against neuroinflammation. *Curr. Med. Chem.* 2022;29(27):4756–75.
6 Blair JM, Webber MA, Baylay AJ, et al. Molecular mechanisms of antibiotic resistance. *Nat. Rev. Microbiol.* 2015;13(1):42–51.
7 Hansch C, Fujita T. p-σ-π analysis. A method for the correlation of biological activity and chemical structure. *J. Am. Chem. Soc.* 1964;86(8):1616–26.

8 Redman J, Willett P, Allen FH, et al. A citation analysis of the Cambridge Crystallographic Data Centre. *J. Appl. Cryst.* 2001;34(3):375–80.

9 Free SM, Wilson JW. A mathematical contribution to structure-activity studies. *J. Med. Chem.* 1964;7(4):395–9.

10 Burley SK, Berman HM, Christie C, et al. RCSB protein data bank: sustaining a living digital data resource that enables breakthroughs in scientific research and biomedical education. *Protein Sci.* 2018;27(1):316–30.

11 Rice CM, Fuchs R, Higgins DG, et al. The EMBL data library. *Nucleic Acids Res.* 1993;21(13):2967.

12 Salmaso V, Moro S. Bridging molecular docking to molecular dynamics in exploring ligand-protein recognition process: an overview. *Front. Pharmacol.* 2018;9:923.

13 Cramer RD, Patterson DE, Bunce JD. Comparative molecular field analysis (CoMFA). 1. Effect of shape on binding of steroids to carrier proteins. *J. Am. Chem. Soc.* 1988;110(18):5959–67.

14 Henderson R, Baldwin JM, Ceska TA, et al. Model for the structure of bacteriorhodopsin based on high-resolution electron cryo-microscopy. *J. Mol. Biol.* 1990;213(4):899–929.

15 Klebe G, Abraham U, Mietzner T. Molecular similarity indices in a comparative analysis (CoMSIA) of drug molecules to correlate and predict their biological activity. *J. Med. Chem.* 1994;37(24):4130–46.

16 Thakur A, Mehta V, Nagu P, et al. *Computer-Aided Drug Design: QSAR, Molecular Docking, Virtual Screening, Homology and Pharmacophore Modeling*. Walter de Gruyter GmbH & Co. KG; 2024.

17 Karplus M, Levitt M, Warshel A. *The Nobel Prize in Chemistry 2013*. Nobel Media AB; 2014.

18 Leach AR. *Molecular Modelling: Principles and Applications*. 2nd ed. Prentice Hall; 2001.

19 Cramer CJ. *Essentials of Computational Chemistry: Theories and Models*. 2nd ed. Wiley; 2013.

20 Jensen F. *Introduction to Computational Chemistry*. 3rd ed. Wiley; 2017.

21 Schlick T. *Molecular Modeling and Simulation: An Interdisciplinary Guide*. Springer; 2010.

22 Nadendla RR. Molecular modeling: a powerful tool for drug design and molecular docking. *Resonance* 2004;9(5):51–60.

23 Trott O, Olson AJ. AutoDock Vina: improving the speed and accuracy of docking with a new scoring function, efficient optimization, and multithreading. *J. Comput. Chem.* 2010;31(2):455–61.

24 Jorgensen WL, Maxwell DS, Tirado-Rives J. Development and testing of the OPLS all-atom force field on conformational energetics and properties of organic liquids. *J. Am. Chem. Soc.* 1996;118(45):11225–36.

25 Frenkel D, Smit B. *Understanding Molecular Simulation: From Algorithms to Applications*. Elsevier; 2023.

26 Young D. *Computational Chemistry: A Practical Guide for Applying Techniques to Real World Problems*. John Wiley & Sons; 2004.

27 Manathunga M, Götz AW, Merz Jr KM. Computer-aided drug design, quantum-mechanical methods for biological problems. *Curr. Opin. Struct. Biol.* 2022;75:102417.

28 McCammon JA, Harvey SC. *Dynamics of Proteins and Nucleic Acids*. Cambridge University Press; 1988.

29 Cheng X, Ivanov I. Molecular dynamics. In: *Computational Toxicology: Volume I*. Humana Press; 2012. pp. 243–85.

30 Kapral R, Ciccotti G. Molecular dynamics: An account of its evolution. In: *Theory and Applications of Computational Chemistry*. Elsevier; 2005. pp. 425–441.

31 Roccatano D. A short introduction to the molecular dynamics simulation of nanomaterials. In: *Micro and Nanomanufacturing Volume II*. Springer; 2018. pp. 123–55.

32 McQuarrie DA. *Quantum Chemistry*. University Science Books; 2008.

33 Atkins PW, Friedman RS. *Molecular Quantum Mechanics*. Oxford University Press; 2011.

34 Friesner RA, Guallar V. Ab initio quantum chemical and mixed quantum mechanics/molecular mechanics (QM/MM) methods for studying enzymatic catalysis. *Annu. Rev. Phys. Chem.* 2005;56(1):389–427.

35 Dewar MJ, Zoebisch EG, Healy EF, Stewart JJ. Development and use of quantum mechanical molecular models. 76. AM1: a new general purpose quantum mechanical molecular model. *J. Am. Chem. Soc.* 1985;107(13):3902–9.

36 Batool M, Ahmad B, Choi S. A structure-based drug discovery paradigm. *Int. J. Mol. Sci.* 2019;20(11):2783.

37 Morris GM, Lim-Wilby M. Molecular docking. In: *Molecular Modeling of Proteins*. Humana Press; 2008. pp. 365–82.

38 Dias R, de Azevedo J, Walter F. Molecular docking algorithms. *Curr. Drug Targets* 2008;9(12):1040–7.

39 Huang S-Y, Grinter SZ, Zou X. Scoring functions and their evaluation methods for protein-ligand docking: recent advances and future directions. *Phys. Chem. Chem. Phys.* 2010;12:12899e12908.

40 Leach ARG, Valerie J. *An Introduction to Chemoinformatics*. Springer; 2007.

41 Wang M, Wang Z, Sun H, et al. Deep learning approaches for de novo drug design: an overview. *Curr. Opin. Struct. Biol.* 2022;72:135–44.

42 Mouchlis VD, Afantitis A, Serra A, et al. Advances in de novo drug design: from conventional to machine learning methods. *Int. J. Mol. Sci.* 2021;22(4):1676.

43 Sharma V, Wakode S, Kumar H. Structure-and ligand-based drug design: concepts, approaches, and challenges. In: *Chemoinformatics and Bioinformatics in the Pharmaceutical Sciences*. Elsevier; 2021. pp. 27–53.

44 Aparoy P, Kumar Reddy K, Reddanna P. Structure and ligand based drug design strategies in the development of novel 5-LOX inhibitors. *Curr. Med. Chem.* 2012;19(22):3763–78.

45 Hendrickson JB. Concepts and applications of molecular similarity. *Science* 1991;252(5009):1189–90.

46 Lipnski CA, Lombardo F, Dominy BW, et al. Experimental and computational approaches to estimate solubility and permeability in drug discovery and development settings. *Adv. Drug Deliv. Rev.* 2001;46:3–26.

47 Daniel FV, Stephen RJ, Hung-Yuan C, et al. Molecular properties that influence the oral bioavailability of drug candidates. *J. Med. Chem.* 2002;45:2616–23.

48 Teague SJ, Davis AM, Leeson PD, et al. The design of leadlike combinatorial libraries. *Angew. Chem.* 1999;38(24):3743–8.

49 Hoque I, Chatterjee A, Bhattacharya S, et al. An approach of computer-aided drug design (CADD) tools for in silico pharmaceutical drug design and development. *Int. J. Adv. Res. Biol. Sci.* 2017;4(2):60–71.

50 Chanin N, Chartchalerm INA, Thanakorn N, et al. Practical overview of quantitative structure-activity relationship. *EXCLI J.* 2009;8:74–88.

51 Gramatica P. Principles of QSAR models validation: internal and external. *QSAR Comb. Sci.* 2007;26:694–701.

52 Wermuth CG, Ganellin CR, Lindberg P, et al. Glossary of terms used in medicinal chemistry (IUPAC Recommendations 1998). *Pure Appl. Chem.* 1998;70(5):1129–43.

53 Anu G, Manish G, Komal S. A practical overview of quantitative structure-activity relationship. *EXCLI J.* 2016;5(2):427–37.

54 Tonde V. Pharmacophore modeling in drug designing. *Adv Pharmacol.* 2025;103:313–324.

55 Anderson AC. The process of structure-based drug design. *Chem. Bio.* 2003;10(9):787–97.

56 Mirzaei H, Zarbafian S, Villar E, et al. Energy minimization on manifolds for docking flexible molecules. *J. Chem. Theory Comput.* 2015;11(3):1063–76.

57 Hillisch A, Pineda LF, Hilgenfeld R. Utility of homology models in the drug discovery process. *Drug Discov. Today* 2004;9(15):659–69.

58 Sharma T, Mohapatra S, Dash R, et al. Recent advances in CADD. In: *Computer Aided Drug Design (CADD): From Ligand-Based Methods to Structure-Based Approaches.* Elsevier; 2022. pp. 231–81.

59 Cui W, Aouidate A, Wang S, et al. Discovering anti-cancer drugs via computational methods. *Front. Pharmacol.* 2020;11:733.

60 Mak KK, Pichika MR. Artificial intelligence in drug development: present status and future prospects. *Drug Discov. Today* 2019;24(3):773–80.

61 Mak KK, Wong YH, Pichika MR. Artificial intelligence in drug discovery and development. In: *Drug Discovery and Evaluation: Safety and Pharmacokinetic Assays.* Springer; 2024. pp. 461–98.

2

Quantitative Structure–Activity Relationships (QSARs): Principles and Applications

*Abhinav[1], Prabhjot Kaur[1], Ajmer Singh Grewal[1], and Mimoza Basholli Salihu[2]**

[1]*Guru Gobind Singh College of Pharmacy, Yamuna Nagar, Haryana, India*
[2]*Faculty of Medicine, University of Prishtina, Prishtina, Kosova*

*Corresponding author: mimoza.basholli@uni-pr.edu

Introduction

The field of quantitative structure–activity relationships (QSARs) modeling was founded by Corwin Hansch [1] 50 years ago. It began as a logical extension of physical organic chemistry and has since expanded and changed, progressing from the analysis of small groups of similar compounds using relatively simple regression techniques to the analysis of very large datasets containing thousands of structurally distinct molecular structures using an extraordinarily wide variety of statistical and machine learning algorithms. QSAR is a sophisticated diagnostic and prediction technique that can identify relationships between biological response and chemical structure. In order to satisfy the needs and ambitions of medicinal chemists to forecast biological reactions, QSAR has developed further [2]. To establish QSAR as one of the recognized modeling approaches for the physical and biological characteristics of chemicals currently in use, more than 50years of constant progress, interdisciplinary breakthroughs, and concerted efforts were required [3]. The European Union Registration, Evaluation, Authorization and Restriction of Chemicals (EU REACH) Legislation [4], the International Council for Harmonization (ICH) M7 guideline [5], and some international regulatory bodies such as the United States Food and Drug Administration (US FDA) [4, 5] have all supported and encouraged the use of QSAR as a tool for regulatory purposes. In particular, the European Union's Enforcement of the Cosmetics Regulation [5] forbids animal testing in the cosmetics industry and requires the use of alternative techniques, such as QSAR. Around the world, QSAR modeling is used in government, business, and academia. Leaders of many bigger organizations of computer-aided drug design (CADD) are now favorably reconsidering the value of statistically based methods to QSAR, despite years of unrelenting structure-method dominance, according to relatively recent accounts. The application of predictive models to regulatory endpoints by government agencies is one of the active, ongoing areas of QSAR development. This allows for the development and testing of a growing, yet still evolving, range of specialized regulatory tools and databases [6]. As far as their initial characterization is concerned, the general objectives of QSAR remain the same and are still centered on the methodology's predictive power and its vulnerability to mechanical interpretation [7]. The foundations of QSARs, Hansch analysis, Free–Wilson (FW) models, and 3D-QSAR techniques are covered in this chapter. The validation approach, statistical procedures, models, and physicochemical descriptors are also presented in this chapter. To show the true value of QSAR-guided drug creation, case studies are presented.

Physiochemical Descriptors of QSAR

Significant information about a QSAR dataset can be extracted using descriptors which are classified in Table 2.1 [8].

Fundamental Principles of QSAR

To put it briefly, a set of compounds is initially screened against a variety of selection assay systems in the QSAR modeling process. The chemicals are calculated in order to calculate the descriptors, and a machine learning technique or suitable data analysis is used to compute the resulting numerical data in order to look for correlation vs. the biological activities. This results in the development of a QSAR model. In order to understand the chemical effects that lead to biological activity and/or to anticipate the biological activity of a group of novel chemicals, the developed

Table 2.1 Physiochemical descriptors of QSAR.

Category	Type of descriptor
Size-related descriptors	Molecular weight, refractivity, volume, surface area. These are pre-computed from tabulated values
Hydrophobicity-related descriptors	Logarithm of octanol–water partition coefficient (log *P*), the *P* fragment constant, the logarithm of molar aqueous solubility
Electronic effects-related descriptors	The tabulated estimated polarizability, Huckel-level estimates of the highest occupied and lowest occupied molecular orbitals
Hydrogen bonding descriptors	The HYBOT, Abraham descriptors, electro-topological descriptors
Topological descriptors	Derived from connectivity matrices

Source: Data from [8].

model must be validated. The chemistry of a group of molecules serves as the data source for QSAR modeling. A theoretical space is defined as the chemical space by the chemistry of such a collection. A molecule's biological activity is determined by where it is located in this space. As a result, it is believed that the QSAR models cover a portion of the entire chemical space, and model prediction in a group of compounds is deemed appropriate [9]. The development of so-called Hammett constants to measure the impact of substituents on organic compounds on chemical reaction rates was one of the most significant developments in the history of QSAR. Various substituents affect a chemical reaction's pace. Hammett developed an empirical equation (Equation 2.1) that shows a reaction constant ρ as a function of the type of reaction and a substituent constant σ for each substituent in the molecule,

$$\log K = \log K_0 + \rho \times \sigma \tag{2.1}$$

where $\log K_0$ and $\log K$ are the dissociation constants of the unsubstituted and substituted benzoic acid, σ is the Hammett constant, and ρ is the reaction constant [10]. The actual dawn of QSAR is considered to be in the early 1960s with the contributions of Hansch and Fujita. It was then that scientists began to correlate the biological activity of compounds more logically with their structural characteristics (now also known as molecular descriptors) [10, 11]. Simply said, QSAR is the process of employing a mathematical model to build a link between a selection of structure-based descriptors and molecular activity (defined as broadly as possible). The structure-based descriptors are employed on the idea that structure perceived by descriptors provides information on activity. It is therefore feasible to alter molecular characteristics and, consequently, activities by logically changing the molecular structure [12]. Depending on the chemical representation from which they are formed, the QSAR descriptors are often categorized as 1D, 2D, or 3D (4D [13], 5D [14], and 6D [15] descriptors have also been proposed).

The 2D QSAR Models

Hansch Analysis

A key component of QSAR research, Hansch analysis is a groundbreaking method that combines statistical modeling with a molecule's properties to clarify the connection between chemical structure and biological activity [16]. The method transformed predictive toxicology and medication design, and it was first termed by chemist Corwin Hansch [17]. In essence, Hansch analysis is the process of creating linear models that relate the biological activity of a group of substances to molecular characteristics or physical descriptors. A collection of characteristics that define a molecule's shape, including electrical, steric, and hydrophobicity aspects, are represented by the descriptors [18]. The derived quantitative equations reveal the structure–activity relationship by characterizing the impact of a descriptor on the quantifiable effect. Since scientists may now create molecular structures that increase or change desired biological activity, Hansch analysis has proven to be of unmatched help in rational drug design [19]. The method also sheds light on the mechanism underlying the compound–biological target interaction. Despite having its roots in the linear model, Hansch analysis has been extended to nonlinear and even higher models, which has improved QSAR models' predictive power [20]. Generally speaking, Hansch analysis is a fundamental technique in the theory of QSAR that enables researchers to decipher the intricate relationship between chemical structure and biological response. Its lasting relevance to computational chemistry is demonstrated by its historical significance and ongoing relevance to drug development and toxicity testing today [21]. During the initial phases of QSAR research and development, researchers attempted to link the drug molecule's biological activity to either its electronic or lipophilicity properties, as indicated by the following equations [19, 20]:

$$\log(1/c) = \alpha \log P + b \tag{2.2}$$

$$\log(1/c) = k_1\sigma + k_2 \tag{2.3}$$

where P is the lipophilicity partition coefficient of the drug molecule, and σ is the Hammett constant determining the electronic alteration of the molecule by different substituents. α, b, k_1, and k_2 are constants. None of these equations could suit the wide variety of molecules and other QSAR models for other biological activities, but each one might be able to fit the biological activity with one of them locally. In 1962, Hansch and Fujito introduced a new development in their work on QSAR [21]. Their work's main idea was to distinguish between the fields of drug administration and pharmacological activity [22, 23]. To get to the site of action, the medicine must pass through a number of phases. Protein and lipid membranes, as well as the division of drug molecules between the two liquid systems, are involved in adsorption and desorption. Once the medicine has reached the area of action, any kind of biological effect requires some kind of interaction with the receptor. The drug–receptor relationship is determined by the characteristics of the electronic drug molecule. Therefore, the electrical and lipophilic properties must be connected to the biological activity in order to characterize any QSAR model. This resulted in the QSAR model that is currently quite popular and is known as Hansch analysis. The following are the starting equations in the format that Hansch suggested:

$$\log(1/c) = k_1\pi + k_2\sigma + k_3\sigma \tag{2.4}$$

$$\log(1/c) = k_1\pi^2 + k_2\pi + k_3\sigma + k_4 \tag{2.5}$$

$$\log(1/c) = k_1\pi + k_2\sigma + k_3E_s + k_4 \tag{2.6}$$

$$\log(1/c) = k_1\pi^2 + k_2\pi + k_3\sigma + k_4E_s + k_5 \tag{2.7}$$

Here, π, σ, and E_s are Hansch hydrophobic constant, Hammett constant, and Tafts steric parameter, respectively. Equation (2.6) is applicable when the study is done on small range values of π and Equation (2.7) is applicable when the study is done on broader range values of π. The constants k_1–k_5 used in Equations (2.4)–(2.7) are derived from the multiple linear regressions (MLRs) [24].

FW Models

The foundation of FW analysis is the idea that comparable changes in potency are additive, while chemical changes within a series of compounds are independent of one another. Because cooperativity between replacement sites is also possible, the additivity assumption is only an estimate. However, the FW approximation will often be accurate in practice, which serves as a foundation for important predictions about compound potency [24].

The 3D QSAR Models

Atomic, group, or molecular descriptors like polarizability, lipophilicity, electronic, and steric parameters (Hansch type of work), or the presence of specific structural features (FW type of work) have all been linked to biological activities in traditional QSAR research. Since the three-dimensional structures of the molecules were taken into consideration in these kinds of studies, it becomes inevitable to use the limitations of these techniques to prepare structurally divergent working new molecules. In order to fully utilize the three-dimensionality of the ligands and forecast biological activity using sophisticated chemometric techniques, 3D-QSAR has developed as a logical extension of the traditional Hansch and FW methods. All of these QSAR techniques that relate macroscopic target characteristics to calculated atom-based descriptors of the spatial arrangement of molecular structure are collectively referred to as 3D-QSAR. In the design of pharmaceuticals and agrochemicals, these techniques have been shown to be effective prediction tools [25, 26].

Comparative Molecular Field Analysis

Cramer et al. created the alignment-dependent, field-based, ligand-based technique known as comparative molecular field analysis (CoMFA), which aids in establishing a quantifiable connection between a molecular structure and its response attribute. The technique primarily deals with ligand characteristics such as electrostatic and steric characteristics, as well as positive and negative receptor–ligand interactions. Since CoMFA is a descriptor-based, alignment-dependent technique, energy is computed by positioning an appropriate probe at each lattice point and arranging all aligned ligands in an energy grid. Each unit fraction's computed energy corresponds to both steric (van der Waals) and electrostatic (coulombic) properties. When creating a model, these computed values are used as descriptors. A robust linear regression technique, such as partial least squares (PLSs), is used to link these descriptor values with biological responses. A significant signal for distinguishing between excellent and bad electrostatic and steric potential is provided by the PLS result, which can also be correlated with biological reactions [27].

GRID

The first specialized software that medicinal chemists had access to as an alternative to the original CoMFA process was called GRID. GRID predicts the energetically advantageous binding locations on the structures of known molecules by computing the interaction energy fields in the molecular field computation. Although it follows the CoMFA's methodology in that it also computes the explicit nonbonded (or noncovalent) contact between a known three-dimensional structure molecule and a probe (i.e., a user-defined arbitrary small-sized chemical group) positioned on a lattice at the sample points by passing and lying through the macromolecule, it surpasses it in two ways: first, it uses a 6–4 potential function to describe the interaction energies rather than the CoMFA's softer 6–12 function of the Lennard-Jones type, and second, it has access to a wide range of probe types [28, 29]. Additionally, the program uses a donor and a "DRY probe" to estimate the hydrophobic and hydrogen bond potentials, as well as normal steric and electrostatic potentials. A water probe was then used to calculate hydrophobic interactions [30, 31]. Methyl groups, amine NH_2 groups, carboxylate groups, and hydroxyl groups are employed as separate probes in addition to water and a DRY probe. Along with the protein structure, the contour surfaces for each of the aforementioned probes are calculated at different energy levels and displayed graphically. Positive energy levels are typically employed to define the target's form, while negative energy levels in the contours indicate areas where a ligand binding would be advantageous [31].

Comparative Molecular Similarity Indices Analysis

One modification of CoMFA is the alignment-dependent, ligand-based, linear 3D-QSAR technique known as comparative molecular similarity indices analysis (CoMSIA). With the exception of the latter method's determination of molecular similarity, CoMFA and CoMSIA procedures are fairly comparable. CoMFA primarily consists of molecule alignment and can result in inaccurate sensitivity to alignment as well as interpretation of electrostatic and steric potential. This is rectified by using Gaussian potentials, which are significantly softer than CoMFA functions, in CoMSIA fields. The same probes are dispersed throughout the grid lattice, and an energy grid box with the typical shape is constructed. Furthermore, the CoMSIA incorporates the solvent-dependent molecular entropy (hydrophobicity) term. A standard probe is introduced and similarity is computed at each grid point in order to examine the property of a dataset molecule. Steric, electrostatic, hydrophobic, and hydrogen-bonding characteristics are the primary subjects of calculations. These characteristics, which are computed at regularly spaced grid points that correspond to a specific descriptor, are crucial in relation to the biological response [32].

Molecular Shape Analysis

Molecular shape analysis (MSA) is a ligand-based 3D-QSAR technique that aims to integrate Hansch classical methods and conformational analysis. The quantitative description, composition, and use of molecular shape in the construction of a QSAR model are of importance to the MSA. Determining the physiologically active conformation at the beginning, without any prior knowledge of the receptor geometry, is the ultimate goal of the MSA. The MSA is an alignment-based protocol that incorporates shape information and conformational flexibility into the 3D-QSAR.

The conformational search technique can be used to calculate the number of conformers for each molecule in a set. All of the research compounds' structures are superimposed pairwise on a conformer of the most active sample molecule, which serves as the shape reference compound. Using graph theory tools, the MSA alignment process must deal with molecules as a collection of points and lines in order to find patterns. The MSA aligns based on the greatest common set of atoms in the shape reference compound that are shared by all structures in the research table. The MSA procedure incorporates the form reference compound and entails a strict fitting of the atom pairs to overlay each structure. Accelrys Inc.'s Cerius2 software can be used to perform MSA [33].

Molecular Descriptors

MSA descriptors are used to define molecular shape similarity. The most common ones are listed as follows:

- Difference volume: refers to the difference between the volume shape of the reference substance and the volume of the single molecule.
- Overlap volume ratio: this refers to the common steric volume descriptor divided by the volume of a single molecule.
- Non-common overlap steric volume: the term used to describe the volume of the separated molecule and common overlap steric volume.
- RMS to shape reference: the root-mean-square deviation of the single molecule from the shape reference substance.
- Steric overlap common volume: this refers to the volume of intersection of each analog and the reference of choice. It defines how much each analog resembles the shape reference in steric shape.
- Reference shape volume: this is the volume of the shape reference compound [34].

Receptor Surface Analysis

Since receptor surface analysis (RSA) allows one to construct a hypothetical receptor site model, it can be a helpful technique when the 3D receptor structure is unknown. In contrast to pharmacophore models, which aim to extract information about the similarity of molecules that bind to a receptor, RSA aims to extract crucial information about the receptor. Important details on the hypothetical receptor site as a three-dimensional surface and associated characteristics, including hydrophobicity, partial charge, electrostatic potential, van der Waals potential, and hydrogen-bonding potential, can be seen in a receptor surface model. Representative 3D data of interaction energies as a function of steric and electrostatic fields at each surface point are succinctly and quantitatively described by receptor surface models [35].

Molecular Descriptors

The interaction energy (also known as surface point energies) between every receptor surface point and every model is described by RSA descriptors. Depending on the size of the drug molecule, their surface point numbers would vary. With the exception of using consideration points on the hypothetical receptor surface in place of a dimensions grid, the methodology is similar to CoMFA. Because they are on a surface that resembles a molecule, at least a surface of a subset of functioning molecules, they will therefore have greater chemical significance than a grid of rectangles. RSA descriptors that are most commonly used are as follows:

- Intra energy: energy internal to the molecule of the receptor.
- Internal energy: nonbonded electrostatic energy between the receptor and the molecule.
- Inter vdW energy: the van der Waals nonbonded energy between a receptor and a molecule.
- Inter energy: total nonbonded energy between the receptor and the molecule.
- Min intra energy: minimum internal energy of the molecule without the receptor.
- Strain energy: the strain energy of the molecule in the receptor, i.e., the difference between the internal energy of the optimized molecule in the receptor model.
- Intra energy and optimized molecule without the receptor model (min intra energy) [34].

Statistical Analysis for QSAR Model Buildups

Statistical or chemometric techniques form the mathematical foundation for building a QSAR model (Figure 2.1) [34].

Linear Regression Models

Linear regression studies are regarded as an easy-to-understand method recommended for QSAR analysis among the growing number of diverse statistical techniques in the literature. In order to describe the relationship between one or more independent variables (**x**) and a dependent explicative variable (**y**), these regression

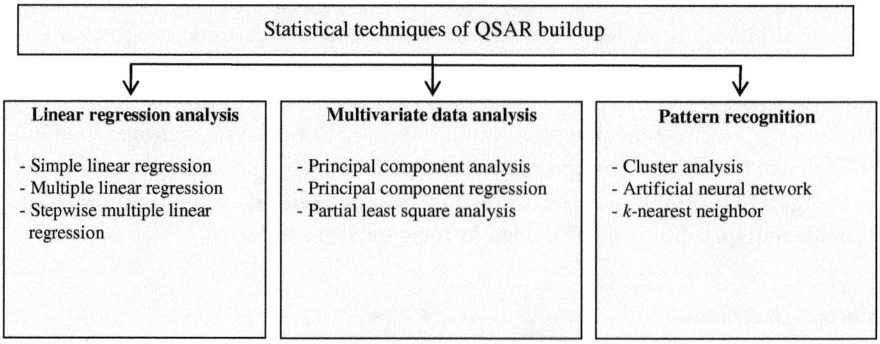

Figure 2.1 Different statistical techniques used for QSAR buildup.

approaches develop a statistical model. Knowing **x** factors, which may be qualitative or quantitative, allows the model to predict **y** [36].

Simple Linear Regression Model

Using a single independent descriptor (**x**) and a dependent variable (**y**), the simple linear regression approach generates a sequence of QSAR equations by doing a typical linear regression analysis. As a result, each independent variable in the descriptor set has its own one-term linear equation. The method works well for analyzing some of the most important characteristics that influence the activity and creating basic structure–activity correlations. The interaction between several descriptors is disregarded, though. The following formula can be used to express linear regression:

$$\mathbf{y} = a + b\mathbf{x} \tag{2.8}$$

where the dependent variable **y** is expressed in terms of the independent variable **x** by means of two parameters: the constant a, also referred to as the intercept, and the regression coefficient b [37].

MLR Model

It is a generalized form of simple linear regression to several dimensions and is also known as the linear free-energy connection technique. In order to describe the pharmacological property reliance on any or all of the descriptors being studied, MLR creates QSAR equations utilizing standard multivariable regression computations. The correlation coefficient (R), Student's t-value, Fisher's ratio, standard deviation (S), and independent verification methods like the leave-one-out (LOO) procedure all support the probability link. According to MLR, each variable is independent and unrelated to others. In the multivariate scenario, where MLR has multiple independent variables, none of these presumptions are valid. The following straightforward multiple-term linear equation represents the correlation in the latter scenario:

$$\mathbf{y} = b_0 + b_1\mathbf{x}_1 + b_2\mathbf{x}_2 + \cdots + b_m\mathbf{x}_m + \mathbf{e} \tag{2.9}$$

the MLR analysis estimates the regression coefficients (b_i) by minimizing the residual error (**e**), which quantifies the deviation of a particular point from the regression line, as in the case of simple linear regression [38].

Stepwise MLR Model

Although not all of the independent variables are included, this is another popular kind of MLR that likewise produces a multiple-term linear equation. In contrast to MLR, each time a new regression is carried out, each independent variable is added to the equation one at a time. Only when the model is significant is the new term retained. When there are several descriptors and it is unknown which descriptors are significant, this regression technique works well [39].

Multivariate Data Analysis

The aforementioned techniques have now been replaced by MLD, which aims to provide the most information possible about the problem while describing a greater number of variables in terms of fewer new latent variables. Large datasets can be seen, categorized, and modeled with the help of these methods, which project multivariate data (MLD) onto a low-dimensional space. MLRs can make advantage of these latent variables because they are orthogonal [40, 41].

Principal Component Analysis and Principal Component Regression

Another data reduction technique that looks for relationships between independent variables but does not produce a QSAR model is principal component analysis (PCA). In terms of decreasing variance, it then generates a new orthogonal set of descriptors known as principal components (PCs), which encapsulate the majority of the information present in the independent variables. As a result, PCA reduces the dimensionality of an MLD collection of descriptors to reflect the true volume of data. Principal component regression (PCR) is a technique used when PCs are used as the independent variables in a linear regression. In other words, PCR creates a multiple-term linear equation by using the PCA decomposition scores as regressors in the QSAR model [42, 43].

PLS Analysis

A large number of initial descriptors are linearly transformed into a small number of new orthogonal terms, known as latent variables, as the result of this recursive regression technique. When the independent variables have a high degree of correlation with one another or exceed the number of observations, PLS provides a statistically stable solution. As a result, PLS can more realistically handle complex structure–activity data and comprehend the relationship between molecular structure and biological activity. One of the common statistical methods used to create predictive 3D-QSAR models is this one [44].

Pattern Recognition

Alternative techniques for conducting qualitative or classification research have been used in QSAR in recent years. To identify the proximity or distance between the vast amount of MLD, the so-called pattern recognition techniques that adhere to the analogy principle are employed. In order to classify newly synthesized compounds as similar or distinct to members of the known classes, it looks for structural characteristics like the presence (or lack) of specified groups, the number of atoms of a given type, or mass spectral fragmentation [45].

Cluster Analysis

It is the statistical identification of patterns used to separate the dataset into classes made up of related items and investigate the relationship between the observations that have reference to multiple variables. In addition to allowing inactive compounds to be included in the analysis, it may be used to examine a large number of substituents and determine which subsets have comparable physical characteristics [46].

Artificial Neural Network

The actual neurons found in an animal's brain serve as the foundation for the artificial neural network (ANN) technique. ANNs are parallel computing systems made up of layers of highly interconnected calculating units called neurons [47]. Each of the neurons in the first layer, sometimes referred to as the input layer, receives information from the user or outside world, signifying one of the independent variables that are used as inputs in QSAR. Then, one or more layers of neurons, collectively referred to as the hidden layers, appear. The output layer is the final layer, and the network's output is handled by its neurons. Every layer has the ability to perform its own calculations and transmit the results to a subsequent layer [48].

The *k*-nearest Neighbor

This method, which is most frequently applied to the classification of novel patterns (like molecules), is arguably the most straightforward machine learning technique. A new or unknown molecule is labeled based on the majority of its *k*-nearest neighbors in the training set using a straightforward distance-based learning technique. A Euclidean distance measure, such as a similarity measure calculated using the molecules' structural descriptors, is used to calculate the proximity [49, 50].

Validation of QSAR

The statistical approach must be selected to produce a definite relationship between the biological activity and the descriptors. Additional checks such as bootstrapping, cross-validation, and activity data scrambling can ensure this. It is based on OECD PRINCIPLE 4. At the 37th Joint Meeting of the Chemicals Committee and Working Party on Chemicals, Pesticides, and Biotechnology, held in Setubal, Portugal, in November 2004 [49], OECD member states adopted these principles.

Cross Validation/Predictive Correlation Coefficient (Q^2)

Q^2, also known as the cross-validation correlation coefficient when LGO (Q^2 LGO) or LOO (Q^2 LOO) cross-validation is applied, is calculated as follows:

$$Q^2 = \frac{1 - \sum\left(y_{\text{pred}} - y_{\text{exp}}\right)^2}{\sum\left(y_{\text{exp}} - y_{\text{mean}}\right)^2} = 1 - \frac{\text{PRESS}}{S_{\text{S}}} \tag{2.10}$$

where y_{pred} is the predicted activity and PRESS is the predictive residual sum of squares. Q^2 values are usually between 0 and the R^2 value. A negative Q^2 indicates a non-predictive model. Values above 0.5 are a sign of a valid model, while above 0.8 are excellent. If $R^2 - Q^2$ differs by more than 0.2–0.3, this may be a sign of chance correlations, irrelevant descriptors, or outliers in the data [49].

The Correlation Coefficient (*r*)

The linearity of the relationship is reflected in it. It represents the variance in the data and serves as a gauge for the model's fit quality. The correlation coefficient should ideally be equal to or near 1, but in reality, any result above 0.9 is acceptable, because biological data is so complicated. A correlation matrix, which illustrates the relationship between one descriptor and another, lists the correlation coefficients of the variables in a data collection. Variables of significance are made orthogonal to one another by the correlation matrix. Unless the new variable is a linear combination of the other variables or a constant of the other variables, in which case it would have no effect, adding each new variable to the model always results in an increase in r. The increase in r because of introducing a new variable reflects overfitting of data [49].

The Coefficient of Multiple Determinations (Pearson's Correlation Coefficient, R^2)

R^2 measures the strength of the linear relationship between predicted and experimental activity values (y). It is calculated using the following equation:

$$R^2 = 1 - \sum (y_{\text{calc}} - y_{\text{exp}})^2 / \sum (y_{\text{exp}} - y_{\text{mean}})^2 = 1 - S_{\text{SR}} / S_{\text{S}} \quad (2.11)$$

where y_{calc} is the predicted value of the model, y_{exp} is the observed value, and y_{mean} is the mean observed value. S_{SR} is the sum of squares of the residuals and S_{S} is the total sum of squares. R^2 values range from 0 to 1, where values closer to 1 indicate a better fit. However, R^2 increases with model complexity, potentially causing overfitting when the number of descriptors exceeds the number of observations [51].

Bootstrapping

It is an additional method that can be used in conjunction with cross-validation to verify the QSAR model's stability and statistical significance. It is carried out by randomly selecting a large number of datasets that are the same size as the original dataset and resampling them [49]. Some objects are included more than once in each PLS execution, while others are left out. Every bootstrap sampling is determined through computation. A measure of the biasness of the initial computations is the difference between the parameters that were calculated from the original dataset and the average of the parameters that were determined by massive bootstrap samplings. Since it entails massive calculation with relatively minor gain compared to cross-validation, the technique is less in demand [50].

Table 2.2 Summary of case studies employing QSAR, where JAK is Janus kinase.

Compound	Target	Targeted disease	QSAR model	Reference
Pyrimidine-4,6-derivatives	JAK3/STAT	Rheumatoid arthritis	3D-QSAR	[53]
1,2,4-Triazine-3(2*H*)-one derivative	Tubulin protein	Breast cancer	2D-QSAR	[54]
Autoimmune disorders inhibitions	PAD4, HMC class II HLA-DQ-ALPHA, TSH	Rheumatoid arthritis, type 1 diabetes, Grave's disease	2D-QSAR	[55]
Imidazo [4,5 − *b*] pyridine derivatives	Aurora kinase (serine threonine kinase)	Breast cancer	3D-QSAR	[56]
Amodiaquine derivatives	Choroquine-resistant antimalarial	Malaria	Genetic function-based QSAR	[57]
Hydroxyquinoline scaffold derivatives	MDA-MB cell lines	Breast cancer	2D-QSAR	[58]
Quinazoline analogs	Epidermal growth factor receptor	Breast cancer	3D-QSAR	[59]

Applications of QSAR Techniques

In drug design and discovery, QSAR is utilized to find chemical compounds with minimal toxicity and strong inhibitory activity against particular targets [52]. The use of QSAR in the development of several therapeutic classes, such as antibacterial and antitumor medicines, by many authors amply demonstrates the technology's value in drug design. Various researchers have carried out earlier studies in this area [53]. Table 2.2 explains many case studies employing QSAR.

Conclusion

Despite all of its flaws, the modern drug discovery community now generally agrees that good statistical foundation-based QSAR is a natural, sensible, and useful medicinal chemistry tool. Its range of applications ranges from providing synthetic guidance toward logical, testable concepts through experimentation to quantitatively and retrospectively rationalizing the structure–activity relations. The assumptions that underpin the 3D-QSAR algorithms will be insightfully criticized with the introduction of more structural bioinformatics data. To prevent misuse and abuse, it would be required to look at the mechanics of the chosen methodology before making any real-world predictions. More accurately, problems with conformational sensitivity and alignment dependency need to be fixed. In a similar vein, whenever target receptor data are available, they must be used when creating models or verifying methods that were developed using just ligand data. Finally, the models must be extensively validated both internally and externally using a stringent cross-validation technique because the models' quality is dependent on their statistical significance and accurate predicting capability of out-of-set-of-training compounds. Aside from that, the QSAR result must be seen as a valuable hypothesis that can be tested using suitable statistical techniques and clarified by further synthesis and biotests of acceptance or rejection. Rightful understanding and flawless execution of such a procedure in the case of QSAR modeling would reward the medicinal chemist for sorting out their experimental attempts and substantially raise the experimental rate of hit.

List of Abbreviations

$\log K_0$ and $\log K$	Dissociation constants
α	Hammett constant
ρ	Reaction constant
α, b, k_1, and k_2	Constants
$\log P$	Partition coefficient
Π	Hansch hydrophobic constant
E_s	Tafts steric parameter
$\mathbf{x}$	Independent variables
$\mathbf{y}$	Dependent variables
R	Correlation coefficient
S	Standard deviation
$\mathbf{e}$	Residual error
Q^2	Predictive correlation coefficient
S_{SR}	Sum of square of residuals
S_S	Total sum of squares
1D	One-dimensional
ANN	Artificial neural network
CADD	Computer-aided drug design
CoMFA	Comparative model field analysis
CoMSIA	Comparative molecular similarity indices analysis
EU REACH	European Union Registration, Evaluation, Authorisation, and Restriction of Chemicals
FW	Free–Wilson
ICH	International Council for Harmonization
JAK	Janus kinase
MLRs	Multiple linear regressions
MSA	Molecular shape analysis
OECD	Organisation for Economic Co-operation and Development

PCA	Principal component analysis
PCA	Principal component regression
PLS	Partial least square
QSAR	Quantitative structure–activity relationship
RSA	Receptor surface analysis
US FDA	United States Food and Drug Administration

References

1 Chatterjee M, Roy K. Quantitative structure-activity relationships (QSARs) in medicinal chemistry. In: *Cheminformatics, QSAR and Machine Learning Applications for Novel Drug Development*. Academic Press; 2023. pp. 3–38.

2 Bajot F. The use of QSAR and computational methods in drug design. In: *Recent Advances in QSAR Studies: Methods and Applications*. Springer; 2010. pp. 261–82.

3 Pisani L, de Candia M, Rullo M, Altomare CD. Hansch-type QSAR models for the rational design of MAO inhibitors: basic principles and methodology. In: *Monoamine Oxidase: Methods and Protocols*. Springer US; 2022. pp. 207–20.

4 Gozalbes R, de Julián-Ortiz JV. Applications of chemoinformatics in predictive toxicology for regulatory purposes, especially in the context of the EU REACH legislation. *Int. J. Quant. Struct.-Prop. Relat.* 2018;3(1):1–24.

5 Myatt GJ, Ahlberg E, Akahori Y, et al. In silico toxicology protocols. *Regul. Toxicol. Pharmacol.* 2018;96:1–7.

6 Cherkasov A, Muratov EN, Fourches D, et al. QSAR modeling: Where have you been? Where are you going to? *J. Med. Chem.* 2014;57(12):4977–5010.

7 Thakur A, Mehta V, Nagu P, Goutam K. *Computer-Aided Drug Design: QSAR, Molecular Docking, Virtual Screening, Homology and Pharmacophore Modeling*. Walter de Gruyter GmbH & Co. KG; 2024.

8 Oprea TI. On the information content of 2D and 3D descriptors for QSAR. *J. Braz. Chem. Soc.* 2002;13:811–5.

9 Dimitrov SD, Diderich R, Sobanski T, et al. QSAR Toolbox – Workflow and major functionalities. *SAR QSAR Environ. Res.* 2016;27(3):203–19.

10 Hitaoka S, Chuman H. Revisiting the Hansch–Fujita approach and development of a fundamental QSAR. *J. Pestic. Sci.* 2013;38(2):60–7.

11 Vasilev B, Atanasova MA. Review of the application of quantitative structure–activity relationship (QSAR) in the prediction of new compounds with anti-breast cancer activity. *Appl. Sci.* 2025;15:1206.

12 Tropsha A, Isayev O, Varnek A, et al. Integrating QSAR modelling and deep learning in drug discovery: the emergence of deep QSAR. *Nat. Rev. Drug Discov.* 2024;23(2):141–55.

13 Fourches D, Ash J. 4D-quantitative structure–activity relationship modeling: making a comeback. *Expert Opin. Drug Discov.* 2019;14(12):1227–35.

14 Vedani A, Dobler M. 5D-QSAR: the key for simulating induced fit? *J. Med. Chem.* 2002;45(11):2139–49.

15 Vedani A, Dobler M, Lill MA. Combining protein modeling and 6D-QSAR. Simulating the binding of structurally diverse ligands to the estrogen receptor. *J. Med. Chem.* 2005;48(11):3700–3.

16 Malik JK, Soni H, Singhai AK. QSAR – Application in drug design. *Int. J. Pharm. Res. Allied Sci.* 2013;2(1):1–13.

17 Benigni R, Andreoli C, Giuliani A. Quantitative structure-activity relationships: principles, and applications to mutagenicity and carcinogenicity. *Mutat. Res. Genet. Toxicol.* 1989;221(3):197–216.

18 Selassie CD, Mekapati SB, Verma RP. QSAR: then and now. *Curr. Top. Med. Chem.* 2002;2(12):1357–79.

19 Hansch C, Hoekman D, Gao H. Comparative QSAR: toward a deeper understanding of chemicobiological interactions. *Chem. Rev.* 1996;96(3):1045–76.

20 Kubinyi H. *QSAR: Hansch Analysis and Related Approaches.* Wiley; 1993.

21 Jhanwar B, Sharma V, Singla RK, Shrivastava B. QSAR: Hansch analysis and related approaches in drug design. *Pharmacologyonline* 2011;1:306–44.

22 Kubinyi H. QSAR and 3D QSAR in drug design: Part 1: methodology. *Drug Discov. Today* 1997;2(11):457–67.

23 Yonchev D, Bajorath J. Integrating computational lead optimization diagnostics with analog design and candidate selection. *Future Sci. OA* 2020;6(3):FSO451.

24 Veerasamy R. QSAR – An important in-silico tool in drug design and discovery. In: *Advances in Computational Modeling and Simulation.* Springer Nature Singapore; 2022. pp. 191–208.

25 Zhao M, Wang L, Zheng L, et al. 2D-QSAR and 3D-QSAR analyses for EGFR inhibitors. *Biomed. Res. Int.* 2017;2017(1):4649191.

26 Zhao X, Chen M, Huang B, et al. Comparative molecular field analysis (CoMFA) and comparative molecular similarity indices analysis (CoMSIA) studies on α_{1A}-adrenergic receptor antagonists based on pharmacophore molecular alignment. *Int. J. Mol. Sci.* 2011;12(10):7022–37.

27 Caron G, Nurisso A, Ermondi G. How to extend the use of grid-based interaction energy maps from chemistry to biotopics. *ChemMedChem* 2009;4(1):29–36.

28 Lozano JJ, Pastor M, Cruciani G, et al. 3D-QSAR methods on the basis of ligand–receptor complexes. Application of COMBINE and GRID/GOLPE methodologies to a series of CYP1A2 ligands. *J. Comput.-Aided Mol. Des.* 2000;14:341–53.

29 Akamatsu M. Current state and perspectives of 3D-QSAR. *Curr. Top. Med. Chem.* 2002;2(12):1381–94.

30 Diudea MV. *QSPR/QSAR Studies by Molecular Descriptors.* Nova Science Publishers; 2001.

31 Sobhia ME, Bharatam PV. Comparative molecular similarity indices analysis (CoMSIA) studies of 1,2-naphthoquinone derivatives as PTP1B inhibitors. *Bioorg. Med. Chem.* 2005;13(6):2331–8.

32 Verma J, Khedkar VM, Coutinho EC. 3D-QSAR in drug design: a review. *Curr. Top. Med. Chem.* 2010;10(1):95–115.

33 Datar PA, Desai PV, Coutinho EC. A 3D-QSAR of angiotensin II (AT1) receptor antagonists based on receptor surface analysis. *J. Chem. Inf. Comput. Sci.* 2004;44(1):210–20.

34 Gramatica P. On the development and validation of QSAR models. In: *Computational Toxicology: Volume II*. Humana Press; 2012. pp. 499–526.

35 Duchowicz PR. Linear regression QSAR models for polo-like kinase-1 inhibitors. *Cells* 2018;7(2):13.

36 Pirhadi S, Shiri F, Ghasemi JB. Multivariate statistical analysis methods in QSAR. *RSC Adv*. 2015;5(127):104635–65.

37 Rajabi M, Shafiei F. QSAR models for predicting aquatic toxicity of esters using genetic algorithm-multiple linear regression methods. *Comb. Chem. High. Throughput Screen*. 2019;22(5):317–25.

38 de Oliveira LH, Cruz JN, Dos Santos CB, de Melo EB. Multivariate QSAR, similarity search and ADMET studies based in a set of methylamine derivatives described as dopamine transporter inhibitors. *Mol. Divers*. 2023;28(5):2931–46.

39 Karelson M, Maran U, Wang Y, Katritzky AR. QSPR and QSAR models derived with CODESSA multipurpose statistical analysis software. In: *Predictive Toxicology of Chemicals: Experiences and Impacts of AI Tools*. Association for the Advancement of Artificial Intelligence; 1999. pp. 12–23.

40 Agarwal A, Shah D, Shen D, Song D. On robustness of principal component regression. In: *Advances in Neural Information Processing Systems*. Curran Associates; 2019. pp. 9893–903.

41 Ul-Saufie AZ, Yahaya AS, Ramli NA, et al. Future daily PM_{10} concentrations prediction by combining regression models and feedforward backpropagation models with principle component analysis (PCA). *Atmos. Environ*. 2013;77:621–30.

42 Abdi H, Williams LJ. Partial least squares methods: partial least squares correlation and partial least square regression. In: *Computational Toxicology: Volume II*. Humana Press; 2012. pp. 549–79.

43 Zhu Z, Rahman Z, Aamir M, et al. Insight into TLR4 receptor inhibitory activity via QSAR for the treatment of *Mycoplasma pneumonia* disease. *RSC Adv*. 2023;13(3):2057–69.

44 Santos-Filho OA, Cherkasov A. Using molecular docking, 3D-QSAR, and cluster analysis for screening structurally diverse data sets of pharmacological interest. *J. Chem. Inf. Model*. 2008;48(10):2054–65.

45 Santos GR, Chiari LP, da Silva AP, et al. A partial least squares and artificial neural network study for a series of arylpiperazines as antidepressant agents. *J. Mol. Model*. 2021;27:1–7.

46 Nabi AE, Pouladvand P, Liu L, et al. Machine learning in drug development for neurological diseases: a review of blood brain barrier permeability prediction models. *Mol. Inf*. 2025;44(3):e202400325.

47 Atasever S. Enhancing HCV NS3 inhibitor classification with optimized molecular fingerprints using random forest. *Int. J. Mol. Sci*. 2025;26(6):2680.

48 Qi Q, Wang Z. Integrating machine learning and nano-QSAR models to predict the oxidative stress potential caused by single and mixed carbon nanomaterials in algal cells. *Environ. Toxicol. Chem*. 2025;44(9):2535–44.

49 Lasfar R, Tóth G. The difference of model robustness assessment using cross-validation and bootstrap methods. *J. Chemom*. 2024;38(6):e3530.

50 Héberger K. Selection of optimal validation methods for quantitative structure–activity relationships and applicability domain. *SAR QSAR Environ. Res.* 2023;34(5):415–34.

51 De P, Kar S, Ambure P, Roy K. Prediction reliability of QSAR models: an overview of various validation tools. *Arch. Toxicol.* 2022;96(5):1279–95.

52 Abdel-Ilah L, Veljović E, Gurbeta L, Badnjević A. Applications of QSAR study in drug design. *Int. J. Eng. Res. Technol.* 2017;6(06):582–7.

53 Faris A, Ibrahim IM, Alnajjar R, et al. QSAR-driven screening uncovers and designs novel pyrimidine-4,6-diamine derivatives as potent JAK3 inhibitors. *J. Biomol. Struct. Dyn.* 2025;43(2):757–86.

54 Moussaoui M, Baammi S, Soufi H, et al. QSAR, ADMET, molecular docking, and dynamics studies of 1,2,4-triazine-3(2H)-one derivatives as tubulin inhibitors for breast cancer therapy. *Sci. Rep.* 2024;14(1):16418.

55 Edache EI, Uzairu A, Mamza PA, et al. Design of some potent non-toxic autoimmune disorder inhibitors based on 2D-QSAR, CoMFA, molecular docking, and molecular dynamics investigations. *Intell. Pharm.* 2024;2(5):688–706.

56 Tian YY, Tong JB, Liu Y, Tian Y. QSAR study, molecular docking and molecular dynamic simulation of Aurora kinase inhibitors derived from imidazo[4, 5-*b*] pyridine derivatives. *Molecules* 2024;29(8):1772.

57 Ibrahim ZY, Abdulfatai U, Ejeh S, et al. Genetic function algorithm (GFA) based QSAR, molecular design, and ADMET screening to assess the antimalarial potential of Amodiaquine derivatives. *The Microbe* 2024;5:100208.

58 Aloui M, Mujwar S, Er-rahmani S, et al. Design of novel potent selective surviving inhibitors using 2D-QSAR modeling, molecular docking, molecular dynamics, and ADMET properties of new MX-106 hydroxyquinoline scaffold derivatives. *Heliyon* 2024;10(19):e38383.

59 Anwar S, Alanazi J, Ahemad N, et al. Deciphering quinazoline derivatives' interactions with EGFR: a computational quest for advanced cancer therapy through 3D-QSAR, virtual screening, and MD simulations. *Front. Pharmacol.* 2024;15:1399372.

3

Molecular Mechanics and Quantum Mechanics in Drug Design and Energy Minimization

Vijay Kotra[1], B. N. B. Vaidehi[2], and A. Sree Gayatri[2]*

[1]*Department of Pharmaceutical Chemistry, Faculty of Pharmacy, Quest International University, Ipoh, Malaysia*
[2]*Faculty of Pharmacy, Aditya University, Surampalem, Andhra Pradesh, India*

*Corresponding author: vijay.kotra@qiu.edu.my

Introduction

Sometimes in molecular mechanics (MM), also known as classical mechanics, the molecular structure can be described through bonded atoms whose ideal geometries deviate due to van der Waals effects alongside coulombic non-bonded interactions [1]. This is the main distinction from quantum chemical models, which have nothing to do with chemical bonds. In order to develop an MM model, geometrical parameters must be highly transferable from one molecule to another and dependent on atomic hybridization. Carbon–carbon single bonds, for example, usually have a length between 1.45 and 1.55 Å, and they get longer as the p-character of the carbon hybrids rises. Therefore, if the specific molecule has a known valence structure, it is possible to make a fairly accurate prediction regarding molecular geometry in terms of bond angles, bond length, and torsion angles. The majority of organic compounds fall under this group. Quantum mechanics (QM) treats molecules as groupings of nuclei and electrons without explicitly addressing chemical bonds. The atomic-level behavior of systems is understood via the QM method [2]. In order to estimate the wave function and solve the Schrödinger equation, which is a mathematical model for the development of physical systems and through which the wave function operating in quantum systems remains crucial, QM approaches employ the laws of QM. The outcome of the Schrödinger equation is in terms of electron motions, which subsequently result in energy and molecule structure, among other observables, as well as bonding information. Electrons are confined to atoms and cannot hold any arbitrary energy, according to QM restrictions.

Molecular Mechanics

Andrews [3] coined the term "molecular mechanics" in 1930 (implemented in 1960) to assess the characteristics and geometry of molecular systems. It utilizes force fields (FFs) to determine a molecule's potential energy (PE) based on its structure and interactions, treating atoms as point masses with fixed charges. FFs are used to calculate the PE of each system as a function of the nuclear coordinates, assuming the validity of the Born–Oppenheimer [4] approximation. MM can be used to study molecular systems of any size and complexity, from enormous material assemblies with dozens to millions of atoms to small biological systems. Potential energies are determined by modifying bond lengths, angles, and twists according to the atoms' hybridizations to achieve equilibrium values. MM processes are usually less complicated than those of QM. They can manage more complex systems, such as enzymes, and are faster. MM computations are commonly used to determine the following [5]:

1. Best stable conformation
2. Desired confirmation
3. Energy disparities across stable conformations

Principle of MM

MM is a computer method that uses the concepts of classical mechanics to model molecular systems. This technique assigns a PE function to every molecule's degree of freedom, such as bond lengthening, valence angle variant, and dihedral rotation. Optimizing all parameters and minimizing energy yields conformers' equilibrium geometry and corresponding energies. Automated search strategies for local energy minima are available for molecules with many conformers. The MM technique involves establishing mathematical formulas and parameters for the PE of a molecule according to its geometry:

- It depicts each atom as a distinct particle.
- Each particle has a radius (often the van der Waals radius), a constant net charge, and polarizability, all of which are often established through tests and/or quantum computations.
- The equilibrium distance of bonded interactions, which are regarded as springs, is equal to the experimental or computed bond length.

Force Field

An FF is a set of variables and mathematical equations that are used to calculate a molecular system's PE based on atom positions. Molecular movement is governed by two types of atomic forces: chemically bonded interactions and nonbonded interactions,

$$E = E_{\text{bonded}} + E_{\text{nonbonded}}$$

$$E_{\text{bonded}} = E_{\text{bond}} + E_{\text{angle}} + E_{\text{dihedral}}$$

$$E_{\text{nonbonded}} = E_{\text{electrostatic}} + E_{\text{van der Waals}}$$

A common FF model uses simple springs (quadratic energy functions) to describe chemical bonds and atomic angles, preventing bond breakdown. A sinusoidal function is used to approximate dihedral angles, or rotations about a bond, due to the great degree of variability in dihedral energy. The energy differences between eclipsed and staggered conformations are roughly represented by this function. To characterize the planarity of aromatic rings, "cross-terms" can be employed to define internal variables like angles and bond lengths. Some FFs additionally define hydrogen bonding explicitly. Nonbonded (or noncovalent) terms provide long-range interactions (van der Waals forces by Lennard-Jones Potential and electrostatic by Coulomb's law),

$$E_{\text{total}} = \sum_{\text{bond}} K_r(r - r_{\text{eq}})^2 + \sum_{\text{angles}} k_\theta(\theta - r\theta_{\text{eq}})^2 + \sum_{\text{dihedral}} \frac{V_n}{2}[1 + \cos(n\phi - \gamma)]$$
$$+ \sum_{i<j} \left[\frac{A_{ij}}{R_{12}} - \frac{B_{ij}}{R_6} + \frac{q_i q_j}{\varepsilon R_{ij}}\right]$$

Bond Stretching (Harmonic Potential)

$$\sum_{\text{bond}} K_r(r - r_{\text{eq}})^2$$

- r = Actual bond length
- r_{eq} = Equilibrium bond length (ideal distance)
- K_r = Bond force constant (stiffness of the bond)

Angle Bending

$$\sum_{\text{angles}} k_\theta(\theta - r\theta_{\text{eq}})^2$$

- θ = Actual bond angle
- θ_{eq} = Equilibrium bond angle
- K_θ = Angle force constant

Dihedral Angles

$$\sum_{\text{dihedral}} \frac{V_n}{2}[1 + \cos(n\phi - \gamma)]$$

- ϕ = Dihedral angle
- V_n = Barrier height
- n = Multiplicity
- γ = Phase offset

Nonbonded Interactions

van der Waals

$$\frac{A_{ij}}{R_{12}} - \frac{B_{ij}}{R_6}$$

- Repulsion – $1/R_{12}$ prevents atoms from overlapping
- Attraction – $1/R_6$ dispersion/London forces
- R_{ij} = Atomic distances I and J
- A_{ij}, B_{ij} = van der Waals constant for the atom pair

Electrostatic

$$\frac{q_i q_j}{\varepsilon R_{ij}}$$

- q_i, q_j = Atomic partial charges i and j
- ε = The value of dielectric constant
- R_{ij} = Distance between charges

Parameters

Physics, chemistry, and/or electronic structure calculations (using QM) are used to determine the FF parameters. For various atom types, chemical bonds, dihedral angles, and other factors, a set of parameters is established. For instance, an FF would have different values for the oxygen atom in a hydroxyl group compared to a carbonyl functional group. A typical parameter set includes the atomic masses, van der Waals radii, and partial charges of individual atoms; the effective spring constants for each potential [6]; and the equilibrium values of bond lengths, bond angles, and dihedral angles for pairs, triplets, and quadruplets of bonded atoms.

Limitations of FF

- There is no polarizability because each atom is fixed and cannot change over time.
- To increase concordance with experimental data and allow for electronic polarizability, nonbonded terms can be scaled using a constant factor. Polarizable FFs seem promising, but they are less user-friendly and computationally costly than their fixed charge counterparts [7].

- FFs are limited in exploring reactivity, as bonds cannot be destroyed or created during simulations.

Types of FFs

The first functional FFs were developed to predict the compounds' molecular structures, vibrational spectra, and enthalpies. They were mostly used to address tiny organic compounds. These technologies are constantly evolving and being used. Allinger's group invented the MM potentials, which provide an excellent example. The application of MM to complicated systems, such as biological molecules, has resulted in the development of more widely applicable functional forms. Dreiding [8] and Universal (UFF) [9] FFs are effective examples. These FFs include parameters for all atoms listed in the periodic table. Some popular FFs include CHARMM [10], AMBE [11], CFF [12], MMFF [13], GROMOS [14], and OPLS [15]. Various versions of these FFs exist owing to ongoing development. The FFs were created by academic research groups. FFs are categorized into three classes: I, II, and III.

Class I: first-generation or class I FFs are those with mostly harmonic or sinusoidal internal terms, e.g., AMBER, CHARMM, COSMIC, DREIDING, ECEPP, GROMOS, MM1, OPLS, UFF, and PFF.

Class II: class II FFs are those that explain about cross-terms. Cross terms describe the relationship between stretching, bending, and torsion. Some FFs, such as Allinger's MM2, MM3, and MM4 have cross terms with multiple internal coordinates, such as angle–torsion–angle and bond–angle–bond. The PE for bonds and angles includes terms to the third and fourth powers (cubic and/or quartic, respectively). Higher order terms are used to accurately reproduce quantum mechanical potential energy surfaces (PES) and experimental features, such as vibrational spectra. Higher terms also include more parameters, such as force constants, making their optimization more challenging, e.g., CFF, CVFF, EFF, MM2, MM3, MM4, and QMFF.

Class III: to accurately simulate drug target systems, it's necessary to explicitly address electrostatics and electronic polarization. Advancements in quantum physics, simulation methods, and computer hardware have led to improved class III FFs. Polarizable FFs were produced. Class III FFs are the second-generation FFs that include terms for bond stretching and angle bending at powers 3 and 4. They emerged in 1990 [16–18], e.g., AMOEBA, Amber, Tinker, Allinger's MM4, and Accelrys CFF.

A comparative overview of commonly used MM FFs and their applications is presented in Table 3.1.

Quantum Mechanics

The characteristics of atoms, molecules, and their constituents, such as protons, neutrons, and electrons, are described by QM. The characteristics are determined by the interactions between particles and electromagnetic radiation, such as light, X-rays,

Table 3.1 Classification of MM FFs with their representations, examples, and typical applications in molecular modeling and drug design.

Type	Description	Example	Application
All atom	Explicitly represent every atom	AMBER (Assisted Model Building and Energy Refinement), OPLS-AA (Optimized Potential for Liquid Simulation), CFF (Consistent FF), MM2	Proteins, DNA Small molecules and macromolecules Wide range of organic compounds For metals and polymers hydrocarbons
United atom	Groups nonpolar H atoms with carbons	GROMOS (GROningen MOlecular Simulation), ECEPP (Empirical Conformational Energy Program for Peptides)	Proteins, nucleotides, sugars, and condensed phase systems Proteins and peptides
Coarse-grained	Groups multiple atoms into beads	MARTINI, SIRAH, VAMM (Virtual Atom Molecular Mechanics), SAFT	Faster simulations for larger systems
reactive	Allow bond breaking/forming	ReaxFF, COMB	Stimulate reactions with bond changes
Machine learning	Set of models	MACE (Multi-atomic Cluster Expansion), ANI (Artificial Narrow Intelligence), FFLUX, TensorMol	Wide range of compounds

and gamma rays. The effects of quantum theory are difficult to comprehend and evaluate because of the often-strange behavior of matter and radiation at the atomic level. A number of factors show why QM is a crucial area of study. In the first place, it illustrates the essential physics [19] technique. Second, it has demonstrated remarkable success in providing precise outcomes in all scenarios in which it has been utilized. An important characteristic of QM is that any measurement of a system invariably disturbs it, though the exact nature and timing of this disturbance are uncertain and widely debated. QM became a powerful force that drew 20th century scientists toward its status as the best intellectual achievement of the period.

Basic Concept of QM

Following Bohr's hypothesis, which postulated that electrons orbited in circles, QM is fundamentally based upon two approaches: (a) matrix mechanics, proposed by Werner Heisenberg, and (b) wave mechanics, given by Erwin Schrödinger. Schrödinger's wave mechanics, which is less abstract and simpler to comprehend than Heisenberg's matrix mechanics, is the basis for the current discussion.

Schrödinger's Wave Mechanics

Schrödinger developed a mathematical tool that removed the need for superfluous assumptions to tackle a variety of physical problems by utilizing de Broglie's theory on matter waves [20]. The mathematical approach to optics motivated his work because he could derive straight-line light ray propagation using wave motion if device dimensions were larger than wavelength scales. Schrödinger described de Broglie's findings about the wave behavior of matter by developing a flexible equation for various physical problems without making any arbitrary assumptions. The mathematical approach to optics motivated his work because he could derive straight-line light ray propagation using wave motion if device dimensions were larger than wavelength scales. Schrödinger established a wave equation for matter to behave as a particle at short wavelengths. When mass-containing particles are subjected to conventional mechanics, force-generated PE $V(x, y, z)$ at the position x, y, z is used. The sum of the $V(x, y, z)$ and kinetic energy $p^2/2m_e$ yields a constant total energy E.

Thus,

$$\frac{p^2}{2m_e} + V(x, y, z) = E$$

Since the particle's energy E is too small to allow it to escape, it is thought to be bound and constrained by the potential to a certain area of space. The wave's momentum and wavelength would also fluctuate since the potential varies with two other factors: position. Schrödinger substituted a differential operator for p (momentum) in the energy equation above, assuming a position-dependent wave function (x, y, z). The partial differential equation is satisfied by the wave function. At this point, QM is established using universal mathematical principles by the time-independent Schrödinger wave equation. It is established QM in a universally applicable mathematical form. Schrödinger's theory provides a key benefit by requiring no addition of arbitrary quantum conditions. Using universally applicable mathematics, Schrödinger developed QM by establishing this time-independent wave equation. The theory of Schrödinger supports an important benefit through its requirement to maintain any additional arbitrary quantum conditions. The necessary quantum results emerge from applying reasonable constraints to the wave function, for example, requiring that it does not diverge to infinity far from the potential's center of gravity.

A precise solution of the Schrödinger equation remains out of reach for any atom containing multiple electrons. In addition to the weaker magnetic forces produced by electron spin and orbital motions, there are electronic interactions between the nucleus and electrons as well as between electrons. Approximation methods resolve these problems by enabling the independent movement of each electron within an average electric field created by nucleus and other electrons. The correlation between electron positions thus remains ignored. The concept of orbitals exists to describe each electron wave function. All of the electron waves in an atom must be calculated using QM while adhering to the exclusion principle. The computed energies are then adjusted based on the magnitudes of the magnetic and electron–electron correlations.

Applications of QM in Drug Discovery

Structure-based drug designing has been using QM approaches for considerable amount of time. QM remains applied to certain medicinally important chemical calculations in order to develop informative descriptors for quantitative structure–activity relationship (QSAR) and generate three-dimensional conformations for ligands. QM generates accurate simulation models for ligands and proteins that MM parametrization does not achieve. Research findings show that QM-MM methods generate inconsistent results regarding protein–ligand binding affinity calculations. When an imprecise ligand structure is given, the QM-MM method appears to be advantageous for X-ray structures with higher resolution. According to the study, there are a number of situations where the usage of precise charges requires improving docking accuracy in your extensive collection of protein data bank complexes. QM applications for drug design mainly compute the optimized ligand structures alongside protein–ligand complexes for accurate docking analysis. QM in MM approaches grants precise solutions for calculating binding energy. However, additional ligand target confirmation sampling using molecular dynamics (MD) simulations is still required for this strategy. Target protein research for HIV1 integrase, trypsin, West Nile virus NS3 serine protease, HIV1 PR, and CDK2 is effectively addressed by the use of QM techniques.

Methods Involved in QM

QM contains a diverse set of analytical approaches that help to evaluate, solve, and understand quantum systems [21, 22]. Different methods exist for analysis dependent on the nature of work between theoretical problems and computational modeling and experimental data analysis [23]. The primary QM research methods can be summarized.

Analytical Methods

These involve solving quantum systems exactly using mathematics [2].

Schrödinger equation (time-dependent and time-independent): QM is a fundamental concept it describes about the time evaluation of a quantum system:

- Central to nonrelativistic QM.
- In addition to the hydrogen atom, the model offers solutions for simple systems that contain a ball of particles and a harmonic oscillator.

Operator methods: different methods exist for analysis dependent on the nature of work between theoretical problems and computational modeling and experimental data analysis. Operators for ladders and relations of commutation form an essential part of the methodology [19]. Quantum mechanical operators function as mathematical constructs that transform the state vector of an experimental system after physical processes complete, such as state changes due to temporal evolution.

These operators show dual functionality as physical system properties that scientists can observe experimentally (such as position, momentum, and energy). A quantum system maintains a separate collection of physical observables for its descriptions. The essential operators exist within the framework of QM. We will discuss two essential types of operators during this presentation because they determine quantum system time evolution and correspond to measurable physical observables in quantum systems. This discussion will explain methods to modify and connect operators between each other. Operators in QM are given below.

Operators as representing observables:

- Operators and wave function evolution
- Linear operators
- Eigenvalues and eigenfunctions
- Hermitian operators

Importance of Operator Methods

Formulating QM: operator techniques are central to the mathematical structure of QM.

Solving quantum problems: they provide a powerful way to solve various quantum mechanical problems, such as calculating energies of atoms and molecules, and understanding scattering processes.

Understanding quantum systems: quantum devices help scientists to comprehend the characteristics of quantum phenomena, which include quantization, superposition, and entanglement.

Matrix mechanics (Heisenberg picture): quantum devices help scientists to comprehend the characteristics of quantum phenomena, which include quantization and superposition and entanglement. In the Schrödinger picture, observables remain constant while state vectors and observables change in tandem. In matrix mechanics, however, both remain constant. The Heisenberg picture serves as a mathematical formulation of QM, which Werner Heisenberg and his collaborators created in 1925 [24].

Time evolution: according to the Heisenberg picture, operators function as the core elements to explain quantum system evolution instead of state vectors.

Heisenberg equation of motion: the Heisenberg equation of motion determines how an observable changes in the Heisenberg picture,

$$\frac{\mathrm{d}A(t)}{\mathrm{d}t} = \left(\frac{1}{ih}\right)[A(t), H] + \frac{\partial A(t)}{\partial t}$$

Here, H is the Hamiltonian, and A denotes the commutator.

State vectors: Heisenberg picture maintains state vectors in a constant time-independent state throughout all times. State vectors display temporal changes within the Schrödinger picture yet not in the Heisenberg picture because they evolve through operators.

Equivalence: the Heisenberg and Schrödinger pictures are mathematically equivalent, meaning they describe the same quantum system and yield the same predictions. Equivalence simply offers different ways of looking at the time evolution of a quantum system.

Applications: the Heisenberg picture is particularly useful when considering quantum time correlation function and the time evolution of operators.

Dirac notation (bra–ket): the mathematical tool called Dirac notation or bra–ket notation functions as a representation system for quantum states together with quantum operations within QM [25]. It was introduced by Paul Dirac and simplifies calculations by representing quantum states as "kets" ($|\psi\rangle$) and their conjugate transposes as "bras" ($\langle\psi|$). Microsoft Quantum reports that this mathematical notation enables simple calculations of both probabilities and amplitudes through linear algebra methods. Key aspects of Dirac notation are as follows:

Kets: the quantum state $|\psi\rangle$ takes the form of a column vector with ψ symbolizing the state.

Bras: a ket vector transforms into a row vector by conjugate transposition and uses the notation $\langle\psi|$ when adopting the same ψ symbol as the ket vector.

Inner product: the quantum state inner product which is expressed as $\langle \text{bra}|\text{ket}\rangle$ joins together the bra state $\langle\psi|$ with the ket state $|\varphi\rangle$.

Operators: quantum operations become easier to represent through bra–ket notation, which serves as the mathematical representation of linear operators. The Dirac notation acts as an effective tool for displaying tensor products because these products help tackle complicated many-particle systems.

Quantum computing: scientists employ the Dirac notation to explain tensor products because the products allow analysis of intricate multiparticle systems.

Benefits of Dirac Notation

Conciseness: a brief and user-friendly system exists to represent quantum state combinations and operation sequences.

Ease of calculation: Microsoft Quantum declares simplified procedures for QM problems.

Approximation Methods

QM uses approximation techniques mainly because researchers can only obtain exact results from basic systems such as the hydrogen atom. For most real-world problems, we need to *approximate* the behavior of quantum systems:

- Many quantum systems have complex potentials, multi-particle interactions, or time-dependent behaviors.
- It is difficult to obtain precise solutions from the Schrödinger equation.
- Approximation methods give **reasonably accurate predictions** with much less effort.

Perturbation theory: perturbation theory in QM provides approximate solutions for the Schrödinger equation after modifying its Hamiltonian operator from a solved system [26].

The total Hamiltonian:

$$\widehat{H} = \widehat{H^0} + \lambda\widehat{H'}$$

$\widehat{H^0}$: the unperturbed Hamiltonian (we know its eigenstates and eigenvalues).
$\widehat{H'}$: the perturbation (a small change).
λ: a small parameter (we often set it to 1 later).

Time-independent perturbation theory: it is a nondegenerate perturbation theory. The modification caused by $\widehat{H'}$ leads to energy level and wave function adjustments.

Dependent-time perturbation theory: it is a powerful method in QM used to study how quantum systems evolve when subject to a time-varying external influence (i.e., a perturbation that changes with time). It's especially useful for analyzing transitions between energy levels due to external fields, such as electromagnetic radiation.

Variational method: in QM, the variational method serves as a fundamental solution technique to estimate ground state system energies (and potentially excited state energies) for problems whose Schrödinger equations become intractable for exact solutions. This mathematical framework employs a basic principle that forms the basis of the variational method. The energy expectation value from any normalized trial wave function ψ trial provides an upper limit to the actual ground state energy E_0,

$$E_0 \leq \left\langle \psi_{\text{trial}} \mid \widehat{H} \mid \psi_{\text{trial}} \right\rangle$$

So, by cleverly choosing and tweaking a trial wave function, you can get very close to the true ground state energy.

WKB approximation: the Wentzel–Kramers–Brillouin (WKB) approximation is a semiclassical method [27] used in QM to approximate the solution of the Schrödinger equation, especially in cases where the potential varies slowly with position:

- WKB can be used with a time-independent Schrödinger equation.
- The rate at which potential $V(x)$ modifies exceeds the wavelength of the particle.
- You're outside the classical turning points or in regions where $E \approx V(x)$ (with care near those turning points).
- Semiclassical method useful for tunneling problems.

Numerical and Computational Methods

Numerical and computational methods in QM involve using computer algorithms to approximate solutions of quantum mechanical problems that are difficult or impossible to solve analytically. These methods, often based on the Schrödinger equation, are essential for studying the behavior of atoms, molecules, and materials at the quantum level.

Density functional theory (DFT): DFT operates as a quantum mechanical computational procedure for the investigation of many-body systems, especially atom-based structures, through electronic structure analysis using spatial functionals [28]. This theory states that functionals, functions that take a function as input and provide a single numerical value as output, can be used to calculate the actual properties of many electron systems. The electron density determines the functional values in DFT, because it functions as the dependent spatial variable. Researchers in condensed-matter physics along with computational physicists and chemists consider DFT to be a highly flexible and widely used method for their field.

MD: this mathematical framework employs a basic principle that forms the basis of the variational method.

Quantum Monte Carlo (QMC): a statistical method that uses random sampling to solve the Schrödinger equation, particularly useful for many-body problems.

Ab initio MD: combines DFT with MD, allowing for simulations that include electronic structure changes during MD.

Quantum Field Theoretical Methods

Quantum field theory (QFT) extends QM by treating particles as excitations of underlying fields [29]. This framework, crucial for understanding fundamental interactions, describes how particles are created, annihilated, and interact. QM, special relativity, and classical field theory are all incorporated within it.

Fields as fundamental: QFT views subatomic particles as quantized excitations of fields that permeate space–time. For example, electrons are seen as localized disturbances in an electron field.

Particle interactions: other particles, which are likewise excitations of underlying fields, are exchanged between particles to explain their interactions. For instance, the electromagnetic force arises from the exchange of photons.

Unification of QM and relativity: the concepts of special relativity (relativity of time and space) and QM (wave–particle duality, quantization) are combined in QFT.

Renormalization: the renormalization method in QFT solves ultraviolet divergences to produce predictive results.

Standard Model of particle physics: presently, the Standard Model functions as a QFT framework to explain both universe fundamental elements and their interactive behavior.

Methods used in QFT

Functional methods: QFT employs functional methods to calculate QFT observables.

Perturbation theory: QFT uses perturbation theory to solve problems by expanding the solution in a series of terms, where each term corresponds to a different interaction.

Feynman diagrams: these diagrams are a visual and mathematical tool used to represent the different ways in which particles can interact and are crucial for calculations in QFT. By combining relativistic effects, allowing for particle creation and annihilation, and building upon the fundamentals of QM, QFT essentially offers a framework for comprehending the behavior of particles and their interactions. Used in high-energy physics and advanced condensed matter.

Second quantization: reformulates QM in terms of creation/annihilation operators.

Path integral formulation (Feynman): the sum of all possible paths in QM.

Feynman diagrams: visual tool for perturbation expansions in quantum electrodynamics (QED)/QFT.

Measurement and Interpretational Frameworks

Copenhagen interpretation: the basic Copenhagen interpretation of QM, which was primarily created by Werner Heisenberg and Niels Bohr, holds that quantum systems "collapse" into a definite state after being measured from a probabilistic state. Despite criticism and the emergence of alternative interpretations, this interpretation is frequently utilized in textbooks and research [30].

Superposition: a quantum system exists in a simultaneous superposition of several possible states prior to measurement. This implies that until a property, such as position or energy, is measured, it lacks a definitive value.

Measurement and collapse: The system "collapses" into a single state during the measurement process, eliminating all other possibilities. There are no gradual steps involved; the change happens immediately.

Indeterminacy: the Copenhagen interpretation allows us to embrace probabilistic outcomes in QM because measurements produce uncertain results, which scientists can only predict through probabilities.

Complementarity: in terms of Bohr's principle of complementarity, two quantities like position and momentum remain complementary, which means scientists cannot determine them simultaneously with complete precision.

Observer dependence: according to certain interpretations of the Copenhagen view, the observer influences both the measurement procedure and the system's final state.

Many-worlds interpretation (MWI) [31]: according to MWI of quantum physics, each quantum measurement causes the universe to divide into several parallel realities, each of which represents a potential consequence of the measurement. Essentially, all possibilities in the quantum wave function are realized in distinct, parallel worlds.

Decoherence theory: decoherence theory in QM explains the transition from quantum behavior, characterized by superposition and entanglement, to classical behavior, where systems exist in a single, definite state. A system's environmental interactions cause the decay of quantum coherence while producing outcomes, which appear classical in nature.

Coherence and superposition: a QM particle simultaneously holds multiple states of existence according to superposition theory. These different states are not simply a mixture of possibilities but rather exist in a coherent way, like a wave with different components adding up to a single, composite wave.

The role of environment: decoherence arises from a quantum system's interaction with its surrounding environment. This interaction can be through subtle forces, like gravity, or by exchanging energy or information with other particles in the environment.

Loss of coherence and emergence of classicality: as the system interacts with the environment, its quantum state becomes entangled with the environment.

Quantum computers: since they can result in computation errors, decoherence is a significant challenge in the development and upkeep of quantum computers.

Schrödinger's cat: the idea that decoherence can give the impression that a quantum system is in a definite state when, in fact, it is in a superposition of states is demonstrated by the Schrödinger's cat thought experiment.

Double-slit experiment: placing a detector behind a single slit in the double-slit experiment leads to the destruction of interference patterns while showing classical behavior.

Implications:

- Decoherence explains how QM can be consistent with classical physics, where objects have definite properties.
- It implies that interactions with the environment drive the quantum-to-classical transition, which is a gradual process rather than a sharp jump.
- It facilitates understanding of the environment's role in quantum measurements as well as the measurement problem in QM.

Quantum information approaches (e.g., entanglement, quantum computing): quantum information theory uses the concepts of quantum physics, particularly superposition and entanglement, to create new technologies for communication and information processing. Quantum computing, quantum cryptography, and quantum communication are the areas that make use of the unique properties of quantum systems.

Quantum superposition: unlike classical bits that are either 0 or 1, quantum bits, or qubits, can exist in a simultaneous superposition of 0 and 1.

Quantum entanglement: this is an important idea: two or more quantum particles can be connected so that, despite their distance from one another, their states are correlated. This field develops computers that can solve some problems exponentially faster than traditional computers by utilizing the concepts of superposition and entanglement.

Quantum communication: this area focuses on using quantum phenomena like entanglement to create secure communication channels.

Quantum cryptography: this involves using QM to develop secure encryption methods.

Quantum error correction: quantum mechanical principles are used to develop secure encryption techniques.

Examples of Quantum Information Applications

Quantum algorithms: these algorithms are made to solve problems more quickly by utilizing quantum phenomena like entanglement and superposition.

Quantum teleportation: this does not actually involve the movement of matter; rather, it refers to the transfer of quantum states from one place to another.

Quantum key distribution (QKD): according to Vaia, this distributes cryptographic keys via secure communication channels made possible by QM [32].

Energy Minimization

Finding a drug's lowest energy conformation is the goal of energy minimization. The basic purpose of MM is to discover a molecule's lowest energy conformation, which is known as energy minimization or the geometry optimization strategy. Energy minimization is the technique of arranging a group of atoms in space so that, in accordance with a computer model of chemical bonding, each atom's net interatomic force is reasonably close to zero and its position on the PES is stationary. The technique's goal is to reduce the system's energy. This method seeks to reduce the system's PE to put it short. The system rotates the atoms multiple times, calculating energy at each point. This method is performed several times to locate the place with the least energy until an overall minimum energy is achieved. Every step maintains the energy low; otherwise, the atom will return to its original position. One complete rotation of an atom makes up the minimization step, sometimes referred to as iteration.

Steps in EM process: the main steps are outlined as follows:

1. Calculation of the PE of the initial geometry.
2. Changing each atom's atomic location and calculating the whole energy of the molecule.
3. If the energy of the new conformer is less than the initial one, adopt this conformation and proceed with the next alteration; otherwise, keep the first one.
4. Repeat the procedure until there is no more loss in PE.

Methods for energy minimization [1, 33]**:** minimization can be done with both derivative and nonderivative approaches:

First derivative methods: steepest descents and conjugate gradient.

Second derivative method: Newton–Raphson.

Nonderivative strategies include reducing energy without using the energy function's derivatives. These techniques find the lowest energy conformation or state by repeatedly adjusting the parameters. Nonderivative approaches eliminate the need to calculate the energy function's gradients or Hessians. High-dimensional problems may take more iterations to converge, making them less efficient than derivative approaches. In molecular modeling, nonderivative approaches are especially helpful when the energy function is difficult or costly to compute.

Steepest descent method: steepest descent, often known as gradient descent, is a first-order optimization method that uses steps proportional to the gradient's positive or negative values to determine a function's local minimum or maximum. The process for energy minimization must include identifying the point that is closest to the initial structure. This is the simplest approach that uses the first derivative ($dE/dX_i = 0$). $R_{i+1} = R_i - \alpha F_i$, where i is iteration number, R_i denotes old coordinates, R_{i+1} denotes new coordinates, F_i is force (energy gradient) on the atoms at the step i, and α is a constant that determines the amount to which force is exerted. When the first derivatives are big, or when the geometry is far from the minimum, this approach converges quickly. When the procedure approaches a minimum, it significantly slows down. It moves so slowly at the minimum that it practically never reaches the bottom.

Conjugate gradient method: this is a first-order minimization strategy that takes into account both the current gradient and the previous search direction. Computing cycles for conjugated gradient calculations are proportional to the number of atoms (N), while duration per cycle is proportional to N^2. Need fewer energy evaluations and gradient calculations. Convergence characterization outperforms steepest gradient; more effective than steepest descent; good for systems that are medium in size.

Newton–Raphson method: this approach employs the inverse of the second derivative matrix (the Hessian). The approach can be implemented in either complete or partial (block diagonal [BDNR]) matrix format. The method has the highest computing cost per step of any method used to perform EM. The equation for updating the geometry is as follows:

$$X_{\text{new}} = X_{\text{old}} - \frac{E'(X_{\text{old}})}{E''(X_{\text{old}})}$$

Notice that the correction term depends on both the first derivative (slope) of the PES at current geometry and also the second derivative (curvature). It is the necessity of calculating these derivatives at each step.

Advantages:

- The reduction may converge in one or two steps.
- This method is highly accurate but expensive.
- Good for small number of systems.

Disadvantage:

- This method needs the calculation of second derivatives.

Simplex method: it is a reliable, nonderivative-based solution that is likely one of the simplest to apply. It merely needs to evaluate functions. This method makes use of a geometric shape (simplex) whose vertices indicate various parameter configurations. The energy is determined for each vertex. The simplex is then changed by reflecting or expanding vertices to minimize the overall energy. The vertices of the original simplex must be generated in order to apply the simplex method. Only one of these vertices corresponds to the system's integration. By gradually adding a constant increment to each coordinate, the remaining points can be found.

The sequential univariate method: this strategy systematically seeks for the minimal energy by changing one parameter at a time while keeping the others constant. It entails testing several parameter values by increasing or decreasing the value of the chosen parameter and determining the energy at each point.

Applications of Energy Minimization

1. Comparison of structures
2. Systematic mapping of *E* space
3. Binding energy evaluation
4. Docking studies
5. Harmonic analysis
6. Comparing/fitting FFs.

Applications of MM

Structure-based drug design (SBDD): MM plays a crucial role in SBDD by supporting molecular docking studies. It assists in predicting how a ligand or therapeutic molecule fits into the active site of a target protein (receptor). MM-based FFs are also employed in scoring functions, where interaction energies are calculated to estimate binding affinity between the ligand and receptor.

MD simulations: MM is extensively used in MD simulations to model the time-dependent behavior of biomolecular systems. These simulations provide insights into protein flexibility, conformational changes, and the stability of drug–receptor complexes under physiological conditions.

Conformational analysis and energy minimization: MM enables detailed conformational analysis by:

- Determining the lowest-energy conformers of pharmaceutical compounds, which is critical for identifying the bioactive conformation that binds to the target.
- Performing energy minimization to refine protein–ligand complexes and drug structures by removing steric clashes and unrealistic geometries.
- Calculating noncovalent interaction energies, such as hydrogen bonding, van der Waals forces, and electrostatic interactions, between drugs and their receptors.

Lead optimization and QSAR studies: MM contributes to lead optimization by improving drug–target interactions and enhancing binding efficiency. It also supports QSAR modeling through the generation of molecular descriptors, including surface area, molecular volume, and charge distribution, which are essential for predicting biological activity.

Pharmacophore modeling and virtual screening: pharmacophore modeling using MM facilitates the identification of key structural features required for biological activity. These models are valuable tools for the virtual screening of large compound libraries to identify potential lead candidates.

Protein and peptide drug design: MM is widely applied in protein and peptide drug design to simulate peptide folding and structural stability. It aids in the rational design and optimization of biologics, including therapeutic peptides, enzymes, and antibodies, with improved efficacy and stability.

Several MM software packages, including AMBER, CHARMM, GROMACS, and NAMD, are widely employed for biomolecular simulations using different FFs (Table 3.2). Quantum mechanical calculations in drug design are routinely performed using software such as Gaussian, ORCA, NWChem, and Q-Chem, which support a range of ab initio and DFT methods (Table 3.3). Energy minimization and

Table 3.2 Commonly used MM software packages and the FFs supported for biomolecular simulations.

Software	Description	FFs supported
AMBER	Primarily for biomolecular simulations (proteins, DNA)	AMBER
CHARMM	Versatile for proteins, lipids, nucleic acids	CHARMM
GROMACS	High-performance MD engine, free and open-source	GROMOS, OPLS-AA
NAMD	Designed for large biomolecular systems	CHARMM, AMBER
TINKER	Good for MM and polarizable FFs	MM2, MM3, AMOEBA

Table 3.3 Widely used QM software and the corresponding electronic structure methods employed in drug design and molecular modeling.

Software	Description	QM methods
Gaussian	Widely used for DFT, semi-empirical, ab initio methods	HF, DFT, MP2
ORCA	Free for academics, excellent for spectroscopy and DFT	DFT, MP2, CCSD
NWChem	Open-source, supports large-scale QM/MM simulations	HF, DFT, MP2
Q-Chem	High-level quantum chemical calculations	DFT, ab initio
GAMESS	General-purpose electronic structure program	HF, DFT, CI

Table 3.4 Software tools commonly used for energy minimization and structural optimization in computational drug design.

Software	Used in	Optimization techniques
Avogadro	Molecular building and simple MM optimization	UFF, MMFF94
Open Babel	Format conversion + basic MM energy minimization	MMFF94
MOE (Chemical Computing Group)	Comprehensive drug design suite	MMFF94x, AMBER
Schrödinger Suite (Maestro, MacroModel)	Advanced modeling, docking, QM/MM	OPLS3e
SYBYL-X	CoMFA, QSAR, minimization	Tripos FF

structural optimization of small molecules and biomolecular complexes are carried out using various computational tools, as summarized in Table 3.4.

Global Minimum and Bioactive Conformation: Relevance in Docking and Structure-based Drug Design

Global minimum conformation: the global minimum conformation represents the lowest-energy structure of a molecule, identified through energy minimization using MM FFs or QM calculations. It corresponds to the most thermodynamically stable state of the ligand in isolation, i.e., in the absence of the biological target.

Relevance and Applications

1. Enables identification of intrinsically stable ligand conformations.
2. Useful for pre-docking conformational analysis and database filtering.
3. Provides a reliable starting point for molecular modeling and simulation studies.

Bioactive conformation: the bioactive conformation refers to the specific three-dimensional arrangement adopted by a ligand upon binding to its biological target, such as the active site of a protein. This conformation may differ from the global minimum and often corresponds to a higher-energy state that is stabilized by receptor–ligand interactions within the binding pocket of the protein.

Relevance and Applications

1. Critical for accurate molecular docking and pharmacophore modeling.
2. Mimics the active binding mode of known ligands, guiding SBDD.
3. Essential for structure–activity relationship and three-dimensional QSAR studies.

Role in Docking

1. Many ligand conformations are usually generated and assessed by docking methods in molecular docking.
2. Using merely the global minimum could lead to the omission of significant bioactive conformations.
3. Consequently, when docking, one should either sample several low-energy conformers (ensemble docking) or use conformationally flexible ligands.

Role in Structure-based design

1. Improving binding affinity in drug design involves creating stiff ligands that resemble the bioactive shape.
2. Use MD simulations or flexible docking to examine induced fit effects, in which a protein conforms to a ligand.
3. Models of pharmacophores ought to depict the bioactive conformation instead of the global minimum [34–36].

Optimization Techniques to Identify Bioactive Conformations

Since the global energy minimum frequently does not match the bioactive conformation, a variety of computational and optimization methods are employed to effectively model or predict it:

1. Using flexible ligands for molecular docking: the goal is to forecast a ligand's most advantageous binding pose or bioactive confirmation at the target site. The way it operates is that docking software looks at several ligand locations and conformations to see which one fits into the binding pocket of the protein. The optimization method makes use of Monte Carlo, simulated annealing, genetic algorithms, and scoring functions. Tools: AutoDock/AutoDock Vina, Glide of Schrödinger, MOE Docking, and CCDC Gold.
2. MD: the aim of MD simulations is to model the evolution of conformational changes and ligand–protein interactions throughout time. How it helps: helps identify transient bioactive states by capturing the dynamic flexibility of the protein and ligand. The optimization method uses MM FFs (such as AMBER and CHARMM) to solve Newton's equations of motion. Instruments: GROMACS, AMBER, and NAMD.
3. The goal of conformational search and sampling: it is to produce a wide range of low-energy conformers in order to increase the likelihood of incorporating the bioactive shape. Techniques: thorough torsion scanning, generation of random conformers, genetic or Monte Carlo algorithms frequently used to provide potential conformations prior to docking. Tools: ETKDG algorithm (RDKit), Open Babel, MOE, Schrödinger's ConfGen, and Balloon (free tool for 3D structure generation).
4. Optimization based on pharmacophores: by matching ligands to established pharmacophore models, the spatial arrangement of characteristics essential for activity, are hoped to identify bioactive conformations. Optimization: MM

reduction is frequently the next step after conformations are graded according to feature alignment. Tools: LigandScout, MOE, and Schrödinger's phase.

5. QM/MM and semiempirical methods: goal: improve protein active site conformations by combining the precision of QM with the effectiveness of MM. Use: particularly beneficial for systems that contain charged residues, metal ions, or proton transfers. Tools: Gaussian (ONIOM method), ORCA, and Q-Chem + AMBER (via interfaces like ChemShell).
6. X-ray crystallography and NMR data can be utilized to limit the use of conformational search. Verify the anticipated bioactive forms. Create ensemble conformations [1, 37, 38].

Energy Calculations

In molecular modeling, energy calculations are essential for comprehending molecule interactions, stability, and reactivity. These computations have applications in material science, drug design, protein folding, and other fields [39].

Types of Energy Calculations

- Molecular mechanics:
  The molecular systems are modeled using classical physics. A sum of variables that take into consideration bond stretching, angle bending, torsional angles, and nonbonded interactions (van der Waals and electrostatics) is used to represent the overall PE. The FFs commonly used are AMBER, CHARMM, OPLS, and GROMOS,

$$E_{\text{total}} = \sum_{\text{bond}} K_r(r - r_{\text{eq}})^2 + \sum_{\text{angles}} k_\theta(\theta - r\theta_{\text{eq}})^2 + \sum_{\text{dihedral}} \frac{V_n}{2}[1 + \cos(n\phi - \gamma)] + \sum_{i<j} \left[\frac{A_{ij}}{R_{12}} - \frac{B_{ij}}{R_6} + \frac{q_i q_j}{\varepsilon R_{ij}} \right]$$

- Quantum mechanics:
  - Schrödinger's equation is used to explicitly treat electrons.
  - It is appropriate for small molecules or the active sites of enzymes.
  - It incorporates techniques such as Hartree–Fock (HF), density functional theory (DFT), and post-HF methods (MP2, CCSD).
- Semiempirical methods:
  - The link between QM and MM.
  - Apply QM concepts while making computations simpler with empirical parameters. For example: AM1, PM3, and MNDO.
- MD and Monte Carlo simulations: utilize probabilistic sampling (MC) or time-dependent behavior (MD) to compute free energy differences, average energies, and conformational flexibility.

The energy minimization techniques like Newton–Raphson, conjugate gradient, and steepest descent algorithms are also used.

Conclusion

MM and QM are foundational to modern drug design, each offering unique strengths for modeling molecular systems. MM enables efficient simulation of large biomolecules and supports key tasks such as energy minimization, docking, and conformational analysis, making it indispensable for structure-based drug discovery. QM provides detailed insights into electronic structure and reaction mechanisms, improving the accuracy of binding energy calculations and ligand optimization. The combination of MM and QM, especially in hybrid approaches, allows researchers to balance speed and precision, expanding the scope of computational drug design. Energy minimization techniques help identify both global and bioactive conformations, which are crucial for predicting biologically relevant molecular interactions. Overall, the synergy between these computational methods accelerates the development of effective therapeutics and deepens our understanding of molecular behavior in biological systems.

List of Abbreviations

AMBER	Assisted Model Building and Energy Refinement
AMOEBA	Atomic Multipole Optimized Energetics for Biomolecular Applications
CDK2	Cyclin-dependent kinase 2
CFF	Consistent force field
CHARMM	Chemistry at HARvard Macromolecular Mechanics
COVID	Corona virus disease
CVFF	Consistent valence force field
DFT	Density functional theory
ECEPP	Empirical Conformational Energy Program for Peptides
EFF	Effective Fragment Potential
EGFR	Epidermal growth factor receptor
EM	Energy minimum
ETKDG	Experimental-torsion-knowledge distance geometry
FDA	Food and Drug Administration
FF	Force field
GROMOS	GROningenen MOlecular Simulation
HER2	Human epidermal growth factor receptor 2
HIV	Human immunodeficiency virus
MACE	Message Passing Atomic Cluster Expansion
MD	Molecular dynamics
MM	Molecular mechanics
MMFF	Merck molecular force field
MWI	Many-worlds interpretation
NF-κB	Nuclear factor kappa-light-chain-enhancer of activated B cells

NMR	Nuclear magnetic resonance
ONIOM	Our own N-layered integrated molecular orbital and molecular mechanics
OPLS	Optimized potential for liquid simulation
PE	Potential energy
PES	Potential energy surface
PFF	Polarizable force field
QED	Quantum electrodynamics
QFT	Quantum field theory
QKD	Quantum key distribution
QM	Quantum mechanics
QMC	Quantum Monte Carlo
QSAR	Quantitative structure–activity relationship
RDKit	Rheinberg Data Kit
ReaxFF	Reactive force field
SARS-CoV-2	Severe acute respiratory syndrome coronavirus 2
SIRAH	South American Initiative for a Rapid and Accurate Hamiltonian
UFF	Universal force field
VAMM	Virtual atom molecular mechanics
WKB	Wentzel–Kramers–Brillouin

References

1 Leach AR. *Molecular Modelling: Principles and Applications*. 2nd ed. Pearson Education Limited; 2001.

2 Szabo A, Ostlund NS. *Modern Quantum Chemistry: Introduction to Advanced Electronic Structure Theory*. Dover Publications; 1996.

3 Andrews DH. The relation between the Raman spectra and the structure of organic molecules. *Phys. Rev.* 1930;36:544.

4 Hanson D. *The Born–Oppenheimer Approximation. Chemistry LibreTexts.* Chemical Education Digital Library; 2022.

5 Pissurlenkar RR, Shaikh MS, Iyer RP, Coutinho EC. Molecular mechanics force fields and their applications in drug design. *Anti-Infect. Agents. Med. Chem.* 2009;8:128–50.

6 Field MJ, Bash PA, Karplus M. Combined quantum mechanical and molecular mechanical potential for molecular dynamics simulations. *J. Comput. Chem.* 1990;11:700–33.

7 Cho AE, Guallar V, Berne BJ, Friesner R. Importance of accurate charges in molecular docking: quantum mechanical/molecular mechanical (QM/MM) approach. *J. Comput. Chem.* 2005;26:915–31.

8 Ermer O, Lifson S. Consistent force field calculations. III. Vibrations, conformations, and heats of hydrogenation of nonconjugated olefins. *J. Am. Chem. Soc.* 1973;95:4121–32.

9 Mayo SL, Olafson BD, Goddard WA. DREIDING: a generic force field for molecular simulations. *J. Phys. Chem.* 1990;94:8897–909.

10 Rappé AK, Casewit CJ, Colwell KS, et al. UFF: a full periodic table force field for molecular mechanics and molecular dynamics simulations. *J. Am. Chem. Soc.* 1992;114:10024–35.

11 MacKerell AD, Bashford D, Bellott M, et al. All-atom empirical potential for molecular modeling and dynamics studies of proteins. *J. Phys. Chem. B.* 1998;102:3586–616.

12 Weiner PK, Kollman PA. AMBER: assisted model building with energy refinement. A general program for modelling molecules and their interactions. *J. Comput. Chem.* 1981;2:287–303.

13 Oostenbrink C, Villa A, Mark AE, van Gunsteren WF. A biomolecular force field based on the free enthalpy of hydration and solvation: the GROMOS parameter sets 53A5 and 53A6. *J. Comput. Chem.* 2004;25:1656–76.

14 Halgren TA. Merck molecular force field. I. Basis, form, scope, parameterization, and performance of MMFF94. *J. Comput. Chem.* 1996;17:490–519.

15 Jorgensen WL, Tirado-Rives J. The OPLS potential functions for proteins: energy minimizations for crystals of cyclic peptides and crambin. *J. Am. Chem. Soc.* 1988;110:1657–66.

16 Huang J, Lopes PEM, Roux B, MacKerell AD. Recent advances in polarizable force fields for macromolecules: microsecond simulations of proteins using the classical Drude oscillator model. *J. Phys. Chem. Lett.* 2014;5:3144–50.

17 Senftle TP, Hong S, Islam MM, et al. The ReaxFF reactive force-field: development, applications and future directions. *npj Comput. Mater.* 2016;2:15011.

18 Leach AR. *Molecular Modelling: Principles and Applications*. Addison Wesley Longman; 1996.

19 Levine IN. *Quantum Chemistry*. 7th ed. Pearson; 2014.

20 Schrödinger E. Quantisierung als eigenwertproblem. *Ann. Phys.* 1926;79:361–76.

21 de Sousa B, Freire F. The role of quantum mechanics in drug discovery: a review. *J. Mol. Model.* 2021;27:300.

22 Senn HM, Thiel W. QM/MM methods for biomolecular systems. *Angew. Chem. Int. Ed. Engl.* 2009;48:1198–229.

23 Jensen F. *Introduction to Computational Chemistry*. 3rd ed. Wiley; 2017.

24 De Lange OL, Raab RE. *Operator Methods in Quantum Mechanics*. Clarendon Press; 1991.

25 Heisenberg W. Über quantentheoretische umdeutung kinematischer und mechanischer beziehungen. *Z. Phys.* 1925;33:879–93.

26 Dirac PAM. A new notation for quantum mechanics. *Math. Proc. Camb. Philos. Soc.* 1939;35:416–18.

27 Wilson AH. Perturbation theory in quantum mechanics. *Proc. R. Soc. Lond. A* 1929;122:589–98.

28 Wentzel G, Kramers HA, Brillouin L. Eine verallgemeinerung der quantenbedingungen für die zwecke der wellenmechanik. *Z. Phys.* 1926;38:518–29.

29 Thijssen J. *Computational Physics: An Introduction*. 2nd ed. Cambridge University Press; 2013.

30 Weisstein EW. *Functional*. Wolfram MathWorld; 2024.
31 Peskin ME, Schroeder DV. *An Introduction to Quantum Field Theory*. Addison-Wesley; 1995.
32 Jammer M. *The Philosophy of Quantum Mechanics*. Wiley-Interscience; 1974.
33 Yadav MR, Murumkar PR. *A Textbook of Drug Design and Development*. Taylor & Francis; 2009.
34 Perola E, Charifson PS. Conformational analysis of drug-like molecules bound to proteins: an extensive study of ligand reorganization upon binding. *J. Med. Chem.* 2004;47:2499–510.
35 Teague SJ. Implications of protein flexibility for drug discovery. *Nat. Rev. Drug Discov*. 2004;2:527–41.
36 Kitchen DB, et al. Docking and scoring in virtual screening for drug discovery: methods and applications. *Nat. Rev. Drug Discov*. 2004;3:935–49.
37 McGann M. FRED and HYBRID docking performance on standardized datasets. *J. Comput. Aided Mol. Des*. 2011;25:937–52.
38 Warren GL, et al. A critical assessment of docking programs and scoring functions. *J. Med. Chem*. 2006;49:5912–31.
39 Brooks BR, et al. CHARMM: the biomolecular simulation program. *J. Comput. Chem*. 2009;30:1545–614.

4

Molecular Docking in Drug Discovery: Techniques and Applications

Azizeh Shadidizaji[1,2], Ahmet Hacımüftüoğlu[3], and Mohamad Warda[4,5]*

[1]*Vaccine Development Application and Research Center, Ataturk University, Erzurum, Türkiye*
[2]*Coordination Office of Computer-Based Modeling and Simulation of Biological Systems, Ataturk University, Erzurum, Türkiye*
[3]*Department of Medical Pharmacology, Faculty of Medicine, Atatürk University, Erzurum, Türkiye*
[4]*Department of Physiology, Faculty of Veterinary Medicine, Atatürk University, Erzurum, Türkiye*
[5]*Department of Biochemistry, Faculty of Veterinary Medicine, Cairo University, Giza, Egypt*

*Corresponding author: shadi.microbiology@gmail.com

Introduction

Molecular docking is a cornerstone computational technique in modern drug discovery, enabling the prediction of binding orientations and affinities between small molecules (ligands) and target biomolecules (receptors). By simulating molecular interactions, docking provides critical insights into protein–ligand binding mechanisms, which are essential for structure-based drug design (SBDD). This approach has revolutionized the identification of potential drug candidates, optimization of lead compounds, and elucidation of protein–ligand interactions [1, 2]. Recent advancements in docking methodologies have introduced sophisticated techniques such as flexible docking, ensemble docking, and extra-precision docking, which account for receptor flexibility and improve prediction accuracy [3]. Scoring functions, a pivotal component of docking algorithms, have evolved to incorporate machine learning and deep learning approaches, enhancing the reliability of binding affinity predictions [1]. Moreover, post-docking refinement using molecular dynamics (MD) simulations has emerged as a powerful tool to validate docking poses and explore dynamic interactions at atomic resolution [4]. This chapter covers the theoretical foundations and practical applications of molecular docking, highlighting its integration with virtual screening, drug repurposing, and lead optimization. It also examines the challenges associated with scoring functions, receptor flexibility, and ligand conformational sampling, offering insights into emerging solutions such as hybrid docking–MD approaches. By bridging computational predictions with experimental validations, molecular docking continues to drive innovation in drug discovery and development. Case studies on drugs targeting mechanistic insights into the *p*-coumaric acid (PCA) protection against bisphenol A (BPA)-induced hepatotoxicity in silico models illustrate the practical applications of docking in drug discovery.

Principles of Molecular Docking

Molecular docking is a computational method that predicts the optimal orientation of a ligand when bound to a receptor/protein to form a stable complex. This preferred orientation can be used to estimate the binding affinity between the ligand and the protein through scoring functions. Molecular docking is widely employed in drug discovery to determine the binding pose of potential drug candidates with target proteins, thereby aiding in the prediction of their biological activity and binding strength. Consequently, molecular docking serves as a critical tool in SBDD and discovery [5].

Definitions, History, and Significance

The primary objective of molecular docking is to simulate molecular recognition computationally, achieving an optimized conformation that minimizes the system's free energy. Developing new drugs remains a highly challenging endeavor. Contemporary drug discovery predominantly relies on an in silico-chemico-biological approach. The application of computer-aided techniques in drug discovery and development has gained significant traction, with increasing implementation and recognition in the field [6]. Since its introduction in the mid-1970s, molecular docking has emerged as a pivotal computational tool for elucidating the interactions between chemical compounds and their biological targets, playing a critical role in drug discovery and development. Indeed, the scientific literature has seen a substantial increase in studies employing docking techniques to (i) identify key structural determinants governing ligand–receptor binding affinity and specificity, and (ii) develop advanced docking methodologies with improved accuracy and predictive power [7]. Molecular docking is a computational process that predicts the optimal spatial arrangement of ligand molecules when bound to a target receptor, maximizing their binding affinity and stabilizing the resulting complex. At the cellular level, this phenomenon occurs dynamically, facilitating rapid and specific molecular interactions that lead to the formation of stable, biologically functional complexes.

Docking Workflow and Methodologies

Preparation

File format conversion (PDB, MOL2, etc.).
Removal of water molecules, cofactors, and ions (if unnecessary).
Adding hydrogen atoms and assigning charges.
Tools: PubChem, PDB RCSB, ZINC, BioChemDraw, and Galaxy.

Prediction of Active Site

Identification of binding pockets (using tools like PyMOL, AutoDock, or CASTp).
Grid generation for docking (defining search space).

Pre-docking Steps (Prepdocking)

Setting up parameters (grid box dimensions, scoring functions).
Converting files to docking-compatible formats (PDBQT for AutoDock Vina).
Tools: Chimera, Schrödinger.

Energy Minimization

Optimization of ligand and receptor structures (using tools like Chimera, GROMACS, or AMBER).
Reducing steric clashes and improving binding conformation.
Tool: Chimera.

Docking

Running the docking simulation (AutoDock Vina, Glide, etc.).
Selecting appropriate algorithms (rigid/flexible docking).
Tools: Schrödinger, Chimera.

Result Docking Analysis

Evaluating binding affinity (docking scores, ΔG).
Analyzing interactions (hydrogen bonds, hydrophobic contacts, and π-π stacking).
Visualization (PyMOL, LigPlot+, and Discovery Studio).
Comparing results with experimental data (if available).

Online Server: PLIP

The study by Shadidizaji et al. provides valuable insights into the interaction between quercetin and the *Melampsora lini*-derived effector AvrM14 using in silico methods. It highlights the role of molecular docking and dynamic simulations in understanding plant defense mechanisms. The research establishes quercetin as a functional inhibitor of AvrM14, suggesting its potential in enhancing plant resistance against pathogens [8]. Recent advancements in molecular docking have provided profound insights into the protective mechanisms of natural compounds against toxic agents. Tekin et al. [9] demonstrated the efficacy of PCA in mitigating BPA-induced hepatotoxicity through both in vivo and in silico models. Their study revealed that PCA interacts with key molecular targets, such as cyclooxygenase-1, via stable hydrogen bonding, thereby attenuating oxidative stress and inflammation pathways. Computational analyses further highlighted PCA's ability to modulate the Kelch-like ECH-associated protein 1 (KEAP1) and nuclear factor erythroid

2-related factor 2 (Nrf2) signaling, showcasing its dual protective effects. These findings underscore the pivotal role of molecular docking in elucidating the mechanistic interactions between bioactive compounds and their targets, paving the way for innovative therapeutic strategies in drug discovery. The study by Tekin et al. [10] also provided molecular and pharmacological insights into the protective effects of PCA against BPA-induced testicular injury, employing a combination of in vivo and in silico approaches. The study utilized molecular docking simulations to investigate PCA's binding interactions with key targets involved in oxidative stress and inflammation, including NF-κB and inducible nitric oxide synthase (iNOS). The results demonstrated significant binding affinity and favorable interaction profiles, which correlated well with the experimental outcomes showing reduced oxidative damage and improved histopathological features in rat testicular tissue. This research further highlights the utility of molecular docking in elucidating the mechanistic basis of natural compound efficacy, reinforcing its role in toxicology-informed drug discovery and therapeutic development. Another recent study by Bolat et al. [11] offers compelling insights into the protective mechanisms of gallic acid against acrylamide-induced pulmonary injury, highlighting both in vivo and in silico approaches. Through molecular docking and dynamic simulation analyses, the study demonstrates how gallic acid interacts with key targets involved in oxidative stress and inflammation, particularly within the Nrf2/HO-1/NFκB signaling pathway. The computational findings align with experimental data, reinforcing gallic acid's potential as a modulatory agent capable of attenuating acrylamide toxicity. This research exemplifies the power of integrative approaches, combining molecular docking with pharmacological validation, in identifying and characterizing bioactive compounds for therapeutic applications. Another recent investigation [12] explored the antidiabetic potential of extra virgin olive oil (EVOO) and its key polyphenol hydroxytyrosol, utilizing both in vivo and in silico methodologies. The study combined biochemical analysis in a streptozotocin-induced diabetic rat model with molecular docking simulations targeting key enzymes in glucose metabolism, such as α-glucosidase and dipeptidyl peptidase-4 (DPP-4). The in silico results revealed strong binding affinities and stable interactions of hydroxytyrosol with these targets, supporting its therapeutic potential. These findings reinforce the value of molecular docking as a predictive tool in nutraceutical-based drug discovery, particularly in metabolic disorder contexts like diabetes mellitus.

Classification of Molecular Docking Approaches

Molecular docking methodologies can be broadly categorized into rigid, flexible, ensemble, and hybrid.

Rigid Docking

In rigid docking simulations, both the ligand and receptor are treated as inflexible entities. This approach involves the spatial optimization of one molecule's position and orientation relative to another in three-dimensional (3D) space, with the objective of identifying the optimal geometric alignment that maximizes complementarity according to predefined scoring metrics. Notably, ligand conformations may be generated either independently or in the context of receptor binding interactions.

Flexible Docking

Flexible docking incorporates molecular conformational flexibility into the computational model, allowing for simultaneous evaluation of both spatial transformations and structural adaptability. This method enables the exploration of viable binding conformations for both receptor and ligand molecules, thereby more accurately representing the dynamic interactions observed in biological complexes [13].

Advanced Techniques (Glide XP, AI Scoring, and MD-refined Docking)

Prior to conducting docking simulations, precise identification of the binding site coordinates represents a critical preparatory step. In a few studies, this was accomplished using the Receptor Grid Generation panel within Maestro (Schrödinger, LLC). The grid parameters were meticulously defined based on the crystallographic coordinates of a ligand–receptor complex, ensuring accurate spatial demarcation of the active site [14]. Following grid generation, molecular docking was performed using the Glide module (Schrödinger, LLC), a sophisticated computational tool for predicting ligand–protein interactions. The Glide module (Schrödinger, LLC) implements three hierarchically organized docking protocols designed to balance computational efficiency with predictive accuracy: High Throughput Virtual Screening (HTVS), Standard Precision (SP), and Extra Precision (XP) [14]. These methodologies exhibit fundamental differences in their sampling algorithms and scoring functions, with HTVS and SP employing similar scoring functions while XP utilizes a more sophisticated scoring paradigm coupled with enhanced conformational sampling [15]. The HTVS protocol employs a rapid but less stringent sampling algorithm optimized for processing large compound libraries, making it suitable for preliminary screening phases. In contrast, both SP and XP implementations utilize advanced hierarchical stochastic search algorithms that systematically explore the receptor's active site through multiple stages of pose refinement [16]. The receptor's structural and physicochemical properties are discretized onto a grid system, with progressively higher resolution field representations enabling more accurate ligand pose scoring in SP and XP modes.

MD Simulation

MD simulation employs Newton's laws of motion to calculate the positions and velocities of atoms within a system. Functioning as a "computational microscope," MD simulations can analyze interactions between all particles at the atomic level, enabling the observation of dynamic changes in protein structures over specified time intervals. MD simulations rely on classical equations of motion to model the system's behavior. The resulting trajectories are often visualized graphically, providing insights into the structural and dynamic properties of the system at specific time points and residues. Additionally, MD simulations can identify distinct protein conformational states at different stages of the simulation. This approach yields valuable data on thermodynamic and kinetic parameters, contributing to a deeper understanding of molecular behavior. MD simulations can elucidate a protein's hidden structural pockets or allosteric binding sites, providing critical insights into its functional mechanisms. These simulations employ various force fields (FFs), such as OPLS and GROMOS, to accurately model molecular interactions. The conformational ensembles generated through MD-derived trajectories are valuable for multi-protein docking studies, enabling the exploration of diverse binding poses. Furthermore, MD simulations serve as a robust validation tool for molecular docking results by assessing the stability of predicted ligand–protein complexes. While docking is a highly flexible computational approach that predicts ligand binding under physiological conditions, MD simulations enhance its reliability by refining dynamic interactions and verifying structural stability over time [17].

Post-docking Refinement Using MD Simulations

Post-docking refinement enhances the accuracy of predicted ligand-binding poses through MD simulations and energy optimization. One such approach, Binding Estimation After Refinement (BEAR), is a computational post-processing method designed to refine ligand-binding modes generated by docking tools (e.g., AutoDock and DOCK) and evaluate binding free energies using molecular mechanics Poisson–Boltzmann surface area (MM-PBSA) and molecular mechanics generalized Born surface area (MM-GBSA) algorithms. The BEAR workflow involves initial preprocessing of docked complexes, followed by binding free energy calculations using MM-GBSA and MM-PBSA. These post-docking refinement techniques, including rescoring and conformational optimization, improve the correlation between computational predictions and experimental data, thereby increasing virtual screening hit rates in drug discovery [18].

Binding Free Energy Estimation Using MM-GBSA and MM-PBSA Approaches

Accurate estimation of binding free energies plays a crucial role in deciphering the affinity between biomolecular complexes such as protein–ligand, protein–peptide, and protein–protein interactions. While absolute free energy calculations are computationally intensive, end-point free energy methods such as MM-GBSA and MM-PBSA offer an efficient compromise between computational demand and predictive accuracy. As reviewed by Genheden and Ryde [19], MM-GBSA/MM-PBSA methods are widely employed post-simulation techniques for estimating the binding affinities of structurally resolved biomolecular complexes. These approaches decompose the binding free energy into key thermodynamic components:

- **Gas-phase molecular mechanical energies**, representing internal interactions (electrostatics and van der Waals).

- **Continuum solvation energies**, accounting for solvent effects via implicit solvent models (GB or PB).
- **Entropic contributions**, often estimated via normal-mode analysis or approximated by empirical terms.

By integrating these components, MM-GBSA and MM-PBSA provide quantitative insights into the molecular basis of binding, enabling efficient screening and prioritization of ligands in rational drug design and molecular biology research.

Scoring Functions in Detail

Scoring functions are essential computational tools used in molecular docking to evaluate protein–ligand interactions. They help predict binding affinity, identify optimal binding poses, and facilitate virtual screening in drug discovery. Over time, various scoring functions have been developed, each with distinct strengths in accuracy and computational efficiency. Knowledge-based scoring functions represent a distinct class of scoring approaches that derive interaction potentials from statistical analysis of experimentally determined protein–ligand structures. These methods are founded on the inverse Boltzmann principle, where interaction energies are inferred from the observed frequency of atom-pair distances in structural databases:

$$w(r) = -k_B T \ln[g(r)]$$

where $g(r) = \rho(r)/\rho^*(r)$ represents the ratio between observed atom-pair density $(\rho(r))$ and their expected distribution in a noninteracting reference state $(\rho^*(r))$.

Theoretical Foundations and Development

The statistical mechanics underpinning these functions originates from mean-force potential theory, though their application to biological systems requires significant adaptation from simple fluid models due to the following:

- atomic connectivity constraints
- excluded volume effects
- system composition heterogeneity

Despite these theoretical approximations, knowledge-based potentials have proven remarkably effective in practice. Their development typically involves the following:

- atom-type classification (typically 13–40 types)
- distance-binning of pairwise interactions
- reference state approximation

Implementation variants and performance: several notable implementations demonstrate the versatility of this approach:

- DrugScore: incorporates both pairwise potentials and surface-dependent terms, showing superior performance to force field methods in flexible docking benchmarks.
- BLEEP: achieves correlation coefficients of 0.71–0.74 with experimental binding data across different test sets.
- SMoG2001: implements a self-consistent reference state determination method.
- PMF/PMF04: introduces ligand volume corrections to the reference state, demonstrating particularly strong performance ($R^2 = 0.61$) across diverse complexes.
- DFIRE: applies a distance-scaled finite ideal-gas reference state, achieving good correlations for various biomolecular interactions.

Reference state challenges and innovations: the fundamental limitation of knowledge-based methods lies in the reference state definition. Traditional approaches use atom-randomized states, while modern variants employ the following:

- volume correction factors (PMF)
- distance-scaling (DFIRE)
- iterative optimization (ITScore)

The ITScore method represents a particularly innovative solution, circumventing explicit reference state calculation through the following:

- iterative potential refinement against native/decoy ensembles
- Boltzmann-weighted sampling of binding landscapes
- recent incorporation of solvation and entropic effects (ITScore/SE)

This approach has demonstrated improved performance in binding mode prediction, affinity estimation, and virtual screening applications.

Comparative advantages: knowledge-based functions offer several practical benefits:

- computational efficiency comparable to empirical methods
- training-set robustness due to large structural databases
- physical interpretability of pairwise interaction terms

However, challenges remain in the following:

- handling novel binding geometries
- accounting for protein flexibility
- improving discrimination between native and decoy poses

These functions continue to evolve through hybrid approaches that combine statistical potentials with machine learning and physical correction terms, promising enhanced accuracy for SBDD applications [20].

Consensus Scoring Strategies in Molecular Docking

Given the inherent limitations of individual scoring functions, consensus scoring has emerged as a powerful approach to improve prediction reliability in molecular docking. This methodology combines multiple scoring functions to leverage their complementary strengths while mitigating individual weaknesses.

Rationale and implementation: the fundamental premise of consensus scoring rests on three key observations:

- Different scoring functions often excel at specific aspects of binding prediction.
- Scoring errors tend to be non-correlated across function types.
- True binders typically perform consistently well across multiple scoring metrics.

Common implementation strategies include the following:

- Vote-by-number: compounds are ranked by how many scoring functions identify them as top candidates.
- Rank-by-number: average ranking positions across multiple functions determine final scores.
- Score averaging: normalized scores from different functions are combined arithmetically.
- Weighted combinations: scores are integrated using empirically determined coefficients.

Practical applications: several successful consensus scoring systems demonstrate this approach's effectiveness:

- MultiScore: combines diverse scoring metrics to improve virtual screening accuracy.
- X-Cscore: integrates results from multiple empirical scoring functions.
- GFscore: utilizes geometric and energetic consensus criteria.
- SCS and SeleX-CS: implement sophisticated selection algorithms for consensus building.

Advantages and considerations: the primary benefits of consensus scoring include the following:

- increased prediction robustness
- reduced false positive rates
- improved ability to identify true binders

However, effective implementation requires careful attention to the following:

- Function selection: optimal combinations of complementary scoring methods.
- Normalization procedures: ensuring score comparability across different functions.
- Weight optimization: determining appropriate contribution levels for each component.

Recent advances have incorporated machine learning techniques to optimize consensus strategies, automatically determining the most effective combinations and weightings for specific target classes or screening scenarios. This data-driven approach has shown particular promise in addressing the challenge of developing universally applicable consensus protocols.

As scoring function development continues to advance, consensus methods remain essential tools for maximizing prediction accuracy in structure-based drug discovery, particularly in challenging cases where individual scoring functions might produce conflicting results [21].

Databases for Evaluating Scoring Functions

Developing an effective scoring function requires a well-constructed training and test dataset. Key considerations for building such datasets include the following:

- The protein–ligand complexes should be high-resolution structures without atomic clashes (e.g., reliable crystal structures).
- The dataset should encompass diverse protein types and a broad range of binding affinities.
- Ligands should resemble drug-like molecules and form noncovalent interactions with proteins.

Several publicly available databases can be used to compile these datasets, such as:

- LPDB – a database of ligand–protein interactions.
- PLD – provides protein–ligand binding data in an accessible format.
- BindingDB – a comprehensive resource for measured binding affinities.
- PDBbind – curates high-quality protein–ligand complexes with binding data.
- Binding MOAD – focuses on biologically relevant ligand–protein structures.
- AffinDB – contains experimentally determined binding affinities.

These databases serve as valuable sources for constructing robust training and test sets in scoring function development [22].

FF-based, Empirical, and ML-based

FF scoring functions estimate binding energies by modeling physical interactions such as van der Waals forces, electrostatic attraction, and bonded molecular mechanics terms. These functions rely on parameters derived from quantum mechanical calculations and experimental data, providing a physically meaningful framework for evaluating protein–ligand interactions. A widely used example is the scoring function in DOCK, which incorporates Lennard-Jones potentials for van der Waals interactions and coulombic electrostatics with a distance-dependent dielectric constant to approximate solvent effects. However, this simplified treatment fails to fully account for desolvation energy, the penalty incurred when polar or charged groups move from water into a protein's binding site. Since desolvation depends on the chemical environment of each atom, neglecting it can bias scoring toward highly charged ligands [23].

Case Studies

In Silico Study of Possible Inhibition of BPA on the Activity of COX-1 (PDB: 3N8Z)

Ligands and Protein Preparation: the identified inhibitory effect of BPA on COX-1 activity prompted us to conduct further in silico exploration into the potential inhibitory action exerted by BPA on the COX-1 active site. Therefore, the 3D structure of BPA (PubChem CID 6623) was obtained from PubChem database [24], and the minimization and optimization of CID 6623 of BPA were then performed by the ChemBioDraw Ultra 14 suite software [25]. The X-ray crystal coordinates of COX-1 (PDB: 3N8Z) were then retrieved from the RCSB database [26]. This, followed by the creation of PDBQT files for both the receptor and ligand, was facilitated by the UCSF Chimera software 1.17.3 [27]. Subsequently, preparation for docking was carried out using the Dunbrack 2010 rotamer library [28]. Additionally, the minimization of the receptor structure (steepest descent steps: 100; conjugate gradient step: 10) was performed using UCSF Chimera 1.17.3 software.

Molecular docking protocol: the active site of COX-1 protein was obtained from the online server [29]. Second, grid box size was fixed at ($20 \times 20 \times 20 \text{Å}^3$), and the grid center (−54,186; 54,440; −0.826) for COX-1 enzyme active site was set according to the setting software. Molecular docking of BPA and COX-1 were run in UCSF Chimera #AutoDock Vina software. The PyMOL software was utilized to generate a 3D representation of the BPA–COX-1 complex in the study [30].

The BPA–COX-1 complex interaction: for this, Protein–Ligand Interaction Profiler (PLIP) server was used [31]. The two-dimensional (2D) image of BPA–COX-1 complex was obtained by Discovery Studio Visualizer (v21.1.0.20298).

Molecular docking simulation: the investigation focused on the active site prediction server's evaluation of the active site of the COX-1 enzyme, revealing cavity_1_KLDPTFRCWGQEYNVHASIM. Notably, our study demonstrated that BPA exhibits specific binding to (THR 322.B, pos 8) positioning itself at a distance of 2.381 Å within the COX-1 active site (Figure 4.1). The complex formed between BPA and COX-1 manifests a single hydrogen bond with the THR 322.B residue within the active site. Our analyses unveiled a root-mean-square deviation (RMSD) of 3.157, accompanied by the formation of five hydrogen bonds. This interaction involves two ligand atoms and two receptor atoms. Importantly, our findings indicate a dock score of −6.2 kcal/mol.

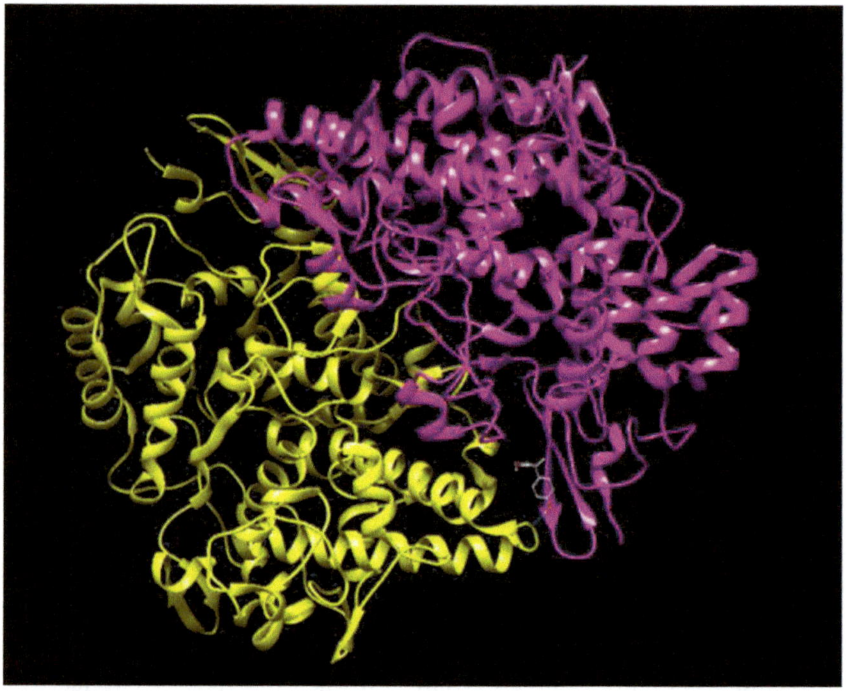

Figure 4.1 BPA–COX-1 complex docking (hydrogen bond with THR 322.B, pos 8, 2.381 Å within the COX-1 active site), 3D image was created by UCSF Chimera software.

Utilizing information presented in Table 4.1 and Figure 4.2, which delineate the hydrogen interactions within the BPA–COX-1 complex, it is discernible that multiple residues (ARG 49A, ALA 133A, and THR 322B) actively participate in forming hydrogen bonds with BPA. The measured distances between hydrogen and acceptor atoms (H–A and D–A) fall within the conventional range for hydrogen bonding, indicating robust and stable interactions. Furthermore, the angles formed by the donor atoms exhibit a configuration conducive to favorable hydrogen bonding geometry, reinforcing the robust nature of these interactions. Notably, ASP 135A and TYR 136A engage in hydrophobic interactions within the BPA–COX-1 complex (Table 4.2). The proximity between the ligand atom and the protein atom conforms to the typical range for hydrophobic interactions, suggesting a durable association. Given their propensity to enhance the stability of protein–ligand complexes, hydrophobic

Table 4.1 Hydrogen interaction between BPA and COX-1 complex.

Index	Residue	AA	Distance H–A	Distance D–A	Donor angle	Protein donor	Side chain
1	49A	ARG	2.50	3.29	136.77	V	V
2	133A	ALA	2.30	3.16	152.04	X	X
3	322B	THR	2.67	3.01	101.36	V	V
4	322B	THR	2.29	3.21	168.82	X	X

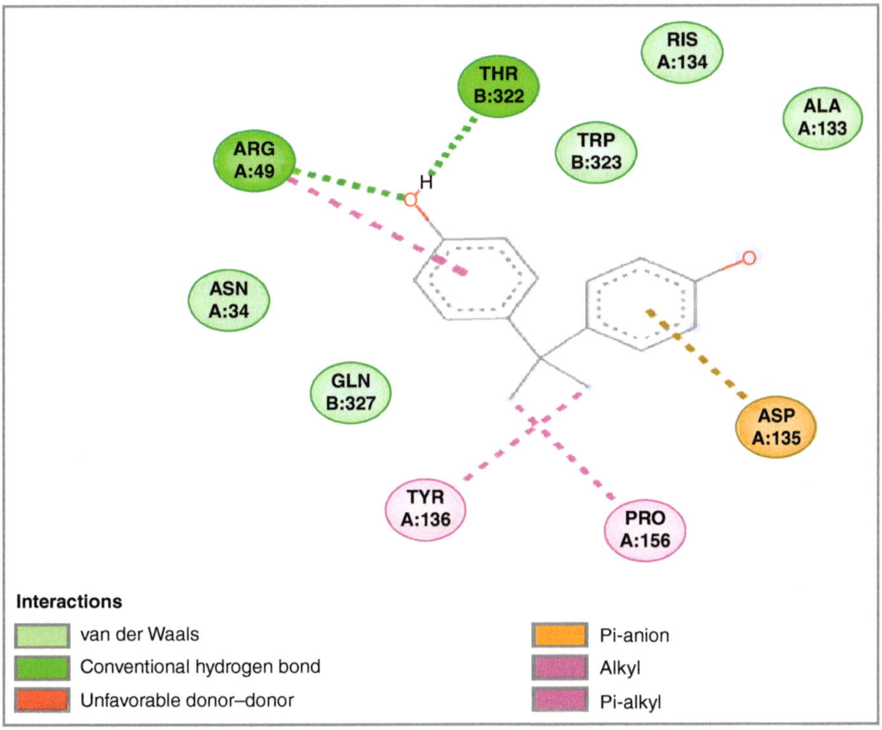

Figure 4.2 Interaction analysis of BPA–COX-1 complex (2D).

Table 4.2 Hydrophobic interaction between BPA and the COX-1 complex.

Index	Residue	AA	Distance H–A	Ligand atom	Protein atom
1	135A	ASP	3.60	8,993	851
2	136A	TYR	3.60	8,993	862
3	136A	TYR	3.70	8,994	864

interactions play a pivotal role in determining the overall binding affinity. These findings offer valuable insights into the molecular interactions and potential inhibitory effect of BPA on the activity of COX-1. In contrast to the other obtaining results of BPA in aggravation of inflammatory mediators, this result sheds light on the potential anti-inflammatory properties of BPA.

In Silico Evaluation of PCA as a Modulator of KEAP1 and TNF Trimer Stability

In line with our investigation of PCA as a bioactive compound with antioxidative and anti-inflammatory potential, we employed a comprehensive suite of computational tools to evaluate its modulatory role on two key molecular targets: KEAP1 and TNF trimer (tumor necrosis factor-α). KEAP1 serves as the cytoplasmic inhibitor of the Nrf2 antioxidant response, whereas TNF is a central mediator in pro-inflammatory signaling.

Protein and Ligand Preparation

- **Ligand (PCA):** 3D structure (SDF format) obtained from PubChem.
- **Proteins:** crystallographic structures of KEAP1 and TNF trimer retrieved from the Protein Data Bank (PDB IDs: 6LRZ and 6OOY, respectively).
- All molecular systems were preprocessed using Schrödinger Maestro 2023/4, ensuring optimized geometries, protonation states, and energy minimization.

Molecular docking and visualization: molecular docking simulations were conducted using the Glide module, assessing binding conformations and docking scores. The docking scores and Glide energies (Table 4.3) indicated significant interactions.

Notably, the more negative values for the PCA–TNF interaction suggest a relatively stronger binding affinity. Visualization tools (2D, 3D, and surface rendering) revealed specific binding pockets and interaction residues. Detailed pharmacophore mapping identified key hydrogen bond acceptors/donors, hydrophobic patches, and aromatic ring interactions (Table 4.4).

Table 4.3 Docking scores and glide energies significant interactions.

Complex	Docking score	Glide energy
PCA–KEAP1	−4.987	−26.192
PCA–TNF trimer	−6.923	−29.609

Table 4.4 Molecular docking analysis: distances between ligand-binding amino acids and atoms.

Docking complex	Bond + amino acid	Atom 1 (receptor)	Atom 2 (ligand)	Distance (Å)
PCA/KEAP1	Hydrogen bond with ARG 415	H:2916	O:4418	1.83
	Hydrogen bond with ARG 415	H:2920	O:4418	1.90
	Salt bridge with ARG 415	N712	O:4418	2.75
	Aromatic H bond with ILE 416	O:716	H:4429	2.53
	Aromatic H bond with VAL 463	O:1082	H:4429	2.74
	Hydrogen bond with ARG 512	H:3639	O:4417	2.79
	Aromatic H bond with LEU 557	O:1822	H:4432	2.68
	Aromatic H bond with LEU 557	O:1822	H:4430	2.77
PCA/TNF trimer	Aromatic H bond with LEU 120	O:790	H:6541	2.67
	Aromatic H bond with TRY 119	H:6035	H:6529	2.65
	Hydrogen bond with TRY 151	H:6267	O:6528	1.95
	Hydrogen bond with TRY 151	O:1039	H:6545	1.97

Binding free energy analysis via MM-GBSA: post-docking MM-GBSA calculations (Prime module, Schrödinger) quantified the thermodynamic stability of the complexes. For the PCA–KEAP1 system:

- **ΔG_bind:** −63.78 kcal/mol, dominated by van der Waals (−1,195.47 kcal/mol), lipophilic (−12,365.56 kcal/mol), and hydrogen bonding (−29.90 kcal/mol) contributions.
- The **electrostatic (Coulomb)** and **solvation** terms also contributed favorably (−37.74 and −35.36 kcal/mol, respectively).

For the PCA–TNF complex:

- **ΔG_bind:** −23.37 kcal/mol, with key contributions from **Coulomb** (−19.44 kcal/mol), **hydrophobic** (−13.11 kcal/mol), and **hydrogen bonding** (−1.37 kcal/mol) interactions.
- These data suggest PCA forms energetically favorable interactions with both proteins, with KEAP1 showing a comparatively higher binding affinity.

Pharmacophore modeling: pharmacophore features, including hydrogen bond donors (D), acceptors (A), hydrophobic centers (H), and aromatic rings (R), were identified and spatially mapped to reveal the structural compatibility of PCA with receptor binding pockets (Table 4.5). These features provide a foundation for rational ligand optimization.

MD simulations: to further validate the stability and behavior of these complexes under physiological conditions, **100 ns MD simulations** were performed using the **Desmond** engine. Key observations include the following:

- **RMSD profiles (Figure 4.3C1, D1):** indicate conformational stability of the complexes.
- **Protein–ligand contact analysis (Figure 4.3C2, D2):** highlighted persistent hydrogen bonds, hydrophobic, ionic, and water-mediated bridges.
- **Root-mean-square fluctuation (RMSF) analysis (Figure 4.3C3, D3):** showed fluctuation patterns within key binding regions, consistent with flexible docking sites.
- **Ligand interaction schematics (Figure 4.3C4, D4):** depicted atomistic details of interactions between PCA and amino acid residues.

Therefore, the in silico assessment presented herein reveals the dual modulatory potential of PCA on the **KEAP1–Nrf2 antioxidative pathway** and **TNF-mediated inflammatory signaling**. The combination of docking, MM-GBSA, pharmacophore modeling, and MD provides a multidimensional view of PCA's interaction mechanisms. These findings warrant further experimental validation and support the therapeutic prospects of PCA as a natural compound targeting oxidative stress and inflammation.

Table 4.5 Pharmacophore mapping analysis results (A = acceptor, D = donor, H = hydrophobic, N = negative ionic, P = positive ionic, and R = aromatic ring).

Features	PCA/KEAP1		PCA/TNF trimer		
Rank	1	2	1	2	3
Feature_label	N3	R4	R4	D2	N3
Score	−1.31	−0.60	−1.34	−0.66	−0.32
X	−90.9677	−91.329	−7.6272	−5.2617	−11.3591
Y	107.3013	102.4028	3.8024	6.0073	−0.6656
Z	−9.0726	−5.8461	−15.1842	−15.9108	−15.9599
Type	N	R	R	D	N
Num	2	3	3	1	2
From_chemscore	0	0	0	0	0
Source	HBond	RingChemscoreHphobe	RingChemscoreHphobe	HBond	HBond

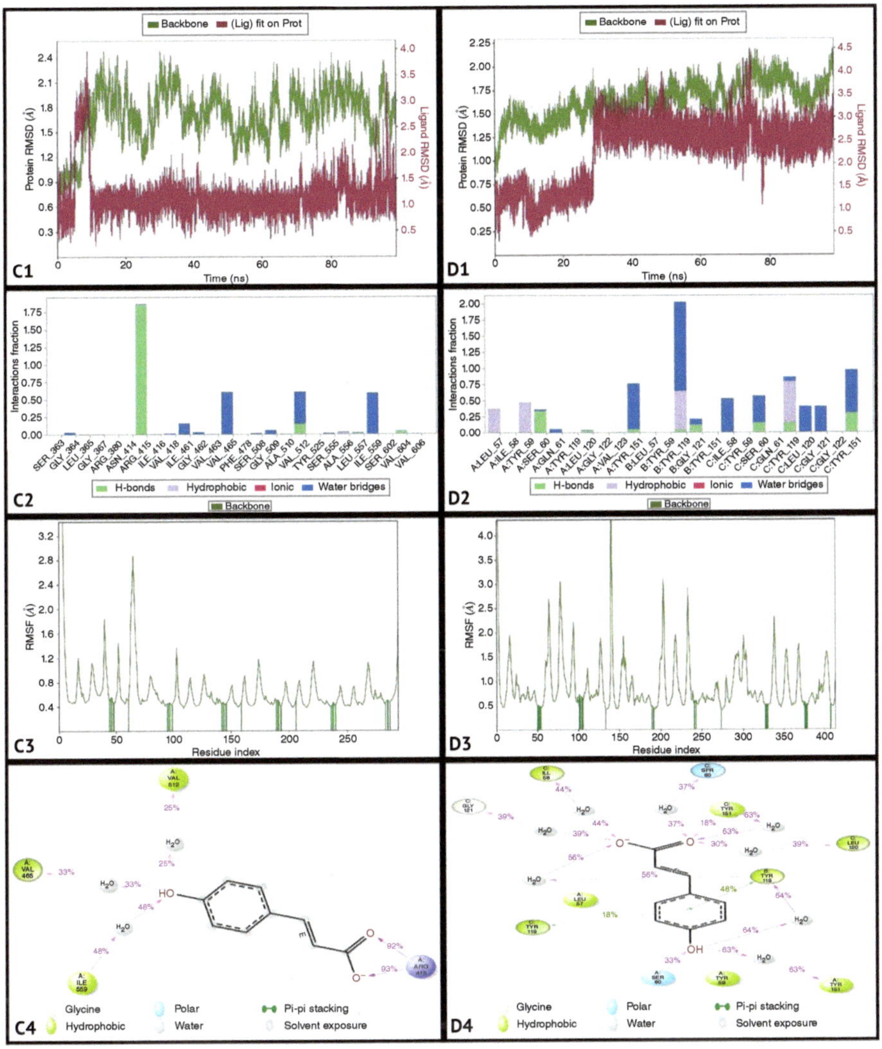

Figure 4.3 **C1:** PCA/NRF2, RMSD; **C2:** PCA/NRF2, protein–ligand interactions (hydrogen bonds, hydrophobic, ionic, and water bridges); **C3:** PCA/NRF2, RMSF; **C4:** PCA/NRF2, schematic of detailed ligand atom interactions with the protein residues; **D1:** PCA/TNF trimer, RMSD; **D2:** PCA/TNF trimer protein–ligand contacts (hydrogen bonds, hydrophobic, ionic, and water bridges); **D3:** PCA/TNF trimer, RMSF; **D4:** PCA/TNF trimer, a schematic of detailed ligand atom interactions with the protein residues. Images made with Schrödinger Maestro 2023/4 licensed version [9].

Challenges and Limitations

Explicitly modeling water molecules (e.g., using free energy perturbation or thermodynamic integration) offers theoretical rigor but is computationally impractical for large-scale virtual screening. Instead, implicit solvent models like PB/SA and GB/SA provide a balance between accuracy and efficiency. These methods approximate water as a dielectric continuum, calculating electrostatic desolvation penalties and hydrophobic contributions based on solvent-accessible surface area. For example, GB/SA-based approaches have been successfully applied in docking programs like SDOCK, where electrostatic, van der Waals, and hydrophobic terms are weighted to match experimental binding data. To further improve speed, multiscale GB methods prioritize accurate calculations for key atoms while approximating others, making them viable for high-throughput screening. A major challenge in FF scoring is combining different energy terms, as each is computed using distinct methods (e.g., coulombic electrostatics vs. Lennard-Jones potentials). Empirical weighting factors are often introduced to balance these contributions, but these coefficients are typically system-dependent and lack universal applicability. Additionally, entropic effects, such as conformational flexibility upon binding, are difficult to model accurately and can significantly impact predictions. Despite these challenges, FF scoring remains valuable, particularly when combined with implicit solvent models. Future improvements may focus on better integration of solvation, entropy, and multi-term weighting schemes to enhance predictive power while maintaining computational efficiency.

Solvent Modeling, Receptor Flexibility, and Entropy

Empirical scoring functions represent a widely used approach for estimating binding affinities in molecular docking studies. These functions employ a linear combination of weighted energy terms to approximate the free energy of binding (ΔG):

$$\Delta G = \Sigma W_{\mathrm{i}} \cdot \Delta G_{\mathrm{i}}$$

where ΔG_{i} corresponds to various interaction components and W_i represents empirically derived weighting coefficients. The energy terms typically include the following:

- van der Waals interactions
- electrostatic contributions
- hydrogen bonding effects
- desolvation penalties
- hydrophobic interactions
- entropic considerations (e.g., ligand flexibility)

Development and Optimization

The development of empirical scoring functions involves regression analysis against experimental binding data from protein–ligand complexes with known three-dimensional structures. Notable examples include the following:

SCORE1: developed by Böhm using 45 complexes, this early function incorporated hydrogen bonding, ionic interactions, lipophilic contacts, and ligand flexibility terms. Subsequent refinements expanded the training set to 82 complexes and introduced additional parameters for binding site electrostatics and aromatic interactions.

ChemScore: introduced by Eldridge et al., this function emphasized hydrogen bonding, metal coordination, lipophilic effects, and ligand flexibility. Validation was performed using independent test sets of 20 and 10 complexes.

X-Score: developed by Wang et al. [32] using an extensive training set of 200 complexes, this function demonstrated improved performance through comprehensive treatment of van der Waals, hydrogen bonding, hydrophobic, and flexibility terms.

Applications and Limitations

These functions have been successfully implemented in popular docking programs such as FlexX and Surflex, offering computational efficiency advantages over FF-based methods. However, several challenges remain, which are as follows:

Double-counting issues: the additive nature of empirical functions may lead to redundant consideration of certain interaction components.

Training set dependence: performance is often limited by the size and diversity of the training data, potentially affecting generalizability.

Term weighting: the empirical determination of coefficient weights may not accurately reflect true physical contributions.

Recent Advances

Modern approaches have begun integrating machine learning techniques with traditional empirical methods:

Knowledge-based quantitative structure–activity relationship (QSAR): incorporates atom-pair interaction potentials derived from structural databases, using machine learning to optimize parametrization.

RF-Score: a random forest-based approach demonstrating exceptional performance ($R = 0.953$) on a large training set of 1,105 complexes, highlighting the potential of data-driven methods.

These hybrid approaches benefit from both physical interpretability and the pattern recognition capabilities of machine learning, offering promising directions for future development of more accurate and transferable scoring functions.

Conclusion and Future Perspectives

Molecular docking has emerged as an indispensable computational tool in drug discovery, enabling the prediction of ligand–protein interactions, virtual screening of compound libraries, and rational drug design. Over the years, advancements in algorithms, scoring functions, and computational power have significantly improved docking accuracy, allowing researchers to identify potential drug candidates more efficiently. Molecular docking will advance through AI and machine learning, improving binding predictions and virtual screening.

Enhanced flexibility modeling will better simulate biological interactions, while quantum computing could revolutionize docking precision and speed. The approach will also expand into multi-target drug design for complex diseases like cancer. Finally, open-source platforms will democratize drug discovery, enabling global collaboration.

List of Abbreviations

AI	Artificial intelligence
BPA	Bisphenol A
COX-1	Cyclooxygenase-1
DPP-4	Dipeptidyl peptidase-4
EVOO	Extra virgin olive oil
GB/SA	Generalized Born surface area
HTVS	High throughput virtual screening
iNOS	Inducible nitric oxide synthase
KEAP1	Kelch-like ECH-associated protein 1
MD	Molecular dynamics
MM-GBSA	Molecular mechanics generalized Born surface area
MM-PBSA	Molecular mechanics Poisson–Boltzmann surface area
NF-κB	Nuclear factor kappa B
Nrf2	Nuclear factor erythroid 2-related factor 2
PCA	*p*-Coumaric acid
PDB	Protein Data Bank
PLIP	Protein–ligand interaction profiler
RMSD	Root-mean-square deviation
RMSF	Root-mean-square fluctuation
SBDD	Structure-based drug design
SP	Standard precision
TNF	Tumor necrosis factor
XP	Extra precision

References

1 Meli R, Morris GM, Biggin PC. Scoring functions for protein-ligand binding affinity prediction using structure-based deep learning: a review. *Front. Bioinform.* 2022;2:885983.

2 Sahu MK, Nayak AK, Hailemeskel B, Eyupoglu OE. Exploring recent updates on molecular docking: Types, method, application, limitation & future prospects. *Int. J. Pharm. Res. Allied Sci.* 2024;13(2):24–40.

3 Bayarsaikhan B, Zsidó BZ, Börzsei R, Hetényi C. Efficient refinement of complex structures of flexible histone peptides using post-docking molecular dynamics protocols. *Int. J. Mol. Sci.* 2024;25(11):5945.

4 Rastelli G, Pinzi L. Refinement and rescoring of virtual screening results. *Front. Chem.* 2019;7:498.

5 Roy K, Das RN, Ambure P, Aher RB. Be aware of error measures. Further studies on validation of predictive QSAR models. *Chemom. Intell. Lab. Syst.* 2016;152:18–33.

6 Agarwal S, Mehrotra R. An overview of molecular docking. *JSM Chem.* 2016;4(2):1024–8.

7 Tripathi A, Misra K. Molecular docking: A structure-based drug designing approach. *JSM Chem.* 2017;5(2):1042–7.

8 Shadidizaji A, Cinisli KT, Warda M, et al. Virtual insights into the quercetin-*Melampsora lini*-derived effector AvrM14 interaction: an in silico exploration of plant defense mechanisms. *Physiol. Mol. Plant Pathol.* 2024;129:102200.

9 Tekin S, Bolat M, Atasever A, et al. Mechanistic insights into the P-coumaric acid protection against bisphenol A-induced hepatotoxicity in in vivo and in silico models. *Sci. Rep.* 2025;15(1):11023.

10 Tekin S, Sengul E, Yildirim S, et al. Molecular insights into the antioxidative and anti-inflammatory effects of P-coumaric acid against bisphenol A-induced testicular injury: In vivo and in silico studies. *Reprod. Toxicol.* 2024;125:108579.

11 Bolat M, Tekin S, Bolat İ, et al. Gallic acid's protective mechanisms against acrylamide-induced pulmonary injury: in vivo and in silico insights into the Nrf-2/HO-1/NFκB pathway modulation. *Naunyn Schmiedebergs Arch. Pharmacol.* 2025;398(9):11821–37.

12 Gamal M, Awad MA, Shadidizaji A, et al. In vivo and in silico insights into the antidiabetic efficacy of EVOO and hydroxytyrosol in a rat model. *J. Nutr. Biochem.* 2025;135:109775.

13 Raval K, Ganatra T. Basics, types and applications of molecular docking: a review. *IP Int. J. Compr. Adv. Pharmacol.* 2022;7(1):12–6.

14 Friesner RA, Murphy RB, Repasky MP, et al. Extra precision glide: Docking and scoring incorporating a model of hydrophobic enclosure for protein–ligand complexes. *J. Med. Chem.* 2006;49(21):6177–96.

15 Dhasmana A, Singh S, Kadian S, Singh L. Skin tissue engineering: principles and advances. *J. Dermatol. Skin Care* 2018;1(1):1–11.

16 Tripathi NM, Bandyopadhyay A. High throughput virtual screening (HTVS) of peptide library: technological advancement in ligand discovery. *Eur. J. Med. Chem.* 2022;243:114766.

17 Hollingsworth SA, Dror RO. Molecular dynamics simulation for all. *Neuron* 2018;99(6):1129–43.

18 Dodaro A, Novello G, Menin S, et al. Post-docking refinement of peptide or protein-RNA complexes using thermal titration molecular dynamics (TTMD): a stability insight. *J. Chem. Inf. Model.* 2025;65(3):1441–52.

19 Genheden S, Ryde U. The MM/PBSA and MM/GBSA methods to estimate ligand-binding affinities. *Expert Opin. Drug Discov.* 2015;10(5):449–61.

20 Li J, Fu A, Zhang L. An overview of scoring functions used for protein–ligand interactions in molecular docking. *Interdiscip. Sci.* 2019;11(2):320–8.

21 Blanes-Mira C, Fernández-Aguado P, de Andrés-López J, et al. Comprehensive survey of consensus docking for high-throughput virtual screening. *Molecules* 2022;28(1):175.

22 Cheng T, Li X, Li Y, Liu Z, Wang R. Comparative assessment of scoring functions on a diverse test set. *J. Chem. Inf. Model.* 2009;49(4):1079–93.

23 Wang J-C, Lin J-H. Scoring functions for prediction of protein-ligand interactions. *Curr. Pharm. Des.* 2013;19(12):2174–82.

24 Kim S, Chen J, Cheng T, et al. PubChem 2023 update. *Nucleic Acids Res.* 2023; 51(D1):D1373–80.

25 Ultra14 C. ChemBioDraw Ultra14: PerkinElmer; 21 Nov. 2018. Available from: https://www.perkinelmer.com/product/ChemDraw-directChemDrawdi.

26 Berman HM, Westbrook J, Feng Z, et al. The protein data bank. *Nucleic Acids Res.* 2000;28(1):235–42.

27 Pettersen EF, Goddard TD, Huang CC, et al. UCSF Chimera – a visualization system for exploratory research and analysis. *J. Comput. Chem.* 2004;25(13):1605–12.

28 Shapovalov MV, Dunbrack RL. A smoothed backbone-dependent rotamer library for proteins derived from adaptive kernel density estimates and regressions. *Structure* 2011;19(6):844–58.

29 Tian W, Chen C, Lei X, et al. CASTp 3.0: computed atlas of surface topography of proteins. *Nucleic Acids Res.* 2018; 46(W1):W363–7.

30 Schrödinger. *Schrödinger Release 2023-1.* Available from: https://www.schrodinger.com/life-science/download/release-notes/release-2023-1/

31 Adasme MF, Linnemann KL, Bolz SN, et al. PLIP 2021: expanding the scope of the protein–ligand interaction profiler to DNA and RNA. *Nucleic Acids Res.* 2021; 49(W1):W530–4.

32 Wang C, Chao C, Yu J, et al. Mechanisms underlying the formation of amylose–lauric acid–β-lactoglobulin complexes: experimental and molecular dynamics studies. *J. Agric. Food Chem.* 2022;70(34):10635–43.

5

Computational Approaches to Drug–Receptor Binding Analysis

*Rahul Pandey[1], Ujjwal Nautiyal[1], Abhishek Chandola[1], and Firoz Anwar[2]**

[1]*School of Pharmaceutical Sciences, Swami Rama Himalayan University, Dehradun, Uttarakhand, India*
[2]*Department of Biochemistry, Faculty of Sciences, King Abdulaziz University, Jeddah, Saudi Arabia*

*Corresponding author: fanwar1@kau.edu.sa

Introduction

Finding and developing new therapeutic molecules for curing a disease is essentially a complex, costly, and time-taking process [1]. Investments worth billions of dollars are required for such processes, which usually span more than a decade [2]. Such processes are associated with a very low success rate. Out of 20,000–30,000 molecules screened, only one molecule comes forward as an acceptable marketed drug [3, 4]. Computational drug design (CDD) evolved as a promising solution to the drug discovery-related pharmaceutical research. Initially conceived as a supporting tool, CDD has become a central component in recognition, enhancement, and evaluation of drug candidate molecules. By leveraging computational power and data-driven methodologies, this field has greatly enhanced our ability to predict drug behavior, assess biological interactions, and optimize chemical structures long before experimental validation. The journey of CDD first covers an early phase of the pre-2000s where basic molecular modeling and structure–activity relationship (SAR) studies were utilized. Researchers were heavily dependent on known compounds, and new drugs were discovered by structural modifications of the older ones, forming the basis of ligand-based (LB) drug design (LBDD). Structural data of the target proteins were very limited and were often treated as rigid structures [5, 6]. After this, transitional phase between 2000 and 2010 followed with the availableness of crystal structures of protein targets, sometimes co-crystallized with ligand leading to the emergence of structure-based drug design (SBDD). Improvements in molecular docking methods and virtual screening tools significantly enhanced the authenticity, speed, and reliability of identifying potential therapeutic candidates, allowing researchers to efficiently prioritize compounds for experimental validation [7, 8]. Broader application of quantitative structure–activity relationships (QSARs) and pharmacophore modeling has enabled the prediction of biological activity by analyzing molecular arrangement, designing, and improvising on new chemical entity with improved efficacy and safety. The modern phase (2015–present) has been marked by the emergence of sophisticated computational techniques in drug discovery integrating AI, ML, and deep learning tools for drug discovery processes. Molecular dynamics (MD) modeling aided in exploring the flexible and time-dependent behavior of receptor–ligand interactions [9, 10]. Cloud computing and high-performance computing (HPC) proved to be of immense support for large-scale screening of drug candidates [11]. AI tools such as AlphaFold have revolutionized prediction of protein structures with remarkable accuracy [12].

Advances in Drug–Receptor Binding Analysis

A central theme in CDD is the analysis of how drug molecules interact with their biological targets, typically proteins. Over time, drug–receptor binding analysis has undergone a major transformation in the sense that earlier techniques relied on rigid docking with simplified scoring functions which were empirical (ChemScore, LUDI), knowledge-based (PMF, DrugScore), and force-field-based (DOCK, Dockthor) [13] but such scoring functions had limited ability to capture conformational changes and induced fit mechanisms [14, 15]. Current practices involved adoption of flexible docking, ensemble docking, induced fit docking, and use of MD simulations to observe binding kinetics, conformational flexibility, and residence time. Enhanced binding affinity prediction is achieved through free energy calculations (MM/Poisson–Boltzmann surface area [PBSA], FEP+) [16]. Nevertheless, AI-based scoring models trained on large datasets ensure improved predictive accuracy during screening.

Evolution of In Silico Techniques

The way computational methods are used to find new medicines has seen significant advancements, particularly in the last decade. Key transitions are covered in Table 5.1.

These improvements have not only sped the procedure of finding innovative medicines but have also led to the precision and reliability of computational predictions.

Table 5.1 Key transitions from conventional in silico drug discovery approaches to AI-enabled and integrative frameworks.

Earlier approaches	Current techniques (2015–2025)
Rule-based QSAR models	AI-driven predictive models (CNNs, GNNs, LSTMs)
Small-scale virtual screening	Large-scale and ultra-large-scale virtual screening
Static molecular docking	MD, ensemble docking, and induced fit methodologies
Isolated ligand or receptor modeling	System-level simulations incorporating network pharmacology
Limited structural data	Protein structure prediction using AI (e.g., AlphaFold)
Ligand-focused modeling	Multi-omics integration (genomics, proteomics, transcriptomics)

CDD has evolved from isolated, structure- or ligand-centric models to comprehensive, system-level, and AI-driven paradigms (Figure 5.1). Today, CDD supports the development of personalized therapies and facilitates drug repositioning, contributing toward rational design of safer and increased effective medications. The overall workflow and integration of modern computational tools within the drug discovery pipeline are depicted in Figure 5.2.

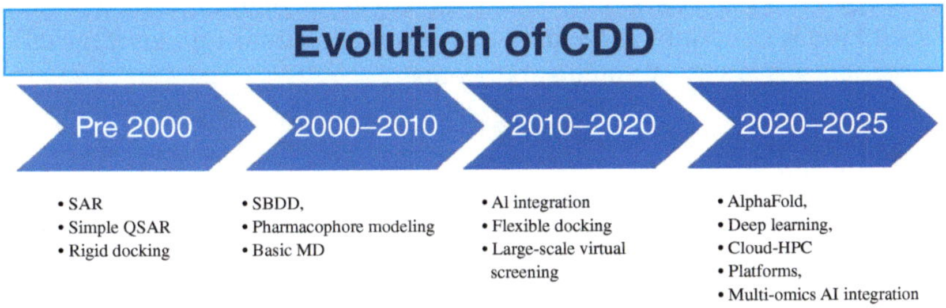

Figure 5.1 Evolution of CDD.

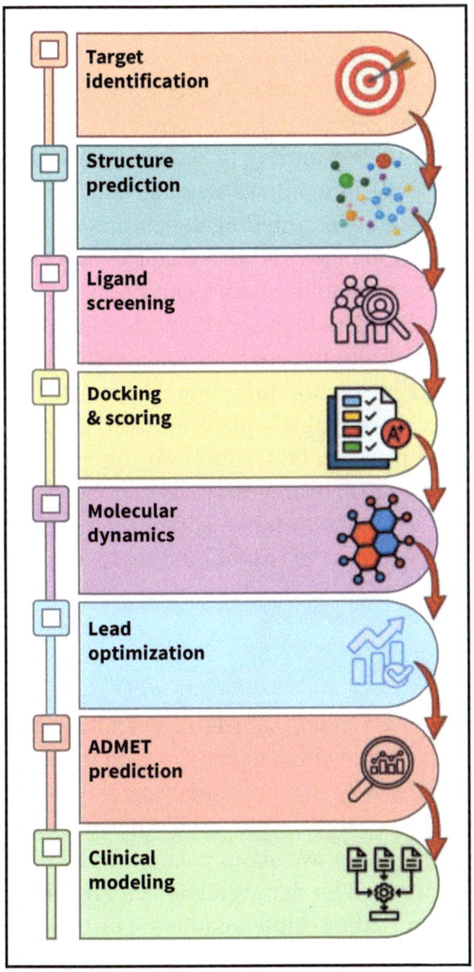

Figure 5.2 CDD process pipeline.

Fundamentals of Drug–Receptor Interactions

Earlier, QSAR methods were applied to understand how drugs interact with their biological targets, which is in fact considered central to rational drug design. However, when the way in which drugs produce their effects is examined, very precise interactions at the molecular level are observed. This is because as a drug gets closer to its target receptor site, changes are triggered at the molecular level [17]. A study of such kind is important as the drug–receptor interactions influence the efficacy, potency, and selectivity of therapeutic agents.

Binding Affinity and Specificity

Binding affinity refers to how strongly a molecule attaches with the action site which is typically a protein receptor. It is quantitatively expressed by K_d. A lower K_d indicates higher affinity which usually translates to effective target engagement at lower drug concentrations. Both labeled [18] and label-free [19] approaches can be applied to measure binding affinity. The labeled approach includes qualitative method like enzyme-linked immunosorbent assay (ELISA). On the other hand, label-free approaches are quantitative in nature and include spectroscopic assays, ITC, or the use of optical biosensors such as SPR, BLI, and GCI. Recently, a variety of innovative affinity-based techniques have been introduced that complement computational predictions and augment our understanding of molecular binding mechanisms. Notably, mass spectrometry (MS) now allows detailed analysis of intact protein–drug complexes, providing insights into binding stoichiometry and kinetics as well. In-cell nuclear magnetic resonance (NMR) enables real-time observation of drug interactions within living cells under native conditions, while surface-enhanced Raman spectroscopy (SERS) offers high sensitivity for detecting binding events. Emerging technologies like affinity magnetic levitation (MagLev) and receptor chromatography provide novel means to study interactions in biomimetic environments, which indeed gives more realistic picture of the process. Additionally, electrochemical and magneto-resistive biosensors are increasingly being used for label-free, real-time monitoring of ligand–receptor binding [20]. Specificity describes the capability of a drug to bind to its target without interacting with other similar proteins. High specificity reduces the risk of binding to another target, which can lead to side effects or toxicity [21].

Types of Molecular Forces

How a drug molecule connects with its intended biological site is typically mediated by noncovalent forces. A prominent one is the hydrogen bond, which is strong and depicts directional interactions between hydrogen donors and acceptors. Ionic interactions happen because opposite electrical charges attract each other, e.g., carboxylate and amine. Hydrophobic interactions play between nonpolar groups which cluster together to minimize contact with water. Van der Waals forces are weak, distance-dependent pulls among atoms. π-π and cation-π interactions are prominent among aromatic rings and positively charged groups [22]. The complementarity of the ligand's functional groups with the receptor's binding site confirms how well such interactions will take place. This will favor effective drug design which aims to maximize favorable interactions while minimizing steric or electrostatic clashes. Pharmacophore modeling is based on these interactions, and implementation of evaluation parameters like scoring functions ascertains degree of binding.

Receptor Flexibility and Allosteric Sites

Earlier, models for drug–receptor interaction study depicted receptors as rigid structures; however, in reality, proteins are rather dynamic and flexible. They can adopt multiple conformations which ultimately influence binding behavior. So, when a ligand approaches a receptor, a conformational change is induced first, while there are other ligands as well that bind selectively to a specific conformation which is already present [23]. Moreover, receptors can also have alternative secondary binding locations, which are separate from the main binding site (called the orthosteric site). Molecules (ligands) can attach to these allosteric sites and change how the receptor works. This happens without the molecule directly blocking the receptor's primary function [24]. Drugs that fall in the category of allosteric modulators are able to augment or lessen the receptor's reaction without directly vying for the primary site with endogenous ligands, thus offering increased specificity and fewer side effects. Allosteric modulation has become a transformative strategy in drug discovery, particularly for complex protein families like GPCRs, kinases, and phospholipases [25]. Recently, MK-2206, an allosteric kinase inhibitor, which selectively targets the PH domain of Akt kinase and locks it in an inactive conformation was developed. Similarly, VU0359595 and VU0364739 are examples of such allosteric inhibitors of phospholipase D isoforms PLD1 and PLD2, respectively. A high isozyme selectivity was observed. These agents demonstrate how allosteric ligands can stabilize inactive protein states, prevent undesired phosphorylation, and provide novel mechanisms for therapeutic intervention. The development of drugs that target these alternative (allosteric) sites has created new possibilities for treating diseases such as cancer, disorders of CNS, and viral infections [26].

Thermodynamic and Kinetic Considerations

Drug binding could be seen from both thermodynamics and kinetics points of view. Thermodynamics specifies whether a ligand would bind to its target or not and, if so, then with what strength? The main parameter is the binding free energy (ΔG), which depends on enthalpy (ΔH) and entropy ($-T\Delta S$). Stronger binding is associated with more negative values of ΔG. Usually, computational approaches are used to calculate this value, including MM/PBSA, MM/generalized Born surface area (GBSA), and free energy perturbation (FEP) [27]. The other consideration is the kinetics of the drug–receptor interaction, which describes how fast the binding–unbinding process will occur. k_{on} and k_{off} characterize the overall binding kinetics and define the length of time a drug occupies its binding site. These parameters are increasingly being appreciated for their importance in predicting in vivo efficacy. Long residence times are often associated with prolonged therapeutic action, thus requiring thermodynamic and kinetic evaluations to be considered in the modern drug discovery paradigm [28]. While docking provides a static view of binding, MD simulations and Markov state models offer insights into kinetic behavior and conformational transitions, helping to assess drug performance more realistically [29].

Molecular Docking

Molecular docking is the mainstay for SBDD, allowing the prediction of interactions between a small molecule (ligand) and a biological macromolecule [30]. The aim of docking is to find the best binding orientation and the concurrent potential of the ligand toward its target to help a researcher prioritize compounds for synthesis and other evaluations. Due to its efficiency and wide applicability, molecular docking is being used extensively in virtual screening, lead optimization, and drug repurposing [31].

Principles and Workflow

The core of molecular docking envisions its role as a model for the prediction of interaction between a ligand and the receptor and determines the best orientation of the ligand and the strength of interaction. Traditionally, the process is based on a mere analogy of a key fitting into a lock of a receptor's binding pocket. Previously considered as rigid, but now some novel concepts of docking include, albeit lightly, the flexibilities of the molecules [32]. More conventional docking processes are conducted in a series of preparatory steps comprising receptor and ligand structure preparation, finding the binding site for the docking simulation by way of some search algorithms that traverse through ligand conformation possibilities, scoring the resulting poses based on predicted binding affinity, and ranking and subsequent analysis of the result sets. This then predicts the binding geometry and estimated binding energy, thereby allowing the prioritization of compounds for further development based on those predictions [33].

Ligand and Receptor Preparation

Accurate and reliable results with molecular docking can only be given through proper preparation of the ligand as well as the receptor. Receptor preparation means the treatment done for cleaning the protein structure with the removal of crystallographic water molecules, the addition of missing hydrogen atoms, the assignment of appropriate protonation states besides clearly defining the binding site whether from experimentally known ligand-bound structures or binding pocket prediction tools [34]. Ligand preparation consists of generation of accurate 3D conformations, molecular geometrical optimization, bond order assignments, and identification of possible tautomeric states and ionization states. A few interesting software tools used in these programs are AutoDock Tools, Schrödinger's LigPrep, and Open Babel. Lack or inadequate preparation of the receptor or ligand will yield faulty docking predictions that, in turn, make virtual screening and lead identification efforts unreliable.

Docking Algorithms and Search Methods

Dockers utilize several algorithms to assess the many possible orientations and conformations that a ligand may adopt in the receptor binding site [35]. These algorithms include systematic searches, in which exhaustive sampling of ligand poses is undertaken; stochastic methods like Monte Carlo simulations and genetic algorithms that rely on random sampling strategies to probe conformational space incrementally; and construction, in which the ligand is built up incrementally in the binding site, as in Flexx [36]. Popular docking programs implementing these strategies are AutoDock and AutoDock Vina, which use a Lamarckian genetic algorithm; Glide (Schrödinger), which puts in place proprietary exhaustive sampling techniques; GOLD, based on genetic algorithms; and DOCK, one of the pioneering tools using rigid-body docking [37]. These algorithms are specifically designed to efficiently traverse the complex and high-dimensional conformational space of ligand–receptor interactions to find binding poses that are geometrically and energetically favorably oriented.

Scoring Functions: Types and Limitations

Scoring functions are mathematical methods that estimate the bond tightness with ligand to its acceptor organ through molecular docking, thus allowing the ranking of predicted binding poses in order of their probable biological relevance [38]. Scoring functions can commonly be grouped into three large categories: force-field-based, which computes interaction energies via principles of molecular mechanics (MM) based on van der Waals and electrostatic forces; empirical, which depend on parameters derived from experimental data such as hydrogen bonding, desolvation, and entropy; and knowledge-based, which apply statistical potentials derived from databases of information containing known protein–ligand complexes [39]. Scoring functions are important for ranking candidate molecules but also have considerable limitations [40]. In particular, scoring functions may simplify complex interactions and fail to properly accommodate protein flexibility, solvent effects, and other variables, sometimes giving a false positive result, while ignoring an active compound. Careful consideration must be employed in interpreting docking and scoring results, particularly in conjunction with validation via accurate methods such as MD simulations or experimental binding assays.

Case Studies Using Docking

From the onset of molecular docking, it has been applied to several areas of therapy, confirming its worth as a tool to generate hypotheses and for early-stage drug discovery. In the case of HIV, very early docking studies aided the development of key protease inhibitors such as ritonavir and indinavir by modeling their interactions with the viral protease enzyme [41]. Docking has been instrumental in the design of EGFR kinase inhibitors such as erlotinib targeting the ATP-binding site of the receptor [42]. In the initial months of the COVID-19 pandemic, molecular docking enabled rapid screening of approved drug libraries against SARS-CoV-2 main protease (Mpro), which facilitated the identification of possible candidates for drug repurposing [43]. Basically, in fighting tuberculosis, virtual screening and docking were used in discovering novel inhibitors of DprE1, whose gene products are indispensable for the survival of *Mycobacterium tuberculosis* [44]. Though some of these cases may ultimately require further validation in light of findings that support biological relevance, they illustrate the versatility and significance of docking in lead identification and drug development acceleration.

MD Simulations

MD simulation is one of the key tools to understand the behavior of biomolecular systems at an atomic level. MD, unlike static modeling methods, provides a view of the movement and interaction of molecules over time and hence becomes a useful tool when assessing drug–receptor interactions [45]. In recent years, the importance of MD simulations in molecular biology and drug discovery has grown tremendously. Growing speed, accuracy, and wealth of software accessibility in combination with an explosion of experimental structural data have further increased the popularity of biomolecular simulation [46].

Basics of MD

An understanding of drug–receptor interaction is dependent on an understanding of how those atoms and molecules behave at the atomic level. Such an understanding is of immense help and enables us to understand how the biomolecule works. One can very hardly observe individual atoms moving in complex molecules such as proteins or DNA directly in real time. Also, manipulating those atoms or controlling them in a desired manner is extremely difficult. In this regard, MD is useful. Rather than attempting to perform such experiments, simulations can be computed that model the behavior of atoms within molecules. It would be similar to an atomic-level virtual microscope. MD simulations, for example, give insight into the detailed motion and interaction of atoms. It can be described as an atomistic simulation of the movements of atoms and molecules with time [47]. That is why it is said that MD produces the "movie" of the molecular system's behavior wherein the regularities hold true since the very atoms are said to be constantly in motion. Hence, the molecular function and the intermolecular interactions would, therefore, depend on the dynamics of the atoms and the molecules involved. The dynamics of the system involve a numerical solution of the Newtonian classical equations of motion for the moving molecules and atoms. MD simulations estimate the movement of each atom within a protein or molecular system over time, using a general physical model that describes how atoms interact with one another [48]. These MD simulations are mainly applied to check the stability of the drug (ligand) bound to the target protein under near-physiological conditions (e.g., in water, at body temperature, with ions). For example, after docking a small molecule inhibitor into the active site of an enzyme, a 100-ns MD simulation could be performed to check for tight binding of the ligand or to observe if it starts to drift away from the binding pocket [49]. During this simulation, important parameters, such as the RMSD and hydrogen bond interactions, are monitored to evaluate the sturdiness of the complex over time. Strong and stable binding would be indicated if the ligand stays and keeps forming important interactions with the protein. Significant movement or detachment of the ligand could indicate weak or adverse binding, thereby providing guidance for the next steps toward experimental validation.

Force Fields and Simulation Protocols

In MD simulations, force fields characterize how atoms and molecules will interact with one another. A force field consists of mathematical equations and the associated parameters that define the potential energy of a molecular system. Such equations describe how the energy of a system depends on the coordinates of its particles. It consists of the interatomic potential energy, $U(r_1, r_2, ..., r_N)$, where r_1–r_N represent the particles in the system, and of a set of parameters entering into this form. Such parameters are usually derived either from purely ab initio or semiempirical quantum mechanical calculations or fitted to experimental data [50]. These values are obtained through hands-on experiments and by using computer simulations that apply the principles of quantum mechanics (QM), and they represent the bonded and nonbonded interactions between atoms. Force fields describe in MD simulations potential energy of a molecular system, considering both bonded and nonbonded interactions. Bonded interactions include bond stretching (an energy change when the lengths of bonds deviate from the equilibrium), angle bending (an energy change when the bond angles deviate), and dihedral or torsional angles (an energy change due to rotation around bonds). Nonbonded interactions include weak attractions or repulsions between molecules, called van der Waals forces, and electrostatic interaction, which refers to the attractive or repulsive force between charged atoms, modeled generally according to Coulomb's law [50, 51]. The most often used force fields in biomolecular simulations are AMBER, CHARMM, GROMOS, and OPLS [52]. Each of these force fields is optimized for a special type of molecules; for instance, AMBER is suitable for proteins and nucleic acids, while the OPLS force field is mainly used to study small and drug-like molecules [53]. Following a well-established workflow synchronizes and ensures the physical and dynamical stability of the system, as well as the reliability of the data generated for further analysis (Figure 5.3). The workflow starts with the preparation of the system, which consists of inputting the three-dimensional model of the protein (taken from the Protein Data Bank), adding the ligand/drug candidates, solvating with water solvents (generally in a cubic or octahedral box), and adding counterions

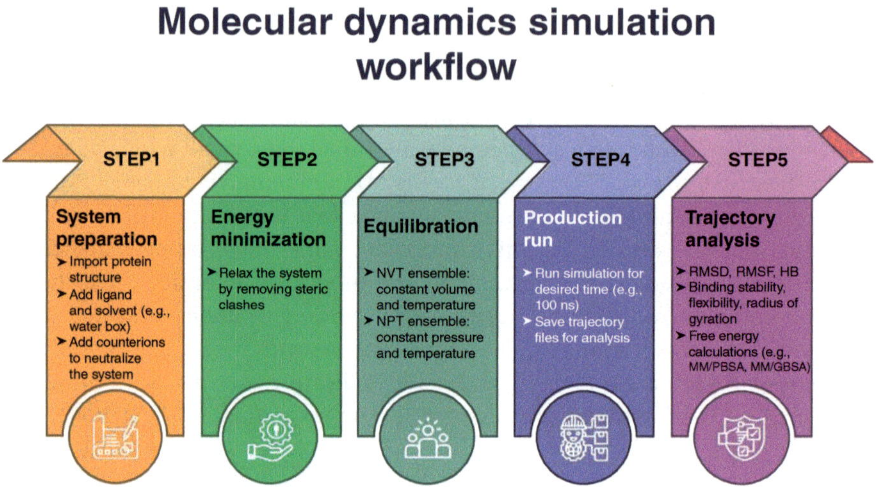

Figure 5.3 MD simulation workflow.

(like Na^+ or Cl^-) to neutralize the system. This is carried out first by energy minimization, a procedure for relaxing the structure by getting rid of steric clashes or unsatisfactory contacts. The next procedure is equilibration, which is performed in two stages: NVT (number of particles, volume, and temperature stay constant) ensemble is used to gradually heat the system to the desired temperature (300 K), while at NPT (constant pressure/temperature) ensemble, the pressure is held constant at 1 atm to allow the system to reach a realistic density. Afterward, the production run is conducted, whereby atomic positions and velocities are updated over a time scale of tens to hundreds of nanoseconds, and resulting trajectories are saved for later analysis. In the final analysis phase, various molecular properties are analyzed, including RMSD to obtain structural stability, RMSF to examine residue flexibility, R_g to check for tightness, and hydrogen bond analysis to unravel key interactions between the drug and receptor [54].

MD simulation runs on many software tools that have their own sets of strengths and idiosyncrasies. GROMACS is a high-performance open-source package renowned for protein–ligand system simulations [49]. Highly regarded for its speed and flexibility, GROMACS has other uses, too. AMBER consists of a comprehensive set of tools for biomolecular simulation with excellent force-field integration and wide applicability to proteins and nucleic acids [55]. For large-system applications in parallel computing environments, nanoscale molecular dynamics (NAMD) is well prepared and is therefore well suited for HPC clusters [56]. Desmond comes as part of Schrödinger Suite; it is well known for its high efficiency and user-friendly interface, especially regarding drug discovery workflows [57]. While these packages differ in terms of their formats and particular workflows, the basic principles of MD simulations, system preparation, energy minimization, equilibration, and production remain the same. With the increase in computational potential, MD simulation is gaining its due place as a pertinent tool in modern drug design, providing relevant atomic-level information about molecular interactions.

Applications in Studying Binding Stability and Flexibility

The power of MD modeling in drug determination application depends on its ability in evaluating the binding strength as well as flexibility of protein–ligand association. MD timing simulations create time-resolved and atomistic grounds for understanding the connection of ligand and target under conditions quite near physiological. Therefore, stability of binding as per time is studied, with how flexible or dynamic the binding site could be as a response to the ligand binding. Today, research activities are particularly effective in using static molecular docking in finding out the potential binding poses, which are snapshots that do not reveal dynamic nature. Binding stability MD simulation studies ascertain whether a drug candidate is continuously bound to target during the time of the simulation. RMSD of the ligand and protein backbone can be evaluated to understand structural stability of the complex [49]. If the ligand RMSD is low and stable, such cases usually indicate a very strong and persistent binding. The contrary is true in the cases where the ligand keeps shifting, drifting outside the pocket of binding, or shows excessive rotation; this may suggest weak or nonspecific binding. Analysis of H-bond is another useful data, which correlates the development and duration of key H-bond between the ligand and active site residues. Stable hydrogen bonding over time increases the probability of biologically relevant interactions. Similarly, interaction energy profiles, calculated during the course of simulation, quantify the affinity of binding in terms of van der Waals and electrostatic contributions [58]. The MD simulations also indicate that the receptor has a certain flexibility, which may have been lost during the determination of crystallographic structures [59]. Proteins should not be understood as rigid entities that either occur in pairs or not at all. Rather, they consist of subtle or occasionally large-scale movements that can affect the parameters of ligands' binding. All of these distinct types of motion can be surveyed using an MD experiment to watch side chains undergoing rotations, loops shifting, and even domain migrations, thus contributing to the whole binding mechanism. Turns out that strong insights can take primacy over allosteric site targets since some of them could probably be hidden or may only show when proteins conform to certain specific fractions. RMSF analysis is another technique to measure the flexibility of a residue from the amino acid level to simulations. From the RMSF values, a high value usually refers to flexible loops or unstructured areas, while a low value indicates rigidly stable regions [60]. Such information is very useful in determining modes of binding through induced fit and may also reveal cryptic pockets that would only emerge under conditions of dynamics. MD has been effectively used in various drug development projects to winnow docking hits, identify false positives, and prioritize molecules for experimental validation. For instance, MD studies were used for kinase inhibitors or GPCR-targeting compounds to gain insight on key water-mediated interactions or rearrangements in binding pocket geometry that could not have been predicted by docking alone [61]. In such cases, MD results guided medicinal chemists to optimize lead compounds with respect to dynamic features of the binding site.

Post-docking Validation and Refinement

Establish durability of docked entities using post-docking MD models to observe how the ligand behaves at the active site during simulation time. A valid binding pose is expected to relatively stabilize throughout the simulation with respect to corresponding key active site residues. To monitor such behavior, RMSD and H-bond analysis are employed [62]. If the ligand shifts significantly or exits the binding pocket during the simulation, that condition is most likely indicative of an unreliable docking prediction. Additionally, MD should expose docks or interactions unfathomable by docking, such as water-mediated hydrogen bonds, access to allosteric sites, and protein flexibility that enables binding. Such evidence would help to identify the best possible binding mode among numerous docked conformations. MD simulations also play their role to somewhat refine the docked ligand conformations beyond validation. While the simulation occurs, the ligand may adjust slightly to form more favorable interactions, while the protein might go through some minor conformational changes to better accommodate the ligand: this is called induced fit. These structures, refined after the previous steps, tend to depict more faithfully the true binding mode and provide a better lead compound in the design strategy [63]. In many cases, after MD simulations, investigators perform binding free-energy calculations based on MM/PBSA or MM/GBSA methods [64]. These methods could have a greater estimation of binding affinity as compared to docking scores alone and might consider solvation and some entropic contributions typically ignored in docking. Although virtual screening provides a long list of potential molecular hits, most potential hits might not bind well in a dynamic sense. MD simulations act as an effective filter for false positives by examining the stability and quality of interactions of each top-scoring compound to substantially increase confidence in hit selection prior to expensive and lengthy experimental validation [65–69]. Hence, MD simulation extends classical docking approaches by allowing the experimentalist to assess the reliability of predicted binding poses and stabilize the accuracy of ligand–receptor models. Post-docking validation and refinement not only mitigate errors in virtual screening but also guide the rational design of more potent and selective drug candidates by complementing the information regarding the dynamic molecular behavior.

QM/MM and Free Energy Calculations

Precise prediction of binding affinities and molecular interactions is a key component in present-day computational drug discovery. Classical methods derived from MM are useful to simulate large biomolecular systems, but may lack the precision necessary to describe processes that involve bond breakage, charge transfer, or electronic polarization. Therefore, hybrid QM/MM methods [70], coupled with free-energy calculation techniques such as FEP [71] and MM/PBSA [72], have arisen to bridge the gap conferred by MM. Such methods penetrate the mechanisms of drug–receptor interactions with much better accuracy and thus aid in rational drug design with high predictive capability.

Introduction to QM/MM Hybrid Methods

QM/MM hybrid methods enable the QM-accrued refinement to be weighed against the classical MM efficiency by offering the possibility for chemical processes to be simulated in very complicated biological systems with reasonable costs [73]. In a normal QM/MM formation, QM calculations are mainly invoked for the region of interest, considered usually as the target site of the enzyme or the docking interface where a drug and receptor interact. This accounts for accurate quantum descriptions, say, during bond formation or cleavage, electron delocalization, and charge transfer. The remainder of the system, which may contain thousands of atoms (like protein surroundings, solvent molecules, and ions), basically is being treated under MM. The QM and MM interaction along the boundary is strategically controlled such that consistency is maintained in energy calculations across the boundary. This approach finds immense value in the study of enzyme mechanisms, reaction pathways, and drug metabolism scenarios, where the capture of electronic detail is very critical. This method is also being applied increasingly for lead optimization when all the classical force fields fail to capture the subtle electronic effects [74].

FEP and MM/PBSA Approaches

Free energy calculations are a very important part of ligand–receptor binding quantification, making it very important for prioritization of the ligands for synthesis and testing. FEP constitutes a process of transferring one molecule to another under the binding site, and in solution, one ligand is usually the other conversion. Critical values for measuring such states directly relate to relative binding affinities; hence, the free energy difference between these two states. Efficiently computing an FEP free energy change needs many computer simulations and very high-quality force-field parameters [75]. MM/PBSA and the MM/GBSA version, generalized Born solvent were both implemented as endpoint methods, where binding free energies could be derived from MD trajectory snapshots. Decomposing the resultant binding energy in this regard were the categories: MM energy terms, solvation energies, and entropic contributions. While costs in computing are lesser than those of FEP, MM/PBSA would still have meaningful application as accurate estimates and are used in virtual screening and lead optimization workflows [76]. These dual approaches supplement each other where a more analytical FEP is found, but MM-PBSA provides the right balance between accuracy and computational expense.

Role in Predicting Accurate Binding Energies

Binding energy prediction is enhanced through QM/MM and free energy methods. This element plays a vital role in both picking out drug prospects and refining them. Unlike docking scores, often semiempirical and qualitative, this quantitative analysis affords high accuracy that reflects realistic biological interaction. They are particularly useful in distinguishing real binders from false positives in virtual screening, in probing energetic contributions of specific residues or ligand modifications, supporting SAR studies, and evaluating how mutations affect binding affinity for resistance prediction. Thus, these methods will be effective to provide a rough estimation on how tightly the drug binds its target and will ultimately lead to better selection of the most promising compounds, thereby reducing trial and error in experimentation and speeding up the drug development [77].

Pharmacophore Modeling and Virtual Screening

Pharmacophore modeling and virtual screening are phases of drug discovery that help to figure out or develop what compounds would likely bind to a biological target [78]. These features serve as the right forms of molecular recognition requirements in linking structural and ligand information for in silico high-throughput screening of large chemical libraries. More importantly, using these methods involves a far lower expenditure than HTS and cut significantly into the time required to experiment and to get lead candidates identified [79].

Pharmacophore Concept and Generation

A pharmacophore is a spatial arrangement of important essential chemical features that is required by a molecule to interact with a particular biological target and exert a desired biological effect [80]. These features usually include hydrogen bond donors and acceptors, aromatic rings, hydrophobic centers, and charged groups, and the concept is based on the functional groups responsible for activity rather than the overall chemical structure [81]. Thus, pharmacophore models are developed by evaluating a number of known active compounds for recognition and its critical sharing for binding through computational algorithms that are generable as a 3D template for designing or screening novel compounds to exhibit biological activity. This includes selecting active ligands or complexes of protein–ligand, feature extraction and alignment, and model refining and validation using active and inactive compounds. Current techniques to enhance pharmacophore model generation, visualization, and virtual screening are the widely known LigandScout, PharmaGist, MOE, and Discovery Studio tools [81–84].

Ligand-based vs. Structure-based Pharmacophore Models

Pharmacophore modeling can be broadly categorized into LB and structure-based (SB) approaches depending on the particular source of input data [85]. Ligand-based pharmacophore (LBP) studies are limited on the analysis of known active compounds due to the unavailability of the target protein; superimposed molecules will be analyzed in order to find common pharmacophoric features, and this is especially useful when there are several active compounds that are structurally diverse. The structure-based pharmacophore (SBP) approach directly extracts pharmacophore features from the ligand-binding site based on the known 3D structure of the specific target protein, obtained through techniques such as X-ray crystallography or AlphaFold predictions [86], giving a better and far more detailed picture of molecular interactions [78]. Each has its own merits: if there is no structural information provided then LBP is suitable, while SBP allows for an understanding of molecular interactions specific to protein–ligand interactions. In many cases, hybrid approaches incorporating both techniques are employed to improve the predictiveness and reliability of models [87].

Screening of Compound Libraries

The established pharmacophore model can be used to perform computational searches of extensive collections of molecules to identify those that possess the specific characteristics described earlier [88]. This pseudo-research typically involves preparing a conformational database of compounds, taking into consideration their tautomers and stereoisomers, then filtering them against the pharmacophore model and scoring them according to fit quality and other pertinent physicochemical parameters. Pharmacophore-based virtual screening is very efficient, narrowing millions of possible candidates to a small focused list of potential hits and further evaluation by molecular docking or MD simulations or experimental validation [89–91]. Pharmacophore modeling has led to the discovery of successful leads in different areas such as oncology, infectious disease, and neurological disorders [2]. Pharmacophore modeling will continue to boost and streamline all these hit-identification and lead-optimization processes as a major part of modern drug discovery pipelines.

Artificial Intelligence and Machine Learning in Binding Analysis

With the integration of AI and ML into drug discovery, binding analyses have been propelled significantly forward by predictive modeling, pattern recognition, and data-driven decision-making [92]. These technologies offer a disruptive way to understand and predict DTIs, recognize binding sites, and calculate binding affinity with greater speed and precision than classical methods [9]. The union of AI with computational biology is thereby changing the paradigm of rational drug design.

Integration of AI in Binding Prediction

Deep understanding-based tools for predicting the interactions between ligands and target proteins are now becoming common using vast databases from chemical and biological repositories. Models are trained on known drug–target interactions and can reveal new potential binders by detecting hidden patterns in the data [93]. For this purpose, many machine learning techniques such as support vector machine (SVM), random forests, and gradient boosting machine (GBM) have been used to classify molecules as binders or nonbinders and rank them according to binding potential [94]. On the other hand, it's all about graph neural networks (GNNs) and some models that clearly borrow inspiration or feature use of natural language processing (NLP) in their architecture [95]. In parallel, these have been revolutionizing the representation of the structural and sequential information of proteins and ligands and creating maps of drug–target interactions to facilitate drug repurposing and identification of off-target effects, which are very important in drug safety. Binding-prediction tools based on AI have now become paramount in the virtual screening platform, which helps in minimizing false positives and prioritizing compounds for further in vitro validation [96–98].

Deep Learning for Binding Site Recognition

Deep learning, being a behavior or an application of AI, replicates the neural networks of a human brain, which have evidenced great success in binding site prediction compared to other methods when there is less explanation on structure or when a lot of noise is present. Such methods are convolutional neural networks (CNNs) and recurrent neural networks (RNNs), among others. For interpreting protein sequences and structures to potentially enable the discovery of binding pockets with high accuracy [99, 100]. AlphaFold, in large part, is a DeepMind AI model that can predict 3D protein structures with extraordinary precision. Structural predictions from AlphaFold are better understood on the uncharacterized proteins and have enabled better identification of druggable binding sites [101]. Some tools like DeepSite, BindSite, and PocketNet use deep networks in searching ligand-binding sites in proteins that have no crystal structures. They use more traditional geometric or energy-based techniques for the site definition, particularly when those models assume that the site will be integrated with the protein–ligand interaction dataset.

Predictive Models for Affinity Estimation

The ability to estimate liking of a drug toward a target is essential for ranking of potential candidates during lead optimization. AI models trained from huge datasets such as PDBbind and BindingDB can pretty well predict K_d or IC_{50} or ΔG values. DeepDTA, KronRLS, and GraphDTA are a few among deep-learning frameworks that integrate ligand descriptors with protein descriptors for predicting quantitative binding affinities [102, 103]. These integrated models generally adopt multilayer neural networks, attention, and feature fusion techniques to deal with diverse types of input and extract their meaningful representation. These predictive models help identify compounds not only based on their ability to bind but also based on interaction strength with a target – an action that informs structure–activity studies and decreases the number of entities to be assessed experimentally.

Applications in Drug Discovery

Computational tools have proved indispensable for drug finding and evolution in the real world since they provide a fast track through which research can be accelerated, costs reduced, and quality enhanced for drugs. Such integrated methods as molecular docking, MD, pharmacophore modeling, AI/ML, and free-energy calculations make it significantly easier to identify, design, and optimize therapeutics.

Lead Identification and Optimization

Computational methods have been widely harnessed in the prima facie discovery of lead compounds that represent molecules that show an early excitement to connect with a specific part of the body. Virtual screening of compound libraries can be aided by either SB or LB methods that help eliminate poor candidates and allow focus on a smaller number of high-potential leads. Subsequent further computational analysis would lead to lead optimization, defined as improvement through enhancement of certain druglike properties of the identified molecules. This involves using tools such as MD modeling, binding free energy calculations, MM/PBSA or FEP, and QSAR modeling for refining binding affinity, selectivity, solubility, and metabolic stability [8]. Computational feedback loops will typically run at rapid cycles of design, prediction, and selection before moving into experimental testing and speed up optimization significantly.

Repurposing of Known Drugs

Drug repurposing (or repositioning) involves the identification of new therapeutic uses for existing, approved drugs. Computational methods become especially powerful in repurposing by taking advantage of knowledge related to drug–target interaction networks, machine learning, and bioinformatics workups. These approaches could match drugs to off-target proteins associated with other diseases, predict interactions through AI models, or correlate drug–disease interactions by transcriptomic and proteomic data. The appeal of computational repurposing is that it promises lower developmental costs and a reduced development timeline and risk, given that the concerned drugs have known safety profiles. A glaring example is remdesivir, an antiviral developed for Ebola that was screened computationally and repurposed for the treatment of COVID-19 early on in the pandemic [104].

Examples in Diseases like HIV, Cancer, and Tuberculosis

Through the integration of various techniques like molecular docking, MD simulations, SB design, and AI-driven screening, computational drug discovery has achieved a huge success in diseases that take a high toll. In cases of HIV/AIDS, these methods have successfully identified inhibitors for the viral enzymes, reverse transcriptase, protease, and integrase, AI being able to discover novel nonnucleoside reverse transcriptase inhibitors (NNRTIs) [9]. On the contrary, genetic variability and complexity pose hurdles in cancer: computational methods have helped discover kinase inhibitors, immune checkpoint modulators, and epigenetic regulators, while AI models are being increasingly employed for patient-specific response prediction-aiding personalized oncology. SB virtual screening and free energy calculations have enabled the rapid discovery of inhibitors designed against critical *Mycobacterium tuberculosis* enzymes like InhA and DprE1 in regard to tuberculosis (TB). Owing to the increased threat of drug-resistant TB, computational drug discovery is now being used for questing new agents and for the repurposing of older ones, thus strengthening its role in the contemporary drug discovery endeavor [105, 106].

Challenges and Future Directions

The remarkable advancement of computational drug discovery notwithstanding, challenges still impede achieving its entire potential. Any computational advancement in the traditional sense would address issues of prediction accuracy, problems of complexity conjoined with system design, and biological variability, thereby enhancing the reliability and applicability of in silico technologies. As the topic develops, however, the synergy between cutting-edge computational tools, personalized data outputs, and biological feedback in real time shall be an important cornerstone for the next great leap into precision medicine. This chapter discusses the most significant present limitations and evaluates some promising pathways for future development [107–110].

Accuracy vs. Computational Cost

Accuracy and computational efficiency are fundamental trade-offs in computational drug discovery. While high-accuracy alternatives such as QM computations, QM/MM simulations, and FEP all provide highly resolved detail about how molecules interact with each other, they are highly expensive computationally and therefore restricted from large-scale uses in virtual screening [111]. In contrast, faster methods like molecular docking, QSAR modeling, or machine learning classifiers are appropriate for screening extensive chemical libraries but typically work on simplified models that can miss important molecular or environmental effects [112]. The use of faster techniques within the framework of initial screening followed by more precise methods for selected candidates constitutes the most common practice but remains a point of ongoing optimization. The future has hybrid approaches, combining speed and accuracy-enhanced force fields by deep learning, transfer from high-accuracy datasets, and adaptive sampling methods focusing computing resources on the most promising candidates.

Receptor Flexibility and Solvent Effects

The incomplete treatment of receptor flexibility and solvent interactions represents one of the major disadvantages of many computational approaches. Most docking algorithms will assume that the protein is rigid or semiflexible, ignoring the well-established fact that the receptors will rearrange to greater or lesser extents when the ligand binds, an event referred to as induced fit [113]. In addition, solvent effects, and especially water-mediated interactions and ionic environments, are extremely significant in ligand recognition and binding stability. Unless these are taken into account, erroneous predictions regarding binding mode and affinity may be generated. Recent developments such as ensemble docking, MD methods, and enhanced sampling techniques stand on the perch of improving our capacity to simulate receptor flexibility and dynamic solvent behavior [114]. The incorporation of explicit water site models and solvation-aware scoring functions would help integrate idealized simulations and physiological realities. However, fully capturing this complexity of biological systems would ultimately remain an exercise in overcoming computational constraints.

Toward Personalized and Predictive Drug Design

The future of drug discovery is bending even more toward personalized and predictive medicine, wherein therapies will be tailored to the genetic, proteomic, and metabolic profile of an organism [115]. To make this vision a reality, one needs not only high-resolution biological data but also computational models for interpreting these in clinically meaningful ways. Advances in AI and machine learning paired with multi-omics integration (genomics, transcriptomics, and proteomics) are setting the stage for prediction of drug response in patients. In this regard, computational pipelines are being developed to show how an individual might respond to a drug depending on his unique molecular makeup. Digital twins are also an interesting avenue: virtual models of patients that could simulate the evolution of diseases and responses to treatment [116]. Such instruments would enable in silico clinical trials, thereby shortening timelines exponentially, saving costs, and improving the way outcomes are predicted. Nonetheless, it is essential to tackle data privacy, model interpretability, standardization, and regulatory approval challenges to unleash the full potential of predictive and personalized drug design.

Conclusion

Computational drug discovery is a paradigm shift in modern pharmaceutical research, bringing powerful tools available to court innovations for accelerating and refining drug development. A few of the revolutionary techniques, such as molecular docking, MD simulations, pharmacophore modeling, QM/MM hybrid methods, and machine learning, enhanced our understanding of drug–receptor interactions at the molecular level. Not only can such methods speed up the lead identification and optimization process but they also reduce experiments that consume huge amounts of time and money for structural evaluation. The combination of artificial intelligence and deep learning further allows binding predictions, custom therapy designs, and large-scale virtual screenings at lightning speed. Progress on these issues is still being done; however, strides in computational power as well as algorithm design overcome speedily a great deal of the issues of accuracy, protein flexibility, and biological complexity. Therefore, as the field matures, computational methods will take an ever more central role in driving innovation in targeted, effective, and exceptionally patient-centric drug development.

List of Abbreviations

K_d	Dissociation constant
K_{off}	Dissociation rate constant
K_{on}	Association rate constant
R_g	Radius of gyration
AI	Artificial intelligence
BLI	Biolayer interferometry
CDD	Computational drug discovery
CNNs	Convolutional neural networks
CNS	Central nervous system
EGFR	Epidermal growth factor receptor
ELISA	Enzyme-linked immunosorbent assay
FEP	Free energy perturbation
GBM	Gradient boosting machines
GCI	Grating-coupled interferometry
GNNs	Graph neural networks
GPCRs	G-protein-coupled receptors
H-bond	Hydrogen bond
HPC	High-performance computing
HTS	High-throughput screening
ITC	Isothermal titration calorimetry
LB	Ligand-based
LBDD	Ligand-based drug design
LBP	Ligand-based pharmacophore
MD	Molecular dynamics
ML	Machine learning
MS	Mass spectrometry
NLP	Natural language processing
QSARs	Quantitative structure–activity relationships
RMSD	Root-mean-square deviation
RMSF	Root-mean-square fluctuation
RNNs	Recurrent neural networks
SB	Structure-based
SBDD	Structure-based drug design
SBP	Structure-based pharmacophore
SERS	Surface-enhanced Raman spectroscopy
SPR	Surface plasmon resonance
SVMs	Support vector machines

References

1 Gangwal A, Lavecchia A. Artificial intelligence in natural product drug discovery: current applications and future perspectives. *J. Med. Chem.* 2025;68:3969. https://doi.org/10.1021/ACS.JMEDCHEM.4C01257/ASSET/IMAGES/LARGE/JM4C01257_0010.JPEG

2 Naithani U, Guleria V. Integrative computational approaches for discovery and evaluation of lead compound for drug design. *Front. Drug Discov.* 2024;4:1362456. https://doi.org/10.3389/FDDSV.2024.1362456

3 Blass BE. *Basic Principles of Drug Discovery and Development.* Academic Press; 2015. https://doi.org/10.1016/C2012-0-06670-7.

4 Smith A. Screening for drugs discovery: The leading question. *Nature* 2002;418:453–9. https://doi.org/10.1038/418453A;KWRD=SCIENCE

5 Lionta E, Spyrou G, Vassilatis DK, Cournia Z. Structure-based virtual screening for drug discovery: principles, applications and recent advances. *Curr. Top Med. Chem.* 2014;14:1923–38. https://doi.org/10.2174/1568026614666140929124445

6 Lounnas V, Ritschel T, Kelder J, et al. Current progress in structure-based rational drug design marks a new mindset in drug discovery. *Comput. Struct. Biotechnol. J.* 2013;5:e201302011. https://doi.org/10.5936/CSBJ.201302011/ASSET/A09942C5-827F-49E7-B0D9-37292F7837A2/MAIN.ASSETS/GR4.JPG

7 Shoichet BK. Virtual screening of chemical libraries. *Nature* 2004;432:862–5. https://doi.org/10.1038/NATURE03197;KWRD=SCIENCE

8 Schneider G. Virtual screening: an endless staircase? *Nat. Rev. Drug Discov.* 2010;9:273–6. https://doi.org/10.1038/NRD3139;SUBJMETA=1435,154,2418,309,606,631;KWRD=COMPUTATIONAL+CHEMISTRY,VIRTUAL+SCREENING

9 Chen H, Engkvist O, Wang Y, et al. The rise of deep learning in drug discovery. *Drug Discov. Today* 2018;23:1241–50. https://doi.org/10.1016/J.DRUDIS.2018.01.039

10 Hollingsworth SA, Dror RO. Molecular dynamics simulation for all. *Neuron* 2018;99:1129–43. https://doi.org/10.1016/J.NEURON.2018.08.011/ASSET/F4A0808D-6E29-4BAA-A1F0-68609B810994/MAIN.ASSETS/GR5.JPG

11 Bonde B. Edge, fog, and cloud against disease: the potential of high-performance cloud computing for pharma drug discovery. *Methods Mol. Biol.* 2024;2716:181–202. https://doi.org/10.1007/978-1-0716-3449-3_8

12 Chang L, Mondal A, Singh B, et al. Revolutionizing peptide-based drug discovery: advances in the post-AlphaFold era. *Wiley Interdiscip. Rev. Comput. Mol. Sci.* 2024;14:e1693. https://doi.org/10.1002/WCMS.1693;CTYPE:STRING:JOURNAL

13 Wang R, Lu Y, Wang S. Comparative evaluation of 11 scoring functions for molecular docking. *J. Med. Chem.* 2003;46:2287–303. https://doi.org/10.1021/JM0203783/ASSET/IMAGES/MEDIUM/JM0203783N00001.GIF

14 David L, Nielsen PA, Hedstrom M, Norden B. Scope and limitation of ligand docking: methods, scoring functions and protein targets. *Curr. Comput. Aided Drug Des.* 2005;1:275–306. https://doi.org/10.2174/1573409054367682

15 Stroud RM, Finer-Moore J (eds.). *Computational and Structural Approaches to Drug Discovery.* RSC Publishing; 2008.

16 Guvench O, MacKerell AD. Computational fragment-based binding site identification by ligand competitive saturation. *PLoS Comput. Biol.* 2009;5: e1000435. https://doi.org/10.1371/JOURNAL.PCBI.1000435

17 Wenke M. Drug-receptor interactions. In: *Fundamentals of Biochemical Pharmacology*. Pergamon; 1971. pp. 367–410. https://doi.org/10.1016/B978-0-08-017775-5.50025-8

18 Fang Y. Ligand–receptor interaction platforms and their applications for drug discovery. *Expert Opin. Drug Discov.* 2012;7:969–88. https://doi.org/10.1517/17460441.2012.715631

19 Fang Y. Label-free receptor assays. *Drug Discov. Today Technol.* 2011;7:e5. https://doi.org/10.1016/J.DDTEC.2010.05.001

20 Zheng X, Zhu H, Zhao X, et al. Emerging affinity methods for protein-drug interaction analysis. *J. Pharm. Biomed. Anal.* 2024;249:116371. https://doi.org/10.1016/J.JPBA.2024.116371

21 Manjari R. Understanding drug-receptor interactions: unlocking the secrets of pharmaceutical action. *J. Med. Org. Chem.* https://doi.org/10.37532/jmoc.2023.6(4).100-103

22 Bissantz C, Kuhn B, Stahl MA medicinal chemist's guide to molecular interactions. *J. Med. Chem.* 2010;53:5061–84. https://doi.org/10.1021/JM100112J

23 Boehr DD, Nussinov R, Wright PE. The role of dynamic conformational ensembles in biomolecular recognition. *Nat. Chem. Biol.* 2009;5:789. https://doi.org/10.1038/NCHEMBIO.232

24 Casadó-Anguera V, Casadó V. Unmasking allosteric-binding sites: novel targets for GPCR drug discovery. *Expert Opin. Drug Discov.* 2022;17:897–923. https://doi.org/10.1080/17460441.2022.2085684

25 Changeux JP, Edelstein SJ. Allosteric receptors after 30 years. *Neuron* 1998;21: 959–80. https://doi.org/10.1016/S0896-6273(00)80616-9/ASSET/2E231A4A-25C7-47FF-A7C9-93718B641DDA/MAIN.ASSETS/GR3.JPG

26 Wenthur CJ, Gentry PR, Mathews TP, Lindsley CW. Drugs for allosteric sites on receptors. *Annu. Rev. Pharmacol. Toxicol.* 2014;54:165–84. https://doi.org/10.1146/annurev-pharmtox-010611-134525

27 Wang E, Sun H, Wang J, et al. End-point binding free energy calculation with MM/PBSA and MM/GBSA: strategies and applications in drug design. *Chem. Rev.* 2019;119:9478–508. https://doi.org/10.1021/ACS.CHEMREV.9B00055/ASSET/IMAGES/MEDIUM/CR-2019-000558_0003.GIF

28 Ganotra GK. *Computational Studies of Drug-Binding Kinetics*. Heidelberg University; 2020. https://doi.org/10.11588/HEIDOK.00027374.

29 Shukla D, Hernández CX, Weber JK, Pande VS. Markov state models provide insights into dynamic modulation of protein function. *Acc. Chem. Res.* 2015;48:414–22. https://doi.org/10.1021/AR5002999/ASSET/IMAGES/LARGE/AR-2014-002999_0006.JPEG

30 Agnihotry S, Pathak RK, Srivastav A, et al. Molecular docking and structure-based drug design. In: *Computer-Aided Drug Design*. Springer; 2020. pp. 115–31. https://doi.org/10.1007/978-981-15-6815-2_6

31 Chowdhury KH, Chowdhury MR, Mahmud S, et al. Drug repurposing approach against novel coronavirus disease (COVID-19) through virtual screening targeting SARS-CoV-2 main protease. *Biology* 2021;10:2. https://doi.org/10.3390/BIOLOGY10010002

32 Le VTT, Nguyen THT, Do P-C. Global ligand-protein docking tools: comparation and case study. In: *Unravelling Molecular Docking*. IntechOpen; 2024. https://www.google.co.in/books/edition/Unravelling_Molecular_Docking_From_Theor/akFLEQAAQBAJ?hl=en&gbpv=1&dq=flexible+docking+recent&pg=PA55&printsec=frontcover

33 Stanzione F, Giangreco I, Cole JC. Use of molecular docking computational tools in drug discovery. *Prog. Med. Chem.* 2021;60:273–343. https://doi.org/10.1016/BS.PMCH.2021.01.004

34 Morris GM, Huey R, Olson AJ. Using AutoDock for ligand-receptor docking. *Curr. Protoc. Bioinform.* 2008;24:8.14.1–8.14.40. https://doi.org/10.1002/0471250953.BI0814S24

35 Dias R, de Azevedo Jr WF. Molecular docking algorithms. *Curr. Drug Targets* 2008;9:1040–7. https://doi.org/10.2174/138945008786949432

36 Atzori A. Conformational analysis of peptides and proteins for drug design using molecular simulations. 2015. https://research.manchester.ac.uk/en/studentTheses/conformational-analysis-of-peptides-and-proteins-for-drug-design-/

37 Mursal M, Ahmad M, Hussain S, et al. Navigating the computational seas: a comprehensive overview of molecular docking software in drug discovery. In: *Unravelling Molecular Docking – From Theory to Practice*. IntechOpen; 2024. https://doi.org/10.5772/INTECHOPEN.1004802

38 Li J, Fu A, Zhang L. An overview of scoring functions used for protein–ligand interactions in molecular docking. *Interdiscip. Sci.: Comput. Life Sci.* 2019;11:320–8. https://doi.org/10.1007/S12539-019-00327-W

39 Jain AN. Scoring functions for protein-ligand docking. *Curr. Protein Pept. Sci.* 2006;7:407–20. https://doi.org/10.2174/138920306778559395

40 Kitchen DB, Decornez H, Furr JR, Bajorath J. Docking and scoring in virtual screening for drug discovery: methods and applications. *Nat. Rev. Drug Discov.* 2004;3:935–49. https://doi.org/10.1038/nrd1549

41 Okafor SN, Angsantikul P, Ahmed H. Discovery of novel HIV protease inhibitors using modern computational techniques. *Int. J. Mol. Sci.* 2022;23:12149. https://doi.org/10.3390/IJMS232012149/S1

42 Sharma VK, Nandekar PP, Sangamwar A, et al. Structure guided design and binding analysis of EGFR inhibiting analogues of erlotinib and AEE788 using ensemble docking, molecular dynamics and MM-GBSA. *RSC Adv.* 2016;6:65725–35. https://doi.org/10.1039/C6RA08517B

43 Hosseini M, Chen W, Xiao D, Wang C. Computational molecular docking and virtual screening revealed promising SARS-CoV-2 drugs. *Precis. Clin. Med.* 2021;4:1–16. https://doi.org/10.1093/PCMEDI/PBAB001

44 Sheikh BA, Bhat BA, Rizvi MA, et al. Computational studies to identify potential inhibitors targeting the DprE1 protein in mycobacterium tuberculosis. *Int. J. Pharm. Invest.* 2022;13:129–38. https://doi.org/10.5530/223097131750

45 Badar MS, Shamsi S, Ahmed J, Alam MdA. Molecular dynamics simulations: concept, methods, and applications. In: *Transdisciplinarity*. Springer; 2022. pp. 131–51. https://doi.org/10.1007/978-3-030-94651-7_7

46 Filipe HAL, Loura LMS. Molecular dynamics simulations: advances and applications. *Molecules* 2022;27:2105. https://doi.org/10.3390/MOLECULES27072105
47 Karplus M, Petsko GA. Molecular dynamics simulations in biology. *Nature* 1990;347:6294. https://doi.org/10.1038/347631a0
48 Karplus M, McCammon JA. Molecular dynamics simulations of biomolecules. *Nat. Struct. Mol. Biol.* 2002;9:646–52. https://doi.org/10.1038/NSB0902-646
49 Shukla R, Tripathi T. Molecular dynamics simulation of protein and protein-ligand complexes. In: *Computer-Aided Drug Design*. Springer; 2020. pp. 133–61. https://doi.org/10.1007/978-981-15-6815-2_7
50 González MA. Force fields and molecular dynamics simulations. *École Thématique de La Société Française de La Neutronique* 2011;12:169–200. https://doi.org/10.1051/SFN/201112009
51 Kong LT, Denniston C, Müser MH, Qi Y. Non-bonded force field for the interaction between metals and organic molecules: a case study of olefins on aluminum. *Phys. Chem. Chem. Phys.* 2009;11:10195–203. https://doi.org/10.1039/B906874K
52 Guvench O, MacKerell AD. Comparison of protein force fields for molecular dynamics simulations. *Methods Mol. Biol.* 2008;443:63–88. https://doi.org/10.1007/978-1-59745-177-2_4
53 Harder E, Damm W, Maple J, et al. OPLS3: A force field providing broad coverage of drug-like small molecules and proteins. *Chem. Theory Comput.* 2016;12:281–96. https://doi.org/10.1021/ACS.JCTC.5B00864/SUPPL_FILE/CT5B00864_SI_001.ZIP
54 Schmitz G, Yönder Ö, Schnieder B, et al. An automatized workflow from molecular dynamic simulation to quantum chemical methods to identify elementary reactions and compute reaction constants. *J. Comput. Chem.* 2021;42:2264–82. https://doi.org/10.1002/JCC.26757
55 Salomon-Ferrer R, Case DA, Walker RC. An overview of the amber biomolecular simulation package. *Wiley Interdiscip. Rev. Comput. Mol. Sci.* 2013;3:198–210. https://doi.org/10.1002/WCMS.1121
56 Phillips JC, Hardy DJ, Maia JDC, et al. Scalable molecular dynamics on CPU and GPU architectures with NAMD. *J. Chem. Phys.* 2020;153:044130. https://doi.org/10.1063/5.0014475/16709547/044130_1_ACCEPTED_MANUSCRIPT.PDF
57 Rustagi V, Gupta SRR, Singh A, Singh IK. Beyond trial and error: leveraging advanced software for therapeutic discovery. *Chem. Biol. Lett.* 2025;12:1251. https://doi.org/10.62110/SCIENCEIN.CBL.2025.V12.1251
58 Durrant JD, McCammon JA. HBonanza: A computer algorithm for molecular-dynamics-trajectory hydrogen-bond analysis. *J. Mol. Graph. Model.* 2011;31:5–9. https://doi.org/10.1016/J.JMGM.2011.07.008
59 Feixas F, Lindert S, Sinko W, McCammon JA. Exploring the role of receptor flexibility in structure-based drug discovery. *Biophys. Chem.* 2014;186:31–45. https://doi.org/10.1016/J.BPC.2013.10.007
60 Martínez L. Automatic identification of mobile and rigid substructures in molecular dynamics simulations and fractional structural fluctuation analysis. *PLoS One* 2015;10:e0119264. https://doi.org/10.1371/JOURNAL.PONE.0119264

61 Heifetz A, James T, Morao I, et al. Guiding lead optimization with GPCR structure modeling and molecular dynamics. *Curr. Opin. Pharmacol.* 2016;30:14–21. https://doi.org/10.1016/J.COPH.2016.06.004

62 Rastelli G, Pinzi L. Refinement and rescoring of virtual screening results. *Front. Chem.* 2019;7:467450. https://doi.org/10.3389/FCHEM.2019.00498/BIBTEX

63 Naqvi AAT, Mohammad T, Hasan GM, Hassan MdI. Advancements in docking and molecular dynamics simulations towards ligand-receptor interactions and structure-function relationships. *Curr. Top. Med. Chem.* 2019;18:1755–68. https://doi.org/10.2174/1568026618666181025114157/CITE/REFWORKS

64 Hou T, Wang J, Li Y, Wang W. Assessing the performance of the MM/PBSA and MM/GBSA methods. 1. The accuracy of binding free energy calculations based on molecular dynamics simulations. *J. Chem. Inf. Model.* 2011;51:69–82. https://doi.org/10.1021/CI100275A/SUPPL_FILE/CI100275A_SI_001.PDF

65 Adeshina YO, Deeds EJ, Karanicolas J. Machine learning classification can reduce false positives in structure-based virtual screening. *Proc. Natl. Acad. Sci. USA* 2020;117:18477–88. https://doi.org/10.1073/PNAS.2000585117/SUPPL_FILE/PNAS.2000585117.SAPP.PDF

66 Awuni Y, Mu Y. Reduction of false positives in structure-based virtual screening when receptor plasticity is considered. *Molecules* 2015; 20:5152–64. https://doi.org/10.3390/MOLECULES20035152

67 Wang L, Shi SH, Li H, et al. Reducing false positive rate of docking-based virtual screening by active learning. *Brief. Bioinform.* 2023;24:1–11. https://doi.org/10.1093/BIB/BBAC626

68 Perola E. Minimizing false positives in kinase virtual screens. *Proteins* 2006;64:422–35. https://doi.org/10.1002/PROT.21002

69 Nada H, Meanwell NA, Gabr MT. Virtual screening: hope, hype, and the fine line in between. *Expert Opin. Drug Discov.* 2025. https://doi.org/10.1080/17460441.2025.2458666

70 Groenhof G. Introduction to QM/MM simulations. In: *Biomolecular Simulations.* Humana Press; 2013. pp. 43–66. https://doi.org/10.1007/978-1-62703-017-5_3

71 Antunes D, Santos LH, Guimarães ACR, Caffarena ER. Free energy perturbation and free-energy calculations applied to drug design. In: *Computer-Aided and Machine Learning-Driven Drug Design.* Springer; 2024. pp. 263–97. https://doi.org/10.1007/978-3-031-76718-0_10

72 Homeyer N, Gohlke H. Free energy calculations by the molecular mechanics Poisson–Boltzmann surface area method. *Mol. Inform.* 2012;31:114–22. https://doi.org/10.1002/MINF.201100135

73 Kar RK. Benefits of hybrid QM/MM over traditional classical mechanics in pharmaceutical systems. *Drug Discov. Today.* 2023;28:103374. https://doi.org/10.1016/J.DRUDIS.2022.103374

74 Senn HM, Thiel W. QM/MM methods for biomolecular systems. *Angew. Chem. Int. Ed.* 2009;48:1198–229. https://doi.org/10.1002/ANIE.200802019

75 Wang L, Chambers J, Abel R. Protein–ligand binding free energy calculations with FEP+. *Methods Mol. Biol.* 2019; 2022:201–32. https://doi.org/10.1007/978-1-4939-9608-7_9

76 Tuccinardi T. What is the current value of MM/PBSA and MM/GBSA methods in drug discovery? *Expert Opin. Drug Discov.* 2021;16:1233–7. https://doi.org/10.1080/17460441.2021.1942836

77 Molani F, Webb S, Cho AE. Combining QM/MM Calculations with classical mining minima to predict protein-ligand binding free energy. *J. Chem. Inf. Model.* 2023;63:2728–34. https://doi.org/10.1021/ACS.JCIM.2C01637/SUPPL_FILE/CI2C01637_SI_002.ZIP

78 Horvath D. Pharmacophore-based virtual screening. methods in molecular biology. In: *Chemoinformatics and Computational Chemical Biology*. Humana Press; 2010. pp. 261–98. https://doi.org/10.1007/978-1-60761-839-3_11

79 Kaserer T, Beck KR, Akram M, et al. Pharmacophore models and pharmacophore-based virtual screening: concepts and applications exemplified on hydroxysteroid dehydrogenases. *Molecules* 2015;20:22799–832. https://doi.org/10.3390/MOLECULES201219880

80 Khedkar SA, Malde AK, Coutinho EC, Srivastava S. Pharmacophore modeling in drug discovery and development: an overview. *Med. Chem.* 2007;3:187–97. https://doi.org/10.2174/157340607780059521

81 Ghose AK, Viswanadhan VN, Wendoloski JJ. The fundamentals of pharmacophore modeling in combinatorial chemistry. *J. Recept. Signal Transduct.* 2001;21:357–75. https://doi.org/10.1081/RRS-100107923/ASSET//CMS/ASSET/AEB60615-88B9-4826-9D1E-A00E42D5AE3D/RRS-100107923.FP.PNG

82 Muhammed MT, Aki-Yalcin E. Pharmacophore modeling in drug discovery: methodology and current status. *J. Turk. Chem. Soc. Sect. A: Chem.* 2021;8:749–62. https://doi.org/10.18596/JOTCSA.927426

83 Vuorinen A, Schuster D. Methods for generating and applying pharmacophore models as virtual screening filters and for bioactivity profiling. *Methods* 2015;71:113–34. https://doi.org/10.1016/J.YMETH.2014.10.013

84 Lu X, Yang H, Chen Y, et al. The development of pharmacophore modeling: generation and recent applications in drug discovery. *Curr. Pharm. Des.* 2018;24:3424–39. https://doi.org/10.2174/1381612824666180810162944/CITE/REFWORKS

85 Qing X, Lee XY, De Raeymaeker J, et al. Pharmacophore modeling: advances, limitations, and current utility in drug discovery. *J. Recept. Ligand Channel Res.* 2014;7:81–92. https://doi.org/10.2147/JRLCR.S46843;WGROUP:STRING:PUBLICATION

86 Gutnik D, Evseev P, Miroshnikov K, Shneider M. Using AlphaFold predictions in viral research. *Curr. Issues Mol. Biol.* 2023;45:3705–32. https://doi.org/10.3390/CIMB45040240

87 Wilson GL, Lill MA. Integrating structure-based and ligand-based approaches for computational drug design. *Future Med. Chem.* 2011;3:735–50. https://doi.org/10.4155/FMC.11.18

88 Gao Q, Yang L, Zhu Y. Pharmacophore based drug design approach as a practical process in drug discovery. *Curr. Comput. Aided Drug Des.* 2010;6:37–49. https://doi.org/10.2174/157340910790980151

89 Giordano D, Biancaniello C, Argenio MA, Facchiano A. Drug design by pharmacophore and virtual screening approach. *Pharmaceuticals* 2022;15:646. https://doi.org/10.3390/PH15050646

90 Kim KH, Kim ND, Seong BL. Pharmacophore-based virtual screening: a review of recent applications. *Expert Opin. Drug Discov.* 2010;5:205–22. https://doi.org/10.1517/17460441003592072

91 Sun H. Pharmacophore-based virtual screening. *Curr. Med. Chem.* 2008;15: 1018–24. https://doi.org/10.2174/092986708784049630/CITE/REFWORKS

92 Uddin MH, Zonder JA, Azmi AS. Exportin 1 inhibition as antiviral therapy. *Drug Discov. Today.* 2020;25:1775–81. https://doi.org/10.1016/J.DRUDIS.2020.06.014

93 Galano-Frutos JJ, Garciá-Cebollada H, Sancho J. Molecular dynamics simulations for genetic interpretation in protein coding regions: where we are, where to go and when. *Brief. Bioinform.* 2021;22:3–19. https://doi.org/10.1093/BIB/BBZ146

94 Lavecchia A. Machine-learning approaches in drug discovery: methods and applications. *Drug Discov. Today* 2015;20:318–31. https://doi.org/10.1016/J.DRUDIS.2014.10.012

95 Lavecchia A. Advancing drug discovery with deep attention neural networks. *Drug Discov. Today* 2024;29:104067. https://doi.org/10.1016/J.DRUDIS.2024.104067

96 Dhakal A, McKay C, Tanner JJ, Cheng J. Artificial intelligence in the prediction of protein–ligand interactions: recent advances and future directions. *Brief. Bioinform.* 2022;23:1–23. https://doi.org/10.1093/BIB/BBAB476

97 Chatterjee A, Walters R, Shafi Z, et al. *AI-Bind: Improving Binding Predictions for Novel Protein Targets and Ligands.* arXiv:2112.13168 [q-bio.QM]. 2021.

98 Chen Z, Hu L, Zhang BT, et al. Artificial intelligence in aptamer–target binding prediction. *Int. J. Mol. Sci.* 2021;22:3605. https://doi.org/10.3390/IJMS22073605

99 Jiménez J, Doerr S, Martínez-Rosell G, et al. DeepSite: protein-binding site predictor using 3D-convolutional neural networks. *Bioinformatics* 2017;33: 3036–42. https://doi.org/10.1093/BIOINFORMATICS/BTX350

100 Mylonas SK, Axenopoulos A, Daras P. DeepSurf: a surface-based deep learning approach for the prediction of ligand binding sites on proteins. *Bioinformatics* 2021;37:1681–90. https://doi.org/10.1093/BIOINFORMATICS/BTAB009

101 Zhao X, Yang VB, Menta AK, et al. AlphaFold's predictive revolution in precision oncology. *AI Precis. Oncol.* 2024;1:160–7. https://doi.org/10.1089/AIPO.2024.0010

102 Öztürk H, Özgür A, Ozkirimli E. DeepDTA: deep drug–target binding affinity prediction. *Bioinformatics* 2018;34:i821–9. https://doi.org/10.1093/BIOINFORMATICS/BTY593

103 Luo J, Zhu Z, Xu Z, et al. GS-DTA: integrating graph and sequence models for predicting drug-target binding affinity. *BMC Genom.* 2025;26:105. https://doi.org/10.1186/S12864-025-11234-4

104 Eastman RT, Roth JS, Brimacombe KR, et al. Remdesivir: a review of its discovery and development leading to emergency use authorization for treatment of COVID-19. *ACS Cent. Sci.* 2020;6:672–83. https://doi.org/10.1021/ACSCENTSCI.0C00489/ASSET/IMAGES/LARGE/OC0C00489_0004.JPEG

105 Hu X-P, Yang L, Chai X, et al. Discovery of novel DprE1 inhibitors via computational bioactivity fingerprints and structure-based virtual screening. *Acta Pharmacol. Sin.* 2022;43:1605–15. https://doi.org/10.1038/S41401-021-00779-1;KWRD=BIOMEDICINE

106 Rathod S, Chavan P, Mahuli D, et al. Exploring biogenic chalcones as DprE1 inhibitors for antitubercular activity via in silico approach. *J. Mol. Model.* 2023;29:1–23. https://doi.org/10.1007/S00894-023-05521-8

107 Tautermann CS. Current and future challenges in modern drug discovery. *Methods Mol. Biol.* 2020;2114:1–17. https://doi.org/10.1007/978-1-0716-0282-9_1

108 Shukla R, Tripathi T. Molecular dynamics simulation in drug discovery: opportunities and challenges. In: *Innovations and Implementations of Computer Aided Drug Discovery Strategies in Rational Drug Design*. Springer Nature; 2021. pp. 295–316. https://doi.org/10.1007/978-981-15-8936-2_12

109 Bano I, Butt UD, Mohsan SAH. New challenges in drug discovery. In: *Novel Platforms for Drug Delivery Applications*. Elsevier Science; 2023. pp. 619–43. https://doi.org/10.1016/B978-0-323-91376-8.00021-5

110 Kumar V, Roy K. Embracing the changes and challenges with modern early drug discovery. *Expert Opin. Drug Discov.* 2025. https://doi.org/10.1080/17460441.2025.2481259;WGROUP:STRING:PUBLICATION

111 Ou-Yang SS, Lu JY, Kong XQ, et al. Computational drug discovery. *Acta Pharmacol. Sin.* 2012;33:1131–40. https://doi.org/10.1038/APS.2012.109;KWRD=BIOMEDICINE

112 Cherkasov A, Muratov EN, Fourches D, et al. QSAR modeling: Where have you been? Where are you going to? *J. Med. Chem.* 2014;57:4977–5010. https://doi.org/10.1021/JM4004285/ASSET/IMAGES/MEDIUM/JM-2013-004285_0009.GIF

113 Vajda S, Weng Z, Rosenfeld R, DeLisi C. Effect of conformational flexibility and solvation on receptor–ligand binding free energies. *Biochemistry* 1994;33:13977–88. https://doi.org/10.1021/BI00251A004/ASSET/BI00251A004.FP.PNG_V03

114 Bozorgpour R, Sheybanikashani S, Mohebi M. Exploring the role of molecular dynamics simulations in most recent cancer research: insights into treatment strategies. arXiv Preprint 2023, arXiv:2310.19950.

115 Serrano DR, Luciano FC, Anaya BJ, et al. Artificial intelligence (AI) applications in drug discovery and drug delivery: revolutionizing personalized medicine. *Pharmaceutics* 2024;16:1328. https://doi.org/10.3390/PHARMACEUTICS16101328

116 Paliwal A, Jain S, Kumar S, et al. Predictive modelling in pharmacokinetics: from in-silico simulations to personalized medicine. *Expert Opin. Drug Metab. Toxicol.* 2024;20:181–95. https://doi.org/10.1080/17425255.2024.2330666

6

Decoding Drug Discovery: Pharmacophore Mapping and Ligand-based Design

Anu Sharma[1], Deepak Kumar Yadav[1], Krishan Kumar Verma[1], Saurabh Agrawal[2], and Amrendra Pratap Singh[3]*

[1]*Metro College of Health Sciences & Research, Greater Noida, Uttar Pradesh, India*
[2]*ABESIT College of Pharmacy, Ghaziabad, Uttar Pradesh, India*
[3]*Morehouse School of Medicines, Atlanta, GA, USA*

*Corresponding author: anukhushi.sharma@gmail.com

Introduction

Drug discovery is likened to the art of lock-and-key fitting. While the single-minded individual hunts for the perfect key to fit a cryptic lock, drug discovery scientists hunt to design molecules that fit and activate, or inhibit, biological targets in the human body. What if the lock structure is unknown, though? Under these circumstances, scientists turn to the exploration of molecules known to work, ligands. This strategy is the cornerstone of ligand-based drug design (LBDD), and the concept of the pharmacophore is around which everything revolves [1]. At the beginning of the 20th century, Paul Ehrlich used the term "pharmacophore" to describe the chemical characteristics that give a drug its biological activity [2]. This has evolved, over time, as a key postulate in computational drug design. A generic representation of the molecular characteristics necessary for a drug molecule's biological activity is called a pharmacophore. They consist of hydrophobic areas, aromatic rings, hydrogen bond donors and acceptors, and ionizable functionalities. Modeling of these features enables the design of new molecules with similar bioactivity to existing drugs, even when structural data on the desired protein is unavailable. Pharmacophore modeling and ligand-based design have become essential tools in modern drug discovery, especially for diseases where protein structures are not known or difficult to determine. In this chapter, the underlying theory, methods, applications, and limitations of pharmacophore mapping and ligand-based design are presented [3, 4].

The Pharmacophore

A pharmacophore is not an actual molecule but a conceptual structure, a map of the minimum necessary features that must be present in a molecule to bind to some specific biological target. It is the collection of steric and electronic characteristics that maximally interact with a particular biological target structure to either activate or inhibit its biological activity.

Key Features of a Pharmacophore

- **Hydrogen bond donors (HBDs):** groups or atoms that are able to donate a hydrogen atom to produce a proton bridge (H bond), e.g., —OH or —NH groups.
- **Hydrogen bond acceptors (HBAs):** groups or atoms that are capable of accepting a hydrogen atom, e.g., oxygen or nitrogen atoms with a lone pair.
- **Aromatic rings:** planar ring systems able to engage in π-π stacking or other noncovalent contact.
- **Hydrophobic areas:** nonpolar groups that engage through van der Waals contacts.
- **Positive or negative ionizable groups:** groups bearing a formal charge under physiological conditions and which are electrostatically interacting.

It is the arrangement in space. Two molecules sharing the same features but with varying 3D arrangements can display very different biological activities [5].

Types of Pharmacophores

- **Ligand-based pharmacophores:** constructed from known active compounds, without requiring target structure knowledge. These models try to define the common elements of active ligands.
- **Structure-based pharmacophores:** derived from the biological structure of the 3D object. To find possible pharmacophoric features, the target binding site is overlaid [6].

Ligand-based Pharmacophores

Legend-driven drug design is a method that centers on utilizing recognized biological interactions and molecular architectures to direct the design of novel drugs. Through learning about how current drugs bind to their sites, researchers have the ability to design new molecules that replicate or improve upon such interactions, resulting in enhanced therapy.

Principles

- **Target identification:** finding the biological target, typically a protein or enzyme linked to a disease, is the first stage in the legend-based drug design process. Designing successful medications requires an understanding of the target's structure and function.

- **Molecular docking:** the interaction between a drug candidate and its target is simulated by this computer method. By analyzing how well a compound fits into the target's active site, researchers can predict its potential efficacy.
- **Structure–activity relationship (SAR):** lead compounds' chemical structures are altered in SAR investigations to determine how these changes affect biological activity. This iterative process helps refine drug candidates for better performance.
- **Pharmacophore modeling:** an abstract depiction of the molecular characteristics required for a medication is called a pharmacophore to interact with its target. By identifying key pharmacophoric elements, researchers can design new compounds that retain essential interactions [7].

Methodologies

- **High-throughput screening (HTS):** by quickly testing thousands of chemicals for activity against a particular target, this method enables researchers to find possible leads for additional development.
- **Computational drug design:** utilizing software tools and algorithms, researchers can model and predict the behavior of drug candidates, optimizing their properties before synthesis.
- **Fragment-based drug design:** this approach involves screening small chemical fragments that bind to the target, which can then be elaborated into more complex molecules [8].

Pharmacophore Mapping Workflow

Pharmacophore modeling involves a number of well-defined steps:

Step 1: Ligand Selection
Choose a group of known active molecules with validated biological activity. Ideally, structurally diverse ones to ensure the model is representative of key features.

Step 2: Conformational Analysis
Create multiple 3D conformations of each ligand using computational resources. This prevents the pharmacophore model from being biased toward a single conformation.

Step 3: Feature Identification
Identify shared pharmacophoric features among the active molecules. Sophisticated software can automatically identify and align the features.

Step 4: Model Generation
Overlay the elements to create a pharmacophore hypothesis. The model represents the bare minimum of characteristics required for biological function.

Step 5: Model Validation
Verify the model with a collection of known active and inactive substances. Measures such as enrichment factor, sensitivity, and specificity are employed.

Step 6: Virtual Screening

Use the verified model to find compounds that fit the pharmacophore model in sizable compound libraries (like ZINC and PubChem). The hits are further assessed using docking and biological assays [9].

Advantages of LBDD

1. No target structural data required.
2. Suitable for HTS.
3. Less expensive and less resource-consuming.
4. Applicable to neglected diseases where target information is not available.

Applications of LBDD

1. Lead compound identification
2. Lead molecule optimization
3. Repurposing current drugs
4. Discovery of multitarget drugs [10, 11]

Case Studies and Real-world Applications

Pharmacophore modeling has become a central method in both research and industrial drug discovery, allowing researchers to find and improve potential drug candidates with speed. The study investigates practical pharmacophore modeling applications and reveals its significant effects across various therapeutic domains, including anticancer drug discovery, HIV protease inhibitors, and natural product drug discovery. We illustrate how pharmacophore modeling can be used in LBDD to help de novo drug creation, virtual screening, and lead optimization through these particular case examples [12].

Anticancer Drug Discovery

Pharmacophore models have made it much easier to find possible kinase inhibitors in the development of anticancer drugs. Scientists managed to cut both the price and duration needed for drug discovery through extensive compound library screening against these models. Understanding specific pharmacophoric features responsible for biological activity creates a targeted method that leads to finding therapeutic agents with precise efficacy [13].

HIV Protease Inhibitors

Pharmacophore modeling approaches have played a major role in the discovery of potent HIV protease inhibitors. Researchers achieved the development of potent and specific inhibitors by mapping crucial pharmacophoric elements such as HBAs and hydrophobic areas. The targeted approach enhances the inhibitors' effectiveness while minimizing side effects, which makes it a vital asset in HIV treatment [14].

Natural Product Drug Discovery

Pharmacophore models are also applied to natural product drug discovery, where scientists seek to discover bioactive lead compounds from a range of plant sources. By creating pharmacophore models from known ligands, researchers enable the discovery of new scaffolds with desired biological activities. This technique not only expedites the search for novel therapeutic leads but also enhances the investigation of the incredible chemical diversity of nature [15].

Challenges and Future Perspectives

Pharmacophore modeling is a very important drug discovery and design tool that acts as a conceptual model with features that depict the key characteristics that are accountable for the biological activity of a drug. This report examines existing challenges in the development of pharmacophores and establishes future directions that may aid its application in medicinal chemistry. Through these challenges and adopting new ideas, the field can evolve toward more efficient drug discovery [16].

Challenges in Pharmacophore Modeling

Pharmacophores are crucial in describing the connection between chemical structure and biological activity. They give a simplified view of molecular attributes required for interaction with a biological target. However, there are a number of difficulties with the creation and application of pharmacophore models, such as data quality, model validation, and combination with other computational approaches. This report seeks to describe these challenges and explore potential future directions that may enhance pharmacophore modeling [17].

Data availability and quality: the quantity and quality of available biological data have a significant impact on the performance of pharmacophore models. The drug development process may be hampered by incorrect models produced by incomplete or flawed datasets. Also, the absence of the standardized databases makes it difficult to retrieve and compare the pharmacophore features.

Model validation: pharmacophore model validation is necessary to determine their predictivity. Lack of a widely agreed set of validation metrics creates a major limitation. Scientists are prone to the use of subjective metrics, and that can contribute to inconsistency in model performance and trustworthiness.

Complexity of biological systems: the inherent complexity of biological systems requires pharmacophores to account for receptor dynamics as well as conformational flexibility and the cellular environment's effects. The significant challenge of integrating these complex factors into a pharmacophore model limits its practical application.

Integration with other computational approaches: pharmacophore modeling's strength is its compatibility with other computational techniques like molecular docking and quantitative structure–activity relationship (QSAR) modeling. Multiple methodologies and different data presentation styles make seamless integration challenging [18].

Future Perspectives

Enhanced data curation and standardization: to address issues of data quality, better curation and standardization of databases are required. Having common guidelines for data sharing and collection will make it easier to create more trustworthy pharmacophore models.

Development of robust validation metrics: the creation of strong and universally agreed validation metrics will increase the validity of pharmacophore models. This may include the creation of benchmark datasets and standardized test protocols to compare model performance.

Incorporating machine learning (ML) techniques: integration of ML methods in pharmacophore modeling is of significant promise. Using massive datasets, ML can find links and patterns that are difficult to recognize using conventional methodologies, ultimately resulting in more precise and predictive models.

Multi-scale modeling approaches: integration of ML methods in pharmacophore modeling is of significant promise. ML, through its utilization of large datasets, can identify patterns and relationships that are not easily recognized using conventional methodologies, ultimately resulting in more precise and predictive models [19].

Future Directions and Emerging Trends

This paper examines the emerging trends and future directions in the development of pharmacophore modeling, an essential element of drug discovery and development. With the changing face of the pharmaceutical industry, the adoption of cutting-edge technologies and methodologies is transforming the process of identifying, validating, and applying pharmacophores. This paper emphasizes some of the significant trends, innovative strategies, and future possibilities that may play a major role in pharmacophore research and its applications [20].

Integration of Artificial Intelligence and ML

One of the most promising trends is the integration of artificial intelligence (AI) and ML into pharmacophore modeling. These tools have the ability to process huge amounts of data to discern patterns and correlations that will not be seen using conventional techniques. By using AI and ML, scientists can improve the precision of pharmacophore predictions and make drug discovery a faster process [21].

Enhanced Virtual Screening Techniques

New virtual screening methods are poised to transform the way pharmacophores are used in drug discovery. Coupling pharmacophore-based virtual screening with molecular docking and dynamics simulations enables a better assessment of candidate drugs. The multifaceted strategy can result in the discovery of new compounds with greater efficacy and less toxicity [22].

Pharmacophore-based Drug Repurposing

Pharmacophore modeling is being used more and more in drug repurposing. By finding already-existing drugs that have similar pharmacophoric properties as new targets, researchers can shorten the process of drug development. This does not only save time and money but also can get new therapeutic possibilities to patients faster [23].

Focus on Polypharmacology

The trend toward polypharmacology, in which drugs are meant to act on several targets, is affecting pharmacophore modeling. Future studies will probably be aimed at preparing pharmacophores that can interact with multiple binding sites and mechanisms of action. This model could result in more effective therapy for complicated diseases like cancer and neurodegenerative diseases [24].

Incorporation of 3D Structural Data

High-quality 3D structural information generated using methods like X-ray crystallography and cryo-electron microscopy is improving pharmacophore modeling. The trend for the future will be increased focus on the incorporation of this structural data to achieve higher accuracy pharmacophore prediction and enable better design of active and selective compounds [25].

Collaborative Platforms and Open Data Initiatives

The creation of collaborative platforms and open data campaigns is driving increasing transparency and openness in pharmacophore research. Through sharing information and methodologies, researchers are able to advance the work of other researchers, achieving quicker developments within the field. This trend looks set to escalate, encouraging cooperation and innovation within the pharmaceutical environment [26].

Structure-based Pharmacophore Modeling

Principles of Structure-based Pharmacophore Modeling

Several key concepts form the foundation of structure-based pharmacophore modeling:

- The target's 3D structure: the presence of high-resolution crystal structures or homology models of the target protein is a prerequisite.
- Ligand binding sites: recognition of the ligand binding site of the target protein is very important for the creation of a correct pharmacophore model.
- Key interactions: determining the pharmacophore features requires an understanding of the interactions between the target and the ligand, such as hydrogen bonds, hydrophobic interactions, and ionic interactions [27].

Methodologies

Data collection: obtaining structural data on the target protein and known ligands is the first step. This may comprise the following:

- The target protein's entries in the Protein Data Bank (PDB).
- Experimental data on ligand binding affinities [28].

Binding site identification: through computer programs, the ligand's binding site in the target protein is determined. This may involve the following:

- molecular docking studies
- site-directed mutagenesis data [29]

Pharmacophore generation: pharmacophore models can be generated using various computational tools. The process typically includes:

- **Feature mapping:** identifying key pharmacophoric features based on the interactions observed in the ligand–target complex.
- **Model construction:** assembling the identified features into a pharmacophore model [30].

Validation: the pharmacophore model created needs to be tested to confirm its predictive capability. This may entail the following:

- Evaluating the model using a collection of recognized active and inactive substances.
- Cross-validation techniques to assess the model's robustness [31].

Applications

Structure-based pharmacophore modeling has numerous applications in drug discovery, including the following:

- **Virtual screening:** rapidly screening large compound libraries to identify potential drug candidates.

- **Lead optimization:** assisting in the design of new compounds with improved efficacy and reduced side effects.
- **Understanding mechanisms of action:** providing insights into how ligands interact with their targets [32].

Challenges

Despite its advantages, structure-based pharmacophore modeling faces several challenges:

- **Quality of structural data:** the quality of the protein structure has a significant impact on the pharmacophore model's accuracy.
- **Dynamic nature of proteins:** proteins are dynamic molecules, and their conformational changes can affect ligand binding.
- **Complexity of biological systems:** biological systems frequently have intricate interactions, making it difficult to predict ligand behavior accurately [33].

Future Directions

Future research in structure-based pharmacophore modeling may focus on:

- integrating ML techniques to enhance predictive capabilities;
- exploring the use of ensemble docking to account for protein flexibility;
- developing more sophisticated algorithms for pharmacophore generation and validation [34].

Core Concepts with Advanced Insights

Having set out the basic tenets of pharmacophore mapping and LBDD, most notably their contribution to the identification of essential molecular features driving biological effect, it is significant to note how these conventional methodologies are being reformed. Initially formulated as rule-based models for the purposes of early-stage drug discovery, pharmacophore-based models have significantly changed with the development of contemporary computational technologies and interdisciplinarity [35]. Although the previous sections have described the central methodologies, such as generation, validation, and application of pharmacophore models, the drug discovery paradigm of today needs to encompass a wider and more comprehensive view. Drug resistance, target complexity, and chemical diversity are the issues that necessitate scientists now bringing together traditional techniques with innovative technologies such as cloud platforms, ML, AI, and quantum modeling [36]. In addition, an integrated drug discovery strategy should also take into account ethical, regulatory, and translational considerations, meaning that computational predictions are not just scientifically sound but also feasibly actionable. As summarized in Table 6.1, the advanced topics discussed in the subsequent sections are systematically linked to the foundational

Table 6.1 Integration of advanced computational topics with core concepts of pharmacophore mapping and LBDD.

Advanced topic	Connection with core concepts
AI and ML	Builds on pharmacophore generation and screening steps by enhancing accuracy and speed using predictive algorithms (e.g., GNNs, DeepChem) [38]
Ligand vs. structure-based design	Adds depth by comparing the LBDD discussed with structure-based methods, showing when and why each is applicable [39]
Emerging trends (dynamic and quantum pharmacophores)	Expands on spatial feature mapping with newer computational layers like molecular dynamics or quantum mechanics [40]
Case studies (success and failure)	Real-world validation of the previously described workflows, giving context and practical insights [41]
Interactive workflow or infographic	Visualizes the stepwise modeling process discussed earlier, making it more digestible and applicable [42]
Ethical and regulatory aspects	Extends the discussion to societal and practical implications, key for real-world implementation of pharmacophore-based hits [43]
Natural products and phytopharmacophores	Directly links back to phytochemical ligands (e.g., phytoestrogens) as input molecules for LBDD [44]
Experimental validation	Bridges virtual predictions to wet-lab techniques, ensuring that theoretical models have practical credibility [45]
Datasets and open resources	Connects with the ligand data collection step, providing high-quality input for pharmacophore modeling [46]
Multitarget drug design	Pushes beyond single-ligand modeling to complex design strategies for diseases needing polypharmacology [47]

concepts of pharmacophore modeling and LBDD. These topics extend the theoretical framework by incorporating real-world case studies, advanced computational paradigms, and forward-looking innovations. This integrative perspective underscores the dynamic nature of pharmacophore modeling and highlights its expanding relevance in precision medicine, multitarget drug design, and natural product-inspired therapeutics [37].

Ethical and Practical Considerations

In drug discovery, ethical practice ensures scientific integrity, protects data rights, and promotes responsible innovation. Practical considerations are crucial for realistic implementation, especially in academic and early-stage research.

Table 6.2 Key differences between open-source and commercial tools in LBDD workflows.

Criteria	Open-source tools	Commercial tools
Cost	Free	Expensive, requires license
Examples	Pharmit, PyRx, UCSF Chimera	MOE, Discovery Studio, Schrödinger
Support	Limited (community-based)	Dedicated customer service
Flexibility	High (modifications allowed)	Limited (based on license agreement)

Intellectual property and licensing: commercial software (like Schrödinger's PHASE or LigandScout) may have restrictions on distribution of models, use for profit, publication without citation, and always disclose tools and respect usage terms.

Open-source vs. commercial tools: a wide range of computational tools are available for pharmacophore modeling, molecular docking, and LBDD, broadly classified into open-source and commercial platforms. Each category offers distinct advantages and limitations in terms of cost, flexibility, technical support, and usability. As summarized in Table 6.2, open-source tools provide high flexibility and accessibility, making them particularly suitable for academic research and methodological development, whereas commercial tools offer comprehensive, integrated environments with dedicated technical support, which are often preferred in industrial and translational settings.

Data transparency and reproducibility: always disclose the following:

- source of ligands;
- tools used;
- conformer settings;
- alignment method;
- share models and workflows via platforms like GitHub, Zenodo, or OSF;
- reproducibility ensures credibility in computational drug design.

FAIR principles in practice: make your pharmacophore data:

- Findable: use DOIs, metadata.
- Accessible: share via open repositories.
- Interoperable: use standard file formats (e.g., MOL2, PDBQT).
- Reusable: provide documentation and licenses.

Example: upload validated pharmacophores to Pharmit or ZINCPharmer.

AI/ML and ethical drug design: if using ML (e.g., to predict activity or generate pharmacophores), ensure the following:

- Model is trained on diverse, unbiased data.
- Interpretability of results is discussed.
- Potential biases are acknowledged.

Conclusion

Pharmacophore mapping and LBDD have become crucial strategies during the early phases of drug discovery, especially if thorough structural data about the biological target is unavailable. These methods are based on the detection and modeling of the spatial relationship between important chemical features required for biological activity and allow for the rapid screening of extensive compound collections and the rational development of new drug candidates. In this chapter, the development of LBDD has been followed, quoting from its classic heritage to incorporating highly sophisticated computational tools, AI, and dynamic modeling methodologies. From the synthesis design of artificial drugs to investigating nature-derived compounds, for example, phytoestrogens, the versatility of pharmacophore models in focused and polypharmacological drug discovery projects has been demonstrated. Pharmacophore modeling is evolving into a more adaptable field thanks to the incorporation of ML techniques, cloud-screening systems, and hybrid approaches that combine ligand-based and structure-based methods, making it a scalable and prophetic tool. Ethical oversight, data reliability, and experiment-based validation guarantee further that such computational models translate smoothly into clinical therapeutic effects in real life. Essentially, pharmacophore-based design is now no longer an isolated method but an integral and dynamic part of a support framework that facilitates innovation, accuracy, and effectiveness in contemporary drug development. Its further development holds the promise of overcoming present challenges in drug discovery and making key contributions toward the discovery of safer, more effective, and targeted medicines.

References

1 Guner O (ed.). *Pharmacophore Perception, Development, and Use in Drug Design*. International University Line; 2000.
2 Wolber G, Langer T. LigandScout: 3-D pharmacophores derived from protein-bound ligands and their use as virtual screening filters. *J. Chem. Inf. Model.* 2005;45(1):160–9.
3 Schuster D, Laggner C, Langer T. Why pharmacophore models fail: Error sources and how to overcome them. *Curr. Pharm. Des.* 2005;11(3):354–64.
4 Yang SY. Pharmacophore modeling and applications in drug discovery: Challenges and recent advances. *Drug Discov. Today* 2010;15 (11–12):444–50.
5 Kellenberger E, Foata N, Rognan D. Ranking targets in structure-based virtual screening of three-dimensional pharmacophores. *J. Med. Chem.* 2008;51(22):7024–36.
6 Leach AR, Gillet VJ. *An Introduction to Chemoinformatics*. Springer; 2007.
7 Baroni M, Cruciani G, Sciabola S, Perruccio F, Mason JS. A common reference framework for analyzing/comparing proteins and ligands. Fingerprints for ligands and proteins (FLAP). *J. Chem. Inf. Model.* 2007;47(2):279–94.

8 Langer T, Krovat EM. Chemical feature based pharmacophores and virtual library screening for discovery of new lead compounds. *Curr. Med. Chem.* 2003;10(9):749–62.

9 Barnickel G, Westen GJ, Labute P, Lemmen C. Combinatorial pharmacophore-based virtual screening: the impact of query flexibility. *J. Chem. Inf. Model.* 2009;49(4):854–61.

10 Kirchmair J, Distinto S, Markt P, et al. How to optimize shape-based virtual screening: choosing the right query and including chemical features. *J. Chem. Inf. Model.* 2009;49(3):678–92.

11 Taminau J, Thijs G, De Winter H. Pharmacophore modeling: advances, limitations, and current utility in drug discovery. *Curr. Med. Chem.* 2008;15(22):2163–80.

12 Lionta E, Spyrou G, Vassilatis DK, Cournia Z. Structure-based virtual screening for drug discovery: principles, applications and recent advances. *Curr. Top. Med. Chem.* 2014;14(16):1923–38.

13 Sliwoski G, Kothiwale S, Meiler J, Lowe Jr EW. Computational methods in drug discovery. *Pharmacol. Rev.* 2014;66(1):334–95.

14 Gohlke H, Klebe G. Approaches to the description and prediction of the binding affinity of small-molecule ligands to macromolecular receptors. *Angew. Chem. Int. Ed. Engl.* 2002;41(15):2644–76.

15 Kitchen DB, Decornez H, Furr JR, Bajorath J. Docking and scoring in virtual screening for drug discovery: methods and applications. *Nat. Rev. Drug Discov.* 2004;3(11):935–49.

16 Bissantz C, Kuhn B, Stahl M. A medicinal chemist's guide to molecular interactions. *J. Med. Chem.* 2010;53(14):5061–84.

17 Li H, Gao Z, Kang L, et al. TarFisDock: a web server for identifying drug targets with docking approach. *Nucleic Acids Res.* 2006;34:W219–24.

18 Sato T, Honma T, Yokoyama S. Combining machine learning and pharmacophore-based interaction fingerprint for in silico screening. *J. Chem. Inf. Model.* 2010;50(1):170–85.

19 Ekins S, Puhl AC, Zorn KM, et al. Exploiting machine learning for end-to-end drug discovery and development. *Nat. Mater.* 2019;18(5):435–41.

20 Schneider G. Automating drug discovery. *Nat. Rev. Drug Discov.* 2018;17(2):97–113.

21 Mullard A. Crowdsourced drug discovery picks up steam. *Nat. Rev. Drug Discov.* 2017;16(12):818.

22 Bajorath J. Integration of virtual and high-throughput screening. *Nat. Rev. Drug Discov.* 2002;1(11):882–94.

23 Ganesan A. The impact of natural products upon modern drug discovery. *Curr. Opin. Chem. Biol.* 2008;12(3):306–17.

24 Schneider G, Fechner U. Computer-based de novo design of drug-like molecules. *Nat. Rev. Drug Discov.* 2005;4(8):649–63.

25 Arimoto R. Computational models for predicting interactions with cytochrome P450 enzyme. *Curr. Top. Med. Chem.* 2006;6(15):1643–57.

26 Nicolaou CA, Brown N. Multi-objective optimization methods in drug design. *Drug Discov. Today Technol.* 2013;10(3):e427–35.

27 Reymond JL. The chemical space project. *Acc. Chem. Res.* 2015;48(3):722–30.

28 Lavecchia A. Machine-learning approaches in drug discovery: methods and applications. *Drug Discov. Today* 2015;20(3):318–31.

29 Chen H, Engkvist O, Wang Y, Olivecrona M, Blaschke T. The rise of deep learning in drug discovery. *Drug Discov. Today* 2018;23(6):1241–50.

30 Gasteiger J, Engel T (eds.). *Chemoinformatics: A Textbook*. 2nd ed. Wiley-VCH; 2003.

31 Sheridan RP. Time-split cross-validation as a method for estimating the goodness of prospective prediction. *J. Chem. Inf. Model.* 2013;53(4):783–90.

32 Hu Y, Bajorath J. Combining machine learning with pharmacophore modeling for improved hit identification. *Future Med. Chem.* 2010;2(4):607–17.

33 DiMasi JA, Grabowski HG, Hansen RW. Innovation in the pharmaceutical industry: New estimates of R&D costs. *J. Health Econ.* 2016;47:20–33.

34 López-Vallejo F, Gil-Redondo R, León LG, et al. The role of databases and data mining in pharmacophore-based virtual screening. *Mini-Rev. Med. Chem.* 2009;9(5):583–94.

35 Mauser H, Guba W. Recent developments in de novo design and scaffold hopping. *Curr. Opin. Drug Discov. Devel.* 2008;11(3):365–74.

36 Lounkine E, Keiser MJ, Whitebread S, et al. Large-scale prediction and testing of drug activity on side-effect targets. *Nature* 2012;486(7403):361–7.

37 Lipinski CA, Lombardo F, Dominy BW, Feeney PJ. Experimental and computational approaches to estimate solubility and permeability in drug discovery and development settings. *Adv. Drug Deliv. Rev.* 2001;46 (1–3):3–26.

38 Veber DF, Johnson SR, Cheng HY, et al. Molecular properties that influence the oral bioavailability of drug candidates. *J. Med. Chem.* 2002;45(12):2615–23.

39 van de Waterbeemd H, Gifford E. ADMET in silico modelling: towards prediction paradise? *Nat. Rev. Drug Discov.* 2003;2(3):192–204.

40 Congreve M, Carr R, Murray C, Jhoti H. A ‘rule of three’ for fragment-based lead discovery? *Drug Discov. Today* 2003;8(19):876–7.

41 Ferreira LG, dos Santos RN, Oliva G, Andricopulo AD. Molecular docking and structure-based drug design strategies. *Molecules* 2015;20(7):13384–421.

42 Kirchmair J, Williamson MJ, Tyzack JD, et al. Computational prediction of metabolism: sites, products, SAR, P450 enzyme dynamics, and mechanisms. *J. Chem. Inf. Model.* 2012;52(3):617–48.

43 Wishart DS, Feunang YD, Guo AC, et al. DrugBank 5.0: a major update to the DrugBank database for 2018. *Nucleic Acids Res.* 2018; 46(D1):D1074–82.

44 Cheng T, Li Q, Zhou Z, et al. Comparative assessment of scoring functions on a diverse test set. *J. Chem. Inf. Model.* 2009;49(4):1079–93.

45 Walters WP, Murcko M. Assessing the impact of generative AI on medicinal chemistry. *Nat. Biotechnol.* 2020;38(2):143–5.

46 Brown N, Fiscato M, Segler MHS, Vaucher AC. GuacaMol: benchmarking models for de novo molecular design. *J. Chem. Inf. Model.* 2019;59(3):1096–108.

47 Elton DC, Boukouvalas Z, Fuge MD, Chung PW. Deep learning for molecular design – a review of the state of the art. *Mol. Syst. Des. Eng.* 2019;4:828–49.

7

Homology Modeling, AI-based Protein Structure Prediction, and 3D Protein Modeling

Pooja Gupta[1], Arsh Chanana[1], Ravindra Pal Singh[1], and Ajay Bilandi[2]*

[1]*NIMS Institute of Pharmacy, NIMS University, Jaipur, Rajasthan, India*
[2]*Bhai Gurdas College of Pharmacy, Sangrur, Punjab, India*

*Corresponding author: ajaybilandi2031@gmail.com

Introduction

Proteins are the cell's functional workhorses, and they carry out virtually all biological functions such as signaling, immune response, molecular transport, and catalysis. The distinct three-dimensional (3D) shape of each protein determines how it will bind to other molecules, how it will carry out its function, and how alterations in structure might result in malfunction or disease. Knowledge of a protein's structure is essential to understanding its biological function and can expose the molecular etiology of disease. For example, prion protein misfolding contributes to Creutzfeldt–Jakob disease, and beta-amyloid aggregation is at the heart of Alzheimer's disease. Structural information on virus proteins, e.g., the SARS-CoV-2 spike protein, facilitated swift vaccine and therapeutic design during the COVID-19 pandemic. Knowledge of protein structures speeds up structure-based drug design (SBDD) in the pharmaceutical industry by enabling the rational design of small molecules that bind with high specificity and affinity to target proteins. Such structural information lowers the trial-and-error aspect of drug discovery and enhances the success rate of therapeutic interventions. Additionally, protein engineering depends greatly on structural data to create enzymes with better stability, activity, or new functions for medical and industrial use [1]. In order to discover protein structures, scientists have traditionally depended on three main experimental methods:

- **X-ray crystallography:** provides atomic-resolution information but involves crystallization of the protein, which may be challenging for membrane proteins or intrinsically disordered proteins (IDPs).
- **Nuclear magnetic resonance (NMR) spectroscopy:** gives structural information on proteins in solution and is appropriate for small- to medium-sized proteins but is size-limited.

- **Cryo-electron microscopy (cryo-EM):** a rapidly developing method that enables visualization of large macromolecular complexes without crystallization, now with near-atomic resolution.

Although they are accurate, these methods tend to be time-consuming, expensive, and not always possible for all proteins of interest. Therefore, computational approaches have become essential, providing scalable, affordable options.

Computational protein structure prediction involves a variety of strategies:

- **Homology (comparative) modeling:** presumes that proteins with similar sequences have similar structures. It utilizes a known structure (template) to model an unknown one (target):
  - **Threading (fold recognition):** tries the target sequence onto a library of known structural folds to determine the optimal fit despite low sequence identity.
  - **Ab initio (de novo) modeling:** structures de novo with physical and chemical principles without utilizing known templates appropriate for novel proteins but resource-intensive.
- **AI-based approaches:** the latest advances in artificial intelligence (AI), particularly deep learning, have radically improved prediction accuracy. Models such as AlphaFold2 (by DeepMind) and RoseTTAFold (by the Baker Lab) have the ability to predict complete 3D atomic-resolution structures with experimental method-level accuracy [2].

These computational techniques have revolutionized structural biology, providing high-confidence structural information for large parts of the proteome, even in organisms where little or no experimental structural information exists. These techniques are now key technologies in functional genomics, systems biology, evolutionary research, and therapeutic discovery.

Accurate protein structure determination is a cornerstone of structure-based drug discovery and rational therapeutic design. A variety of computational, experimental, and validation approaches are currently employed to predict, refine, and assess protein structures, each offering distinct advantages and limitations (Table 7.1).

Homology Modeling

Principles and Workflow

Homology modeling is among the most popular protein structure prediction methods because it is simple, reliable, and biologically relevant when a good template is available. It is based on the evolutionary concept that protein structure is more conserved than sequence. Even if sequence identity is moderate, the 3D protein conformation can be almost identical [3]. The homology modeling workflow generally involves five fundamental steps:

- identification of an appropriate template structure;
- target–template sequence alignment;

Table 7.1 Comparative overview of protein structure prediction and validation methods.

Method/tool	Type	Strengths	Limitations	Common use cases
Homology modeling	Template-based	Fast, accurate with good templates, widely supported	Limited by template quality and sequence identity	Predicting structures of known protein families
AlphaFold2	AI-based	High accuracy, good for novel folds, no need for templates	May miss dynamics, interactions, or PTMs	De novo prediction, structural genomics
RoseTTAFold	AI-based	Fast, accessible, good for multi-domain proteins	Slightly less accurate than AlphaFold2 for some targets	Functional annotation, education, initial modeling
X-ray crystallography	Experimental	High-resolution (atomic level), gold standard	Requires crystallization, time-consuming	Drug design, structure validation
Cryo-EM	Experimental	Suitable for large complexes, membrane proteins	Lower resolution for small proteins, expensive	Virus structures, protein complexes
NMR spectroscopy	Experimental	Studies proteins in solution, captures dynamics	Limited to small proteins, complex data interpretation	Dynamics, flexible regions, small proteins
SAXS	Experimental	Fast, provides shape and size in solution	Low resolution, not suitable for atomic detail	Conformational studies, flexible systems
Molecular dynamics (MD)	Simulation	Explores dynamics, folding, binding, and stability	Computationally intensive, depends on initial model accuracy	Loop refinement, drug binding, stability testing
MolProbity, ProSA, etc.	Validation	Assess stereochemistry, geometry, packing	Cannot confirm biological function	Model quality control and refinement

- model building from the alignment;
- model refinement (loop modeling, side-chain placement);
- model evaluation and validation.

Identification of a Suitable Template Structure

This is the initial and most important step in homology modeling. A template is a protein with a known 3D structure that is sequence similar to the target protein.

Objective: identify a homologous protein structure in databases such as the Protein Data Bank (PDB).

Tools: BLAST, PSI-BLAST, and HHPred.

Criteria: high sequence identity (typically >30%), good resolution, and biological relevance [4].

Target–Template Sequence Alignment

After a template is chosen, the target sequence is aligned with the template to identify the corresponding residues.

Significance: proper alignment guarantees proper structural mapping.

Tools: Clustal Omega, MUSCLE, and T-Coffee.

Challenges: insertions/deletions (indels) and low-conservation regions can compromise accuracy.

Model Building Based on the Alignment

In this step, a 3D model of the target protein is created by transferring the backbone coordinates from the template to the target, based on the alignment.

Tools: MODELLER, SWISS-MODEL, and I-TASSER.

Output: a rough 3D structure with a solid core but potentially erroneous loop and side-chain conformations [5].

Model Refinement (Loop Modeling, Side-chain Placement)

Refinement enhances structure details in variable or poorly ordered areas such as loops and side chains.

Loop modeling: forecasts conformations in areas not preserved between template and target.

Side-chain placement: puts side chains in place based on rotamer libraries or optimization based on energy.

Tools: MODELLER (loop optimization), Rosetta Relax, SCWRL, and MD simulations.

Model Evaluation and Validation

The generated model is checked to identify structural accuracy and credibility.

Validation criteria: geometric quality (bond angles, torsions), steric clashes, and fold correctness.

Tools:

- **PROCHECK:** Ramachandran plots for backbone geometry.
- **Protein structure analysis (ProSA):** *Z*-scores for global model quality.
- **MolProbity:** side chain and packing quality.
- **Verify3D:** environment profile compatibility [6].

Careful consideration and optimization must be done at every step since inaccuracies at earlier steps can have a drastic effect on the quality of the final structure. This strategy is especially valuable when experimental structure determination is impossible because of time, expense, or technical constraints. Though models of increasing precision demand increased sequence identity, productive models can still be built with 25–30% identity using advanced alignment and refinement methods.

Template Selection and Sequence Alignment

The choice of template starts with a search for sequence-similar structurally characterized proteins (usually in the PDB) for the target protein. BLAST, HHpred, PSI-BLAST, or Foldseek are some common programs used to carry out these searches [7]. Key considerations in the choice of an ideal template are the following:

Sequence identity: preferably >50% for good-quality models, although models of 25–30% identity may still prove useful.

- **Coverage:** the fraction of the target sequence that matches the template.
- **Template structure resolution:** templates of higher resolution produce better models.
- **Biological relevance:** structural representations of functional forms (e.g., bound/unbound forms) are preferred.

Sequence alignment is a sensitive procedure that aligns the target onto the chosen template. Even slight misalignments, particularly in loops or active sites, can lead to huge structural errors. Sophisticated multiple sequence alignment (MSA) tools like T-Coffee, PROMALS3D, and MAFFT can enhance alignment quality by using structural information and evolutionary conservation. In reality, multiple templates can be employed for modeling distinct regions of a protein (multi-template modeling), particularly for large or multi-domain proteins.

Model Building and Validation

Model building is followed after the completion of alignment by transferring atomic coordinates from the template into the aligned residues of the target. Conserved regions are simply copied, while loops, gaps, or insertions are rebuilt utilizing fragment libraries, energy-based approaches, or loop databases. Side-chain modeling is another important step, particularly for residues that are not the same in the target and template. Side chains are usually modeled with rotamer libraries and optimized to reduce steric clashes and enhance hydrogen bonding [8]. Important software tools employed for homology modeling include the following:

- **MODELLER:** a very popular tool that performs all steps of homology modeling automatically.
- **SWISS-MODEL:** a web server providing high-quality modeling with in-built validation.
- **I-TASSER:** it combines threading, ab initio modeling, and refinement for high-confidence predictions.
- **Phyre2:** it relies on fold recognition and hidden Markov models (HMMs) to identify remote homologs.
- Once constructed, the model is required to be validated to authenticate structural accuracy.

Validation tools in common use include:

- Ramachandran plot analysis (by PROCHECK or MolProbity) to gauge backbone angles.
- Verify3D and ERRAT to check how well the 3D structure fits its one-dimensional (1D) sequence.
- Qualitative Model Energy Analysis (QMEAN) or ProSA for predicting overall and per-residue quality [9].

Refinement using MD simulations or energy minimization (e.g., through GROMACS or AMBER) can enhance geometry, especially in flexible or poorly modeled areas.

Applications and Limitations

Homology modeling has general applicability across biomedical and life sciences:

- **Structure-based drug discovery:** binding pocket prediction or receptor–ligand modeling when structural information is not available experimentally.
- **Protein function prediction:** prediction of biochemical function from similarity to previously characterized homologs.
- **Rational protein design:** designing proteins with improved properties (e.g., thermostability, substrate specificity).
- **Understanding disease mutations:** modeling the effect of point mutations in genetic disease or cancer (e.g., BRCA1, p53).
- **Antibody modeling:** homology modeling is crucial in building antibody variable regions for therapeutic design.

But it also has its limitations:

- **Dependence on appropriate templates:** no good model can be constructed without a related structure.
- **Poor accuracy in variable regions:** loops, insertions, and disordered regions are difficult to model accurately.
- **Side-chain errors:** particularly for non-conserved residues or active site modeling.
- **Cannot model new protein folds:** de novo methods or AI tools are needed in such situations [10].

Homology modeling is still a powerful, affordable, and feasible method despite these shortcomings, particularly when combined with other predictive methods or as a starting point for more sophisticated modeling.

AI-based Protein Structure Prediction

Development of AI in Structural Biology

The merging of AI and molecular biology has ushered in a new age in protein science. Earlier computational models depended strongly on manually crafted features, stiff scoring functions, and template dependence. Although helpful for predicting the structure of highly conserved families of proteins, they were deficient in generalization capability for complex or new folds. The introduction of deep learning and transformer models, initially developed for natural language processing (NLP), introduced high-performance tools for biological sequence modeling. Proteins, similar to language, have grammar (rules of folding), motifs (keywords), and context (evolutionary relationships), and thus are amenable to attention-based models. Today's AI models don't merely "predict" structure; they're building representations of protein space, reasoning about intricate patterns in evolutionary, physical, and structural data with capabilities far more powerful than the best human intuition [11].

Deep Learning Models: AlphaFold, RoseTTAFold, and Oncoming Tools

AlphaFold2 (DeepMind)

- AlphaFold2's structure changed the scene by eliminating modular modeling requirements. It employs:
  - attention-based transformer networks to ingest MSAs and template data;
  - a recycling approach in which the prediction loop is cycled, structurally refining with each iteration;
  - confidence scores such as predicted local distance difference test (pLDDT) and predicted aligned error (PAE), enabling users to judge the accuracy of the prediction at residue-level and domain-level scales.

Influence of AlphaFold2

- Predicted more than 200 million structures in the AlphaFold Protein Structure Database (AlphaFold DB).
- Enabled studies on human health, for example, simulating disease mutations in proteins like BRCA1, CFTR, and p53.
- Generated structures of orphan disease targets and environmental enzymes, advancing research on bioremediation, vaccine development, and synthetic biology.

RoseTTAFold

- RoseTTAFold introduced a new three-track neural network that simultaneously learns amino acid sequence-to-sequence, distance map, and atomic coordinate relationships.
- Strengths of RoseTTAFold:
  - It performs well with protein–protein complexes.
  - It is faster and computationally lighter compared to AlphaFold2.
  - Supports integrative modeling: integrating AI predictions with electron density maps, crosslinking data, and small-angle X-ray scattering (SAXS) data [12].

Applications

- Applied to predict binding interfaces for antibody–antigen complexes.
- Modeled viral proteins (e.g., from SARS-CoV-2, HIV) to assist vaccine and therapeutic design.

Emerging Models and Tools

The field of computational protein modeling has undergone a revolutionary change with the incorporation of AI and deep learning. Whereas older structure prediction methods relied on sequence homology or physics-based simulations, recent approaches utilize machine learning frameworks, protein language models (PLMs), and generative AI. These methods not only predict protein structures with high accuracy but are also being increasingly applied to protein design, inverse folding, and function prediction. OmegaFold, ESMFold, ProteinMPNN, RFdiffusion, and graph neural networks (GNNs) are employed in Atom3D. Each makes a distinct contribution to various steps of the protein modeling and design pipeline [13].

OmegaFold (MSA-free structure prediction): OmegaFold is a deep-learning-based protein structure prediction program that stands out in that it eliminates the use of MSAs, a normal but computationally expensive step used in AlphaFold2 and the like.

Key Features

- Single-sequence prediction: handles naked amino acid sequences without requiring the use of homologous sequence databases.
- Speed and scalability: much faster than MSA-based approaches, suitable for high-throughput pipelines.
- Transformer-based architecture: learns residue interactions directly from the native sequence using attention mechanisms.

Applications

OmegaFold is especially beneficial for the following:

- predicting structures for orphan proteins with no known homologs;
- conducting large-scale proteome scans;
- quickly modeling short peptides or engineered sequences.

Limitations: OmegaFold might exhibit slightly reduced accuracy relative to AlphaFold2 in the presence of rich evolutionary information. However, its accuracy-to-speed ratio makes it invaluable in most applied contexts.

ESMFold (large PLMs for fast prediction): ESMFold, created by Meta AI, belongs to the evolutionary scale modeling (ESM) family. It relies on large-scale PLMs that have been trained on millions of sequences to estimate protein structure without direct MSAs or templates [14].

Architecture

- Designed based on transformers like the ones utilized in NLP (e.g., GPT, BERT).
- Taught to learn sequence context and evolutionary constraints implicitly from large datasets.

Performance

- Forecasts structures in seconds using consumer-grade GPUs.
- Delivers accuracy reaching AlphaFold2 for most targets, particularly if sequence data are sparse.
- Offers confidence measures akin to AlphaFold's pLDDT.

Benefits

- Best used for real-time applications, for example, annotating structure for metagenomics studies.
- Highly parallelized and scalable to deploy on cloud.

Applications

- Forecasting viral proteins, antibodies, and synthetic peptides.
- Large-scale structural genomics initiatives.
- Integration into structure–function pipelines.

ProteinMPNN (sequence design for given structures): in contrast to structure prediction software, ProteinMPNN (Message Passing Neural Network) addresses the reverse protein folding problem, i.e., sequence design that will fold into a target 3D structure.

Functionality

- Given a backbone structure, ProteinMPNN produces amino acid sequences that are likely to adopt that fold.
- Employs a graph-based neural network to represent interactions between residues and the spatial backbone environment.

Strengths

- It is faster and more diverse than traditional sequence design software (e.g., RosettaDesign).
- Generates sequences with good sequence recovery and thermodynamic stability.
- De novo design, antibody engineering, and enzyme redesign are compatible.

Applications

- Synthetic biology: enzyme design, scaffold or self-assembling protein design.
- Drug discovery: development of binders or therapeutic scaffolds.
- Structural biology: enhancing crystallization success through optimized sequences [15].

RFdiffusion (generative modeling for protein structures): RFdiffusion is a generative AI system that utilizes diffusion models a type of deep generative model made popular by software such as DALL·E and Stable Diffusion, to generate new protein backbones and designs.

Functionality

- Begins with random noise and slowly diffuses toward a structured protein shape, guided by a neural network that has been trained on actual structures.
- May be conditioned on motifs, symmetries, or other structural restraints.

Key Capabilities

- Produces new protein folds never observed in nature.
- Allows for functional site grafting, multimeric design, and scaffold construction.
- Aids in inverse folding through combination with sequence prediction tools such as ProteinMPNN.

Significance: RFdiffusion is a new concept in rational protein design, abandoning template-based methodologies for unconstrained searching of structure space. It holds the key to revolutionizing enzyme design, biosensors, and even protein therapeutics [16].

GNNs in atom-level modeling: GNNs are deep learning models that can deal with graph-structured data. In protein modeling, residues or atoms are considered as nodes, and their interactions (e.g., spatial proximity, bonds) create edges. The graph-based method enables spatially informed learning of molecular shape.

Atom3D Framework: Atom3D is a benchmark suite and dataset that can be used to train and test GNNs for atomic-level protein data. It supports tasks such as the following:

- ligand binding prediction
- protein–protein interface prediction
- side-chain placement
- mutation impact prediction

GNN Capabilities

- Deal with non-Euclidean geometry, perfect for non-regular molecular structures.
- Learn the spatial relations over multiple scales (atoms to domains).
- May be able to combine chemical, geometric, and evolutionary information under one architecture.

Real-world Tools Utilizing GNNs

- **GraphSite:** binding site prediction from 3D structure.
- **Geometric Vector Perceptron (GVP):** enhances GNNs with direction.
- **DimeNet++, EGNN:** high-end GNN architectures for molecular learning.

Advantages

- May be used in both structure prediction and drug discovery.
- Easily plug into multimodal models with sequence and structure data.
- Well suited for multi-task learning in structure–function pipelines.

The rise of deep learning has not only improved protein structure prediction but also facilitated creative and generative modeling for protein design. Applications such as OmegaFold and ESMFold speed up prediction by skipping time-consuming MSAs, whereas models such as ProteinMPNN and RFdiffusion allow de novo protein engineering from the ground up. At the same time, GNNs provide a flexible platform for encoding the rich 3D geometry of proteins and spatially aware predictions.

These technologies are redefining how we tackle structural biology, drug discovery, and synthetic protein design, making the discipline faster, smarter, and more innovative than ever before. These newer capabilities are broadening AI's applications not only to structure prediction but also to protein design, functional annotation, and drug–target interaction modeling [17].

Accuracy, Advantages, and Challenges

Accuracy and benchmarks: AI-based methods have rewritten what's possible computationally:

- AlphaFold2: provides sub-angstrom accuracy for most proteins in blind tests.
- CASP14 results showed median GDT_TS scores >90 for many targets.
- AI predictions were as good as or better than NMR or medium-res cryo-EM resolution for some proteins.

Accuracy tends to be greater for structured, well-conserved domains, and poorer for the following:

- disordered or flexible parts
- proteins with complex allosteric states
- membrane proteins with meager MSAs [18]

Benefits

- **Speed and scale:** predict millions of structures in hours or days.
- **No template requirement:** can predict new folds or orphan proteins.
- **Accessibility:** web tools (e.g., ColabFold) provide access to advanced modeling for nonexperts.
- **Confidence scoring:** provides in-built validation through pLDDT and PAE.
- **Broad utility:** all-around useful from proteomics, synthetic biology, and systems biology, to medicine.

Challenges and Limitations

- **Lack of dynamics:** AI predictions provide one static model, whereas proteins are dynamic by nature.
- **Limited complex prediction:** AlphaFold-Multimer and RoseTTAFold still cannot predict large or transient assemblies well.
- **Uncertainty in unstructured regions:** intrinsically disordered regions (IDRs) remain poorly modeled.

- **Interpretability:** neural networks are "black boxes" and difficult to interpret mechanistically.
- **Computational cost:** resource-intensive high-resolution predictions for big proteins or multimeric complexes [5].

Applications and Future Directions

- **Drug discovery:** ligand docking, structure-based virtual screening, pharmacophore modeling.
- **Mutation impact analysis:** predicting functional and structural impacts of single nucleotide polymorphisms (SNPs) and disease mutations.
- **Synthetic biology:** design de novo enzymes, biosensors, and scaffolds.
- **Structural genomics:** modeling full proteomes of model organisms or pathogens.
- **Personalized medicine:** tailoring treatments based on patient-specific protein variants.

Future Directions

- **Integration with dynamics:** combining AI with MD simulations to predict conformational changes and allosteric effects.
- **Multi-structure prediction:** modeling alternate conformations, ligand-induced shifts, or protein folding intermediates.
- **Protein design revolution:** inverse folding models are opening pathways to rationally create new proteins from scratch.
- **Interdisciplinary convergence:** AI will increasingly integrate with cryo-EM, mass spectrometry, and genomics, creating a holistic view of biological systems.

AI-based protein structure prediction has fundamentally changed our way of thinking in structural biology. Software such as AlphaFold2 and RoseTTAFold have made precise structural modeling possible, scalable, and quick, unlocking new horizons in biomedical science, drug discovery, and protein design. In spite of continued challenges in modeling dynamics, complexes, and disordered proteins, the rate of progress implies that a future of complete integration, real-time, and design-oriented structural prediction is at our doorstep [19].

3D Protein Modeling and Visualization

Computational 3D Protein Modeling: Bridging the Sequence–Structure Gap

Proteins are the molecular workhorses of biology, carrying out an incredible range of cellular functions that rely intimately on their 3D structure. Establishing the 3D structure of a protein is central to our comprehension of its function, interactions, dynamics, and value as a therapeutic target. Although experimental techniques like

X-ray crystallography, cryo-EM, and NMR spectroscopy have yielded tremendous information regarding protein structures, these methodologies are time-consuming, expensive, and not always practical, particularly for membrane proteins, large assemblies, or IDPs. To overcome these limitations, computational 3D protein modeling has emerged as a powerful alternative. It aims to predict a protein's tertiary structure based on its amino acid sequence using algorithms, biophysical principles, and, increasingly, AI. Three main approaches dominate this field: homology modeling, ab initio modeling, and AI-based prediction tools [20].

Homology modeling (leveraging evolutionary relationships): homology modeling, also known as comparative modeling, is one of the most established and widely used methods in computational structural biology. It is based on the premise that evolutionarily related proteins (homologs) will possess similar structures. Thus, if the structure of one protein is known, it can be used as a template for predicting the structure of a related protein with an unknown structure [21].

Workflow and principles: the homology modeling procedure commonly consists of five central steps:

- Template selection: a structural database search (e.g., the PDB) is performed to locate homologous proteins with a known structure to the target sequence.
- Sequence alignment: the target sequence and template sequence are aligned to determine conserved regions and to direct the transfer of structural details.
- Model building: the 3D structure of the target protein is built by duplicating backbone coordinates from the template and modeling variable segments.
- Model refinement: loops, side chains, and gaps are refined with energy minimization or fragment insertion techniques.
- Model validation: the model is tested for structure using structural analysis tools (e.g., Ramachandran plots, ProSA, and Verify3D) to demonstrate physical feasibility.

Strengths and limitations: homology modeling is relatively fast and efficient when the target–template sequence identity is above 30%. However, its accuracy greatly reduces when good-quality templates are unavailable. Loops and non-conserved areas are especially tough to model with high accuracy, which can result in functional misinterpretations.

Ab initio modeling (predicting structure from first principles): ab initio (Latin for "from the beginning") modeling tries to predict protein structure only from its amino acid sequence, without using templates. Instead, it uses physics-based principles, energy functions, and fragment-based assembly strategies to sample the conformational space and identify the most stable structure [22].

Underlying Concepts

- Its basic assumption is that a protein's native conformation is equal to the global free energy minimum of its folding landscape.
- Algorithms model folding routes through repeated sampling of conformations, employing force fields (e.g., AMBER, CHARMM) and scoring functions to measure structure stability.

Tools and Techniques

- Rosetta – employs fragment libraries and Monte Carlo sampling.
- QUARK – makes use of threading and deep learning-scoring.
- I-TASSER – mixes threading with ab initio methods for unaligned segments.

Challenges: ab initio modeling is computationally intensive, particularly for large proteins (>150 residues), owing to the vast conformational space. Although promising, ab initio models are less accurate than homology models unless supported by other information such as co-evolutionary data or experimental restraints.

Use Cases

- Ab initio modeling proves especially useful when:
  - No homologous templates are available.
  - The protein is small or has new folds.
  - Structural motifs must be inferred de novo for synthetic biology [23].

AI-based prediction tools (the AlphaFold revolution): the field of protein structure prediction changed in 2020 with the release of AlphaFold2 by DeepMind. For the first time, a machine learning system obtained near-experimental accuracy within the biennial Critical Assessment of Structure Prediction (CASP) challenge. This opened a new frontier of AI-assisted modeling, where deep learning models can predict very accurate 3D structures, even when no homologous templates were available.

How it works: AlphaFold2 and its alternatives employ the following:

- Neural networks that were trained on thousands of known protein structures.
- MSAs to make predictions about evolutionary relationships.
- Attention mechanisms to learn long-range residue–residue dependencies.
- End-to-end differentiable models that predict atomic coordinates directly [24].

Key Tools

- AlphaFold2: produces atomic-level accuracy for most proteins.
- RoseTTAFold: a lighter and more modular alternative from the Baker Lab.
- ESMFold and OmegaFold: new tools with faster inference times and similar accuracy.

Strengths

- Unprecedented accuracy for single-chain proteins.
- Functional even in the absence of homologous templates.
- Available for free through Google ColabFold and EBI AlphaFold Database.

Limitations

- Less precise in predicting:
  - flexible or disordered regions;

  - protein complexes and multimeric assemblies;
  - effect of ligands or mutations.
- Does not have integrated dynamic simulation or biological validation function [25].

Typically, these approaches are not mutually distinct. For example, I-TASSER combines threading, ab initio, and fragment modeling. Similarly, hybrid methods apply AlphaFold models as templates for homology pipelines or optimize AI-predicted structures through MD simulations. Computational 3D protein modeling is an intersection of biology, physics, and AI. Homology modeling gives a sound basis when appropriate templates are available, but ab initio modeling allows flexibility for new or disordered proteins. The emergence of AI-based software such as AlphaFold and RoseTTAFold has revolutionized the field to the extent of matching accuracy with experimental methods previously believed to be unique to them. As we proceed, joining these methods with experimental validation, dynamics simulation, and functional analysis will hold the key to gaining deeper insights into protein function and facilitating progress in drug discovery, enzyme design, personalized medicine, and synthetic biology [26].

Model Building: From Sequence to Structure

After having a protein sequence, the 3D modeling protocol typically entails:

Structural Template Identification

Knowledge of the 3D structure of a protein is important for exposing its function, interactions, and potential as a drug target. As genomic data have grown exponentially, the divide between protein sequences known and structures determined experimentally grows larger by the day. Computational prediction of protein structure offers a bridge across this divide in the form of an approach to estimate a protein's structure directly from its amino acid sequence. There are two major classes of computational protein structure prediction:

- Homology-based modeling, which takes advantage of known structures of evolutionarily related proteins.
- Template-free or ab initio modeling, which predicts structure from first-principles or by deep learning without requiring known homologous templates.

Both methods have their strengths and limitations and are appropriate to different biological contexts. Here, we discuss both approaches in detail, including the tools that are most typically used in each [27].

Homology-based structure prediction: homology-based modeling, or comparative modeling, is grounded on the principle that proteins with comparable sequences generally have similar structures. The technique takes advantage of the fact that there are experimentally determined structures available in databases like the PDB. If a protein whose structure is unknown (the target) has considerable sequence similarity with one whose structure is known (the template), the template's structural information can be transferred to construct a model of the target.

Workflow

- Template recognition: homology protein searches through databases to find structures with known proteins.
- Sequence comparison: comparing the target sequence to the chosen template(s).
- Model construction: employing the template skeleton to construct the 3D structure of the target.
- Model improvement and checking: loops, side-chain optimization, and checking of geometry.

Tools in Homology-based Prediction

Basic Local Alignment Search Tool (BLAST): BLAST is one of the most popular tools for detecting homologous sequences. It searches a query sequence against a sequence database to detect local similarity regions and is suitable for:

- rapid detection of potential templates for modeling;
- estimating the level of homology by sequence identity and E-values;
- removing unrelated sequences.

BLAST is efficient and quick but can miss low sequence identity distant homologs.

HHpred: HHpred is a more sensitive remote homolog detector that utilizes profile–profile comparisons rather than raw sequence alignment. It translates sequences into HMMs and aligns them with an HMM database containing known structures. HHpred performs best at the following:

- detecting distant evolutionary relationships;
- identifying structural homologs that BLAST misses;
- delivering good quality alignments for modeling.

It is broadly adopted in contemporary homology modeling pipelines because it is more sensitive.

Foldseek: Foldseek is a newer, extremely fast tool that searches for structural homologs by comparing 3D protein structures instead of sequences. It translates structures into low-dimensional representations to allow for fast structural alignment. Foldseek is especially helpful when:

- the user already has a predicted model (e.g., from AlphaFold2);
- the aim is to identify structurally similar proteins even without sequence similarity;
- exploring structural motifs or conserved folds in different sequences.

Foldseek widens the arsenal of homology-based methods by enabling comparisons of structures to structures beyond simple sequence analysis.

Strengths of Homology-based Modeling

- Very high accuracy for structures with >30% sequence identity.
- Highly established workflow with numerous stable tools and databases.
- Applicable to a large number of proteins with known relatives.

Limitations

- Poorly performing in the absence of close homologs.
- Loop regions, insertions, or disordered domains inaccuracy.
- May inherit template-specific mistakes or prejudices [28].

Template-free (ab initio) structure prediction: template-free modeling, or ab initio or de novo structure prediction, is employed when no appropriate homologs or structural templates are found. It tries to predict the 3D conformation of the protein from its primary sequence by physical, statistical, or AI-based approaches. Ab initio modeling has been difficult in the past because of the vast conformational space a protein can take. But with recent developments in machine learning and AI, this method has been transformed.

Key Approaches

- Physics-based simulations: employ MD, force fields, and energy minimization to determine the most stable conformation.
- Fragment assembly: assemble small structural fragments according to statistical probabilities.
- Deep learning: make predictions of inter-residue distances, contact maps, and finally 3D coordinates.

Tools in Template-free Prediction

AlphaFold2: AlphaFold2, created by DeepMind, is a breakthrough in protein structure prediction. It won the CASP14 competition by producing near-experimental accuracy for most tested proteins.

How AlphaFold2 Works

- Leverages MSAs and deep learning to learn evolutionary constraints.
- Uses attention-based neural networks to make residue–residue distance and orientation predictions.
- Constructs a 3D model from these predictions with end-to-end training.

Key Features

- Does not need close homologs, but it does benefit from co-evolutionary data.
- Makes per-residue confidence scores (pLDDT) predictions to determine model reliability.
- It is made available as open-source (AlphaFold2) and web-based (ColabFold, EBI AlphaFold Database).

AlphaFold2 can model proteins with low or no sequence similarity to known structures, and it is particularly useful for the following:

- uncharacterized or orphan proteins;
- disordered or flexible segments;
- large protein families with no known structures.

RoseTTAFold: by the Baker Lab, RoseTTAFold is another deep learning tool that predicts protein structures from a three-track network that jointly combines sequence, distance, and structural information.

Advantages

- Faster and more computationally accessible compared to AlphaFold2.
- Combines sequence and structure constraints at multiple levels.
- Has demonstrated robust performance in predicting multimeric complexes.

Both AlphaFold2 and RoseTTAFold have obfuscated the distinction between homology-based and ab initio prediction, as they are able to take advantage of patterns in large structural databases without copying template coordinates [29].

Both template-based and template-free (ab initio) approaches have made crucial contributions to protein structure prediction. Homology modeling is quick and accurate provided that there are appropriate templates; however, it is limited by sequence similarity requirements. BLAST, HHpred, and Foldseek have become increasingly sensitive and precise in detecting templates, thereby allowing the application of homology modeling to be expanded. Conversely, template-free prediction, previously constrained by low accuracy and computational complexity, has undergone a revolution with AI. Applications such as AlphaFold2 and RoseTTAFold have proved that precise predictions are possible without structural templates, creating new chances for the investigation of uncharacterized proteins and novel structural folds. The future of protein modeling will be hybrid integrated approaches, where homology, AI, and experimental information come together to construct accurate, high-resolution models, driving discoveries in structural biology, drug design, and synthetic biology.

Model Construction

- Homology models are built by transferring backbone coordinates from the template, followed by modeling side chains and loops.
- AI-based predictions output atomic coordinates directly from learned sequence–structure relationships.

Model Refinement

To provide stereochemical accuracy, models typically undergo the following:

- force field energy minimization (e.g., CHARMM, AMBER);
- loop modeling for insertions or deletions;
- rotamer library side-chain optimization (e.g., SCWRL);
- MD simulation to relax and equilibrate [30].

Validation and Quality Assessment (QA)

Validation of a 3D model is necessary prior to downstream use. Popular tools and metrics are the following:

Structure Quality Tools

Protein Structure Validation: Tools and Their Applications

Computational prediction of protein structure is an essential tool in contemporary biology, allowing scientists to make inferences about the 3D conformation of proteins when experimental information is not available. Nevertheless, the accuracy of a predicted model should be thoroughly tested to guarantee that it correctly reflects the physical and chemical restraints of actual proteins. In the absence of proper validation, downstream applications like molecular docking, drug design, or functional annotation can result in false or erroneous conclusions. Protein structure validation entails the evaluation of geometric, stereochemical, and energetic parameters to detect errors or inconsistencies in the model. Tools have been developed for this purpose that address different protein quality aspects. Some of the most frequently used include the Ramachandran plot (through PROCHECK), MolProbity, ProSA-web, QMEAN, and Verify3D. These programs provide both quantitative scores and graphical diagnostics that assist modelers in refining and validating predicted structures prior to their use for biological interpretation [31].

Ramachandran plot (PROCHECK): Ramachandran plot, named after G. N. Ramachandran, who proposed it in 1963, is the most essential of the protein structure validation tools. It analyzes the dihedral angles of the backbone, phi (φ) and psi (ψ), of every amino acid residue in a protein. These are the angles that give the conformational space that the residue will take up without strain or steric clash. PROCHECK is a popular package of software that automatically generates and analyzes Ramachandran plots. It classifies residues into the following:

- Favored regions: sterically permitted and energetically favorable conformations.
- Allowed regions: acceptable but less frequently seen in high-quality crystal structures.
- Disallowed regions: uncommon conformations that can be an indication of errors or strained geometry.

Application: a high-quality model will normally have more than 90% of residues in the preferred areas, and relatively few in the disallowed regions. Deviation can indicate regions that are out of alignment, bad template matches, or loop/turn incorrect modeling. For this reason, the Ramachandran plot is an essential first checkpoint to assess model quality [32].

MolProbity: MolProbity is a general all-atom structure validation program from the Richardson Laboratory at Duke University. It extends beyond backbone angles to judge a range of structural aspects, including the following:

- Clashes: van der Waals overlaps between atoms.
- Rotamers: side-chain conformations, judged against a rotamer library.
- Backbone geometry: bond lengths and angles in comparison with high-resolution crystal structures.

Features: MolProbity offers several outputs:

- Clashscore: number of significant atomic overlaps per 1,000 atoms – lower values are favorable.
- Ramachandran statistics: integrated.
- All-atom contact analysis: helpful in detecting buried clashes not seen in backbone-only structures.
- Hydrogen addition: For hydrogen-bonding assessment.

Significance: the tool is most helpful in side-chain conformation refinement, enhancing rotamer outliers, and steric problems that may compromise molecular simulations or docking. MolProbity's synergy of geometric, steric, and chemical checks makes it among the most stringent validators out there [33].

ProSA-web: ProSA is an online tool that evaluates the overall quality of a protein model based on statistical potentials derived from known structures. It provides both global and local error estimates.

Key Features

- *Z*-score: indicates how far the input structure deviates from the average score of experimentally determined structures of similar size. A *Z*-score outside the range of native proteins indicates problematic models.
- Energy profile: computes a residue-level energy score from a knowledge-based potential. Negative values are typically good, and peaks in positive energy point to local structural issues.

Advantages: ProSA is of greatest use in detecting misfolded areas and determining if a model falls within the range of the known protein structures. Its easy-to-use web interface as well as graphical output make it a favorite among structural biologists [34].

QMEAN: QMEAN is a sophisticated scoring function that merges several structural descriptors into one composite quality score. It was created by the Swiss Institute of Bioinformatics and compares the target model to high-resolution structures in the PDB.

Scoring Components

QMEAN assesses the following:

- Torsion angle potential: quality of local geometry.
- Solvation potential: hydrophobic/hydrophilic residue exposure and burial.
- Secondary structure agreement: agreement with predicted secondary elements.
- Packing quality: amount of optimal atomic packing.

The QMEAN *Z*-score is normalized to high-resolution X-ray structures, and it gives a comparative measure of model quality.

Applications: QMEAN is particularly helpful for ranking several models, checking models from ab initio or AI-based predictions, and combining with tools such as SWISS-MODEL. It provides a balanced evaluation that includes both local and global quality measures [35].

Verify3D: Verify3D is a validation tool for structures that checks each residue in a model against ideal profiles obtained from high-quality experimental structures.

How it works: every residue in a protein model is categorized on the basis of the following:

- solvent exposure (buried or exposed);
- secondary structure context (helix, strand, coil);
- polarity and environment (polar or nonpolar environments).

These features are then matched to expected patterns for that residue type in known structures. A residue score ≥0.2 is generally considered acceptable. The tool reports the percentage of residues with acceptable profiles, and models scoring ≥80% pass the assessment [36].

Strengths: Verify3D works very well in detecting local inaccuracies, particularly those resulting from erroneous loop building or side-chain positioning. It's also worthwhile in detecting mismatches between sequence-based expectations and 3D context.

Ramachandran plots detect backbone strain and misfolding:

- MolProbity deals with atom-level clashes and side-chain conformations.
- ProSA and QMEAN provide statistical and energetic assessments.
- Verify3D verifies environment compatibility between sequence and structure [37].

The use of these tools together guarantees that both global model accuracy and local structural detail are checked. This is especially crucial in areas such as drug discovery, where faulty geometry or local misfolding can result in erroneous docking predictions or interaction analyses. Protein structure validation is a required step in any computational modeling pipeline. Independent verification of model quality, irrespective of whether the model originates from homology modeling, ab initio prediction, or AI-based software such as AlphaFold, is necessary to validate reliability for further applications. PROCHECK (Ramachandran plots), MolProbity, ProSA-web, QMEAN, and Verify3D are important quality control checkpoints for assessing geometric, stereochemical, and environmental aspects of a protein model. As structural bioinformatics advances, the incorporation of multi-tool validation approaches will be crucial to the assurance that computational models reach the precision and usefulness of experimental structures, leading the way toward progress in functional genomics, structural biology, and rational drug design [38].

AI-based Confidence Metrics

- LDDT (AlphaFold): residue-wise confidence scores.
- PAE: domain-specific confidence for the overall structure.

A good model must demonstrate >90% residues in favored Ramachandran areas and few steric clashes or aberrant bond angles [39].

Visualization Tools and Techniques

Effective visualization is key to interpreting, analyzing, and presenting protein structures. Modern tools offer real-time, interactive 3D environments for exploring atomic details, highlighting functional sites, and designing mutations [40].

Popular Visualization Tools

Tool	Features	Platform
PyMOL	High-quality rendering, publication-ready images, scripting, plugins	Desktop
Chimera/ ChimeraX	Advanced analysis, density map fitting, morphing animations	Desktop
VMD	Visualization of MD trajectories	Desktop
Mol	Web-based, PDB-integrated, interactive interface	Web
Jmol/JSmol	Lightweight Java-based viewer, suitable for education	Web/Standalone
iCn3D	NCBI's web-based tool with annotations and alignment	Web

Visualization Capabilities

- Secondary structure display: helices, sheets, and loops.
- Surface and electrostatic maps: highlight active sites or ligand-binding regions.
- Color-coding: by hydrophobicity, charge, conservation, or confidence.
- Animations: morphing between conformations or showing MD.

For AI-predicted structures, visualization also helps assess the reliability of flexible regions (low pLDDT) and domain interfaces (high PAE) [41].

Applications of 3D protein models: 3D models support a vast variety of applications across fields:

Drug Design and Docking

- Identification of ligand-binding pockets.
- Virtual screening of small molecules.
- Docking and scoring ligands with the help of tools such as AutoDock, GOLD, SwissDock, and Schrödinger Glide [42].

Protein–Protein Interactions

- Prediction of interaction interfaces by docking tools (e.g., ClusPro, HADDOCK).
- Prediction of complex formation and co-evolution.

Mutational Analysis

- Inspection of disease-related mutations.
- Prediction of structural destabilization or disruption of function.
- Investigation of mutational tolerance and epistasis [43].

Synthetic Biology and Enzyme Engineering

- Synthesizing new enzymes with a customized active site.
- Synthesizing fusion proteins or artificial scaffolds.
- Predicting substrate specificity and catalytic mechanisms.

Education and Public Engagement

- 3D interactive models in classrooms or on the web.
- Illustrating molecular processes and structural transitions [44].

Limitations and Future Perspectives

Current Limitations

- Static models: most modeling programs don't compensate for protein flexibility or allosteric transitions.
- Low-confidence regions: loops, IDRs, and termini are less reliable.
- Complex environments: modeling proteins in membranes or large assemblies remains challenging.
- The post-close modifications: phosphorylation, glycosylation, and disulfide bridges are usually ignored or inaccurately predicted [45].

Future Directions

- Hybrid modeling: incorporating data from cryo-EM, SAXS, crosslinking, and AI predictions.
- Automated pipelines: using end-to-end systems that construct, optimize, and validate models (e.g., GalaxyWEB, I-TASSER Suite).
- Time-resolved modeling: merging MD simulations with AI to describe conformational dynamics.
- Community resources: development of databases such as AlphaFold DB and the Protein Ensemble Database (PED) [46].

Integration with MD Simulations

Although static 3D models are useful snapshots of protein conformation, biological systems are dynamic. Proteins fluctuate continuously in their conformation, which is essential for the function, ligand binding, and allosteric control. These dynamics can be accessed by MD simulations, which calculate the motion of atoms over time based on physical laws.

Why Combine MD with Modeling?

- Refinement: stereochemical mistakes or strained loops may be present in AI-based models. MD can rectify them by energy minimization and relaxation.
- Exploring conformational space: a variety of conformations, including open/closed states of enzymes or transporter cycles, can be sampled through MD trajectories.

- Simulating environmental effects: MD enables proteins to be simulated in realistic environments, solvated, in membranes, or at various pH/ionic strengths [47].

Software such as GROMACS, NAMD, and AMBER are popular and frequently complemented with visualizers such as VMD or ChimeraX for the analysis and interpretation of dynamics. MD-augmented modeling is particularly helpful for the following:

- membrane-bound proteins (e.g., GPCRs, ion channels);
- protein–ligand binding/unbinding processes;
- folding/unfolding studies [48].

Advanced Model QA

With proteins modeled routinely, maintaining quality and reproducibility becomes essential. While conventional tools are sufficient, recent AI-driven QA tools provide better precision:

- ModFOLD: uses machine learning and statistical scoring to predict global/local accuracy.
- DeepAccNet: estimates residue-level accuracy by using deep learning on input models.
- VoroMQA: uses Voronoi tessellation and statistical potentials to predict model quality.

These programs assist researchers in ranking several predicted models, detecting unreliable regions, and ordering models for further investigation or experimental verification. In addition, benchmarking initiatives like that of CAMEO (Continuous Automated Model Evaluation) continually compare the performance of models in real time against experimentally solved structures updated in real time [49].

3D Modeling of Complexes and Assemblies

It is a significant challenge to predict the structure of complex multimeric proteins, including homodimers, heterooligomers, or large molecular machines. AI has progressed here to a great extent:

- AlphaFold-Multimer: a dedicated version of AlphaFold2 for protein–protein complex prediction. It works well for stable, well-known interactions but might fail to capture transient or weak interactions.
- Docking tools: software such as HADDOCK, ClusPro, and ZDOCK remain in use to model binary or multi-chain complexes using rigid-body or flexible docking [50].

Interface Analysis and Hotspot Prediction

- Tasks such as PDBePISA, HotSpot Wizard, and InterProSurf recognize crucial residues that take part in interactions and binding.

- Mapping binding hotspots aids drug design, antibody engineering, and synthetic interaction networks.

In structural systems biology, these intricate models can be incorporated into macromolecular machines, like the ribosome, proteasome, or nuclear pore complex, frequently merging AI predictions with cryo-EM densities [51].

Databases and Web Portals' Role

With the rampant increase in protein models, various databases and web portals have appeared to store, visualize, and annotate 3D structures:

- AlphaFold Protein Structure Database: holds more than 200 million predicted structures for a variety of species [52].
- Swiss-Model Repository: reliable homology models with reliability annotations.
- ModelArchive: a PDB-like model archive for storing and sharing predicted or computed models.
- Moland iCn3D: facilitates direct structure visualization in a browser, suitable for collaborative and educational uses.

These websites are open science friendly, allowing researchers all over the world to access, download, and utilize high-quality models without having to use high-performance computing resources [53].

Outreach and Education Applications

3D protein structures are not only potent research tools, yet they also provide captivating learning and science outreach resources:

- Protein shapes, folds, and mutations can be accessed by students through PyMOL, Jmol, or web-based viewers.
- AR/VR environments (e.g., Nanome, FoldIt VR) are now being applied to classrooms and public displays to interact with molecular models in virtual space.
- Games of citizen science such as Foldit apply gamification to engage the public in actual scientific discovery, for example, protein puzzle-solving or enzyme design.

With the democratization of access to protein structures by AI, this convergence of science, technology, and society becomes increasingly effective [54].

Convergence of Modeling Tools and Real-world Case Studies

Whereas individual software such as homology modeling, AI predictors, and visualizers are strong by themselves, the potential of protein modeling is fully utilized when these software tools are incorporated into comprehensive pipelines for addressing sophisticated biological challenges [55].

Constructing an Integrated Protein Modeling Pipeline

A complete modeling pipeline often consists of several tools and steps, each being an essential part of structure prediction, refinement, validation, and application:

Typical Workflow

- Input: gene or protein sequence of interest.
- Template search (BLAST, HHpred) if homology modeling is possible.
- Structure prediction (AlphaFold2, RoseTTAFold, and SWISS-MODEL).
- Model refinement (energy minimization, loop modeling, and MD simulation).
- Model validation (Ramachandran plot, MolProbity, and ProSA).
- Visualize and annotate (PyMOL, ChimeraX, Mol*, and iCn3D).
- Functional analysis (binding site prediction, mutational mapping).
- Docking or dynamics (AutoDock, HADDOCK, and GROMACS).
- Iterative refinement with experimental feedback (cryo-EM, SAXS, and mutagenesis).

Current platforms such as I-TASSER, Phyre2, and GalaxyWEB provide unified web-based interfaces, whereas local workflows might integrate tools in tailored pipelines through Python, R, or shell scripting [56].

Case Study 1: COVID-19 Spike Protein Modeling and Vaccine Development

Background: early in the COVID-19 pandemic, structural data regarding the SARS-CoV-2 spike (S) protein were urgently required.

Modeling Method

- Homology models based on the SARS-CoV spike protein were developed initially.
- With data available, AlphaFold2 and cryo-EM structures were incorporated for high-resolution insights.
- Effects of mutational variants (e.g., Delta, Omicron) were studied using 3D mapping on spike models.
- MD simulations and docking were employed to investigate antibody binding, receptor interactions (ACE2), and neutralization.

Outcome

- Structure-guided vaccine design resulted in stabilized prefusion spike constructs.
- Assisted in rapid mRNA vaccine development (Moderna, Pfizer-BioNTech) and neutralizing antibody therapies [57].

Case Study 2: Cancer Mutations in p53 Tumor Suppressor

Background: the tumor suppressor protein p53 is mutated in ~50% of human cancers. Elucidation of structural consequences of these mutations is essential for cancer therapy.

Modeling Strategy

- Full-length p53 (missing in crystal structures) was reconstructed using homology models and AlphaFold2 predictions.
- Mutational hotspots (e.g., R175H, R273H) were positioned on the 3D structure.
- Stability prediction programs (e.g., FoldX, DynaMut) determined how mutations affect folding or DNA binding.

Outcome

- Assisted identification of rescue mutants and the design of small molecules to stabilize mutant p53.
- Directed drug discovery efforts to reactivation of mutant p53 [58].

Case Study 3: Designing De Novo Enzymes

Background: designing enzymes to catalyze reactions that don't occur naturally is a significant objective of synthetic biology [59].

Modeling Strategy

- RoseTTA and ProteinMPNN were employed to design new enzyme scaffolds.
- 3D models were constructed de novo or edited from available templates.
- AI-assisted validation verified folding potential.
- Active sites were optimized by ligand docking and transition-state modeling [60].

Outcome

- New enzymes were designed to catalyze industrially important reactions (e.g., esterification, carbon capture).
- Some were experimentally confirmed and reported in *Nature* and *Science* [61].

Case Study 4: Interpretation of Rare Disease Variants

Background: rare disease patients frequently carry mutations of uncertain significance. Structural modeling aids in interpretation of these variants [62].

Modeling Approach

- Protein models were constructed using AlphaFold2, especially when no template was available.
- Tools such as MutPred, PolyPhen, and Missense3D mapped variants to structures.
- Structural analysis identified if mutations were found in active sites, core domains, or interface interfaces [63].

Outcome

- Clinicians may categorize variants as pathogenic, likely benign, or uncertain.
- Structural information substantiated genetic counseling and individualized treatment decisions [64].

Future Prospects in Integrated Protein Modeling

- Automated pipelines: complete integrated modeling-to-docking platforms are unfolding for high-throughput drug screening.
- Cloud computing and APIs: tools such as Google ColabFold, ESM Metagenomics, and RCSB tools are rendering modeling accessible worldwide.
- Multi-omics integration: integrating modeling with transcriptomics, proteomics, and metabolomics for systems-level analysis.
- Interactive AI assistants: large language models (LLMs) (like this one) will increasingly assist in designing, interpreting, and optimizing protein models in real time [65].

Experimental Validation of Predicted Protein Structures

Introduction

While computational models, especially those from AI tools like AlphaFold2, have revolutionized our ability to predict protein structures rapidly and at scale, experimental validation remains essential. Predicted models must be confirmed to ensure the following:

- accuracy of atomic coordinates;
- correct folding and functional domains;
- biological relevance under physiological conditions.

Experimental validation creates ground truth that computational methods seek to approximate. This chapter considers the key experimental techniques, model confirmation strategies, and some examples where modeling and experimentation integration have led to biological discovery [66].

Key Experimental Techniques for Protein Structure Determination

X-ray Crystallography

- Resolution: high (up to ~1 Å).
- Optimal for: stable, crystallizable proteins.
- Mechanism: proteins are crystallized, and X-ray diffraction patterns are examined to calculate electron density maps, from which atomic models are constructed.
- Validation use: compare predicted structure with crystal-derived coordinates; ensure root-mean-square deviation (RMSD), secondary structure accuracy, and side-chain packing [67].

Cryo-electron Microscopy

- Resolution: medium to high (~2–5 Å; some near-atomic).
- Best suited for: large complexes, membrane proteins, and flexible assemblies.

- Process: proteins are frozen in an instant and imaged with an electron beam to produce 2D projections that are reconstructed to form 3D structures.
- Validation use: superimpose predicted models upon experimental density maps; determine flexibility and domain orientation [68].

NMR Spectroscopy

- Resolution: medium (~2–3 Å), optimal for small proteins (<40 kDa).
- Best for: dynamic, soluble proteins; solution-state analysis.
- How it works: measures interactions between atomic nuclei using magnetic fields; produces distance constraints for structure calculation.
- Validation use: test consistency of model prediction with NOE (Nuclear Overhauser Effect) patterns and dihedral angle constraints [69].

Small-angle X-ray Scattering

- Resolution: low (~10–30 Å).
- Best for: shape and conformational studies.
- How it works: measures X-ray scattering in solution to determine global shape and size.
- Validation use: test if the predicted model is within the SAXS envelope [70].

Structural Comparison and Validation Metrics

Experimental data are compared to predicted models using quantitative measures:

- RMSD: quantifies average distance between atoms (typically backbone Cα atoms).
- Template modeling score (TM-score): assesses global fold similarity; scores >0.5 typically indicate correct topology.
- Global distance test-total score (GDT-TS): quantifies structural similarity by superimposition [71].
- MolProbity scores: evaluate stereochemistry, clashes, and rotamer outliers.
- Fit-to-density metrics (cryo-EM): cross-correlation between model and density map.
- R-free/R-work (crystallography): measures how well the model agrees with experimental diffraction data.

These validation tools are critical in model refinement and confidence evaluation in AI predictions [72].

Wet Lab Functional Validation Strategies

At times, structural predictions cannot be solved experimentally directly. Rather, functional validation is employed to deduce the correctness of a model:

- Site-directed mutagenesis: make mutations at predicted functional residues and measure the effect on activity or stability.
- Enzyme assays: validate catalytic activity in predicted active sites.

- Ligand binding studies: use ITC, SPR, or fluorescence to challenge binding predictions.
- Thermal shift assays (TSA): observe changes in stability due to predicted mutations or ligand interactions.
- Crosslinking and mass spectrometry: map contact points between residues and confirm predicted tertiary structure [73].

Hybrid Modeling and Validation Strategies

Hybrid strategies merge computational and experimental methods to construct or update models:

- AlphaFold-predicted domains are docked into low-resolution EM maps to assemble entire complexes.
- Filter or narrow multiple computational models based on SAXS data, then use homology models.
- Validate or improve AlphaFold/RoseTTAFold predictions with NMR-derived constraints.

These multicomponent methods commonly occur in structural systems biology, membrane protein structure prediction, and drug–target structure investigations [74].

Successful Validation Examples

AlphaFold2 and CASP14: AlphaFold2 reached near-experimental accuracy in the 14th CASP competition. Follow-up experimental comparisons revealed:

- RMSDs of less than 1 Å for most globular proteins.
- Correct prediction of flexible loops and disordered regions (when confident).
- Excellent consistency with newly determined X-ray and cryo-EM structures.

p53 and cancer variants: predicted p53 mutants were confirmed through crystallography and functional assays. Structural modeling accurately predicted loss-of-function mutants and informed stabilizing drug design.

SARS-CoV-2 protein complexes: AI-predicted spike and accessory protein structures were confirmed by cryo-EM, SAXS, and functional binding assays. The models were important for therapeutic interventions worldwide [75].

Challenges and Limitations

- Dynamic zones and IDRs: challenging to confirm due to intrinsic flexibility.
- Multimeric complexes: modeling and validation of large complexes accurately is still challenging.
- Post-translational modifications: AI models tend to overlook PTMs that can affect structure.
- Model-biology discrepancies: a "correct" model can still be wrong biologically if expression systems, cofactors, or environmental conditions are different [76].

Experimental validation is a key step to closing the prediction–reality gap. Although computational models can offer near-native accuracy, only experimental data can verify structural correctness and biological function. The future belongs to hybrid approaches, where AI-based models are employed as starting points and improved with experimental inputs, boosting structure-based discovery like never before. As methods of validation grow faster, cheaper, and more accessible, the prediction–validation–application loop will increasingly become seamless [77–79].

Summary and Future Directions

Present Challenges and Limitations

Despite phenomenal progress, there are a number of challenges still outstanding:

- Disordered and flexible regions: AI models continue to be challenged by IDPs, loops, and mobile domains.
- Protein complexes and dynamics: predicting multimeric assemblies and conformational transitions (e.g., allosteric shifts) remains inaccurate.
- Post-translational modifications: most modeling pipelines lack precise treatment of PTMs such as phosphorylation, glycosylation, and methylation.
- Context-dependence: the predictions can fail to capture structural changes induced by ligands, pH, ionic conditions, or cofactors [80–82].
- Experimental integration: seamless integration of computation and experiment is not yet mature; there are hybrid methods, but this is an emerging area.

To overcome these challenges will involve interdisciplinarity and developments in data gathering, algorithm development, and shared platforms.

Future Trends and Technologies

The prospects for protein structure modeling in the future are very exciting, with promising directions already emerging:

- AI-enhanced refinement: technologies such as AlphaFold3 (expected) and hybrid ML–MD architectures will enable better modeling of dynamics, interactions, and environmental factors.
- Personalized modeling: coupling with genomics will make structure-based precision medicine possible, predicting the effect of single mutations on protein structure and function.
- Cloud and collaborative platforms: open-access tools (e.g., Google ColabFold, ESM Atlas) are making modeling accessible to researchers globally.
- Protein design generative models: methods such as ProteinMPNN, ESM-IF, and Chroma allow for the de novo design of proteins, transforming synthetic biology and enzyme engineering.
- LLMs and AI assistants: language models will help scientists ever more by eliminating pipeline stages, interpreting results, and even suggesting mutations or designs from structure–function paradigms [83, 84].

Conclusion

This chapter showcases the tremendous progress made in protein structure prediction from classical homology modeling to modern 3D modeling enabled by the application of AI. Though homology modeling is still very useful if suitable templates are available, major advances have recently been achieved in the application of AI, in the form of deep learning models, to protein structure prediction. The integration of AI predictive models with refinement, validation steps, and MD simulations has increased the biological accuracy of computational modeling and made them useful tools for drug discovery, functional annotation, and protein engineering. The ever-increasing convergence between computational and experimental techniques is leading to a situation where protein modeling is expected to be a flagship area of future structural biology endeavors. The development of more hybrid and integrative tools is anticipated to improve our capability to model protein structures with high accuracy.

List of Abbreviations

AI	Artificial intelligence
ANN	Artificial neural network
CASP	Critical Assessment of Structure Prediction
CNN	Convolutional neural network
CPU	Central processing unit
Cryo-EM	Cryogenic electron microscopy
DL	Deep learning
DNA	Deoxyribonucleic acid
GPU	Graphics processing unit
GNN	Graph neural network
HMM	Hidden Markov model
MD	Molecular dynamics
ML	Machine learning
MSA	Multiple sequence alignment
NMR	Nuclear magnetic resonance
PDB	Protein Data Bank
PSI-BLAST	Position-specific Iterated Basic Local Alignment Search Tool
RMSD	Root-mean-square deviation
RNA	Ribonucleic acid
SAXS	Small-angle X-ray scattering
SGD	Stochastic gradient descent
SVM	Support vector machine
TM-Score	Template Modeling Score
X-ray	X-ray crystallography

References

1 Chen L, Li Q, Nasif KF, et al. AI-driven deep learning techniques in protein structure prediction. *Int. J. Mol. Sci.* 2024;25(15):8426.

2 Schauperl M, Denny RA. AI-based protein structure prediction in drug discovery: impacts and challenges. *J. Chem. Inf. Model.* 2022;62(13):3142–56.

3 Klebe G. Protein modeling and structure-based drug design. In: *Drug Design: From Structure and Mode-of-Action to Rational Design Concepts*. Springer; 2025. pp. 309–21.

4 Kumar H, Kim P. Artificial intelligence in fusion protein three-dimensional structure prediction: review and perspective. *Clin. Transl. Med.* 2024;14(8):e1789.

5 Chung J, Hahn H, Flores-Espinoza E, Thomsen AR. Artificial intelligence: a new tool for structure-based g protein-coupled receptor drug discovery. *Biomolecules* 2025;15(3):423.

6 Yu CH, Buehler MJ. Sonification based *de novo* protein design using artificial intelligence, structure prediction, and analysis using molecular modeling. *APL Bioeng.* 2020;4(1):016108.

7 Goel D, Kumar R, Kumar S. AI-assisted methods for protein structure prediction and analysis. In: *Microbial Data Intelligence and Computational Techniques for Sustainable Computing*. Springer Nature Singapore; 2024. pp. 365–91.

8 Acharya R, Shetty SS, Pavan G, et al. AI-based homology modelling of fatty acid transport protein 1 using AlphaFold: structural elucidation and molecular dynamics exploration. *Biomolecules* 2023;13(11):1670.

9 Nam KH. AI-based protein models enhance the accuracy of experimentally determined protein crystal structures. *Front. Mol. Biosci.* 2023;10:1208810.

10 Si D, Nakamura A, Tang R, et al. Artificial intelligence advances for de novo molecular structure modeling in cryo-electron microscopy. *Wiley Interdiscip. Rev. Comput. Mol. Sci.* 2022;12(2):e1542.

11 Tsuchiya Y, Yamamori Y, Tomii K. Protein–protein interaction prediction methods: from docking-based to AI-based approaches. *Biophys. Rev.* 2022;14(6):1341–8.

12 Guo SQ, Li YD, Chen P, et al. AI protein structure prediction-based modeling and mutagenesis of a protostome receptor and peptide ligands reveal key residues for their interaction. *J. Biol. Chem.* 2022;298(10):102440.

13 Sebastiano MR, Ermondi G, Hadano S, Caron G. AI-based protein structure databases have the potential to accelerate rare diseases research: AlphaFoldDB and the case of IAHSP/Alsin. *Drug Discov. Today* 2022;27(6):1652–60.

14 Versini R, Sritharan S, Aykac Fas B, et al. A perspective on the prospective use of AI in protein structure prediction. *J. Chem. Inf. Model.* 2023;64(1):26–41.

15 Binbay FA, Rathod DC, George AA, Imhof D. Quality assessment of selected protein structures derived from homology modeling and AlphaFold. *Pharmaceuticals* 2023;16(12):1662.

16 Qiu X, Li H, Ver Steeg G, Godzik A. Advances in AI for protein structure prediction: implications for cancer drug discovery and development. *Biomolecules* 2024;14(3):339.

17 Zhao N, Wu T, Wang W, et al. Review and comparative analysis of methods and advancements in predicting protein complex structure. *Interdiscip. Sci.: Comput. Life Sci.* 2024;16(2):261–88.

18 Cheng J, Liang T, Xie XQ, et al. A new era of antibody discovery: an in-depth review of AI-driven approaches. *Drug Discov. Today* 2024;29(6):103984.

19 Chandra A, Goyal N, Behera A, et al. Artificial intelligence-based protein structure prediction and systems biology-guided smart drug screening. In: *Biochemical and Molecular Pharmacology in Drug Discovery*. Elsevier; 2024. pp. 323–352.

20 Muhammed MT, Aki-Yalcin E. Up-to-date developments in homology modeling. In: *Applied Computer-Aided Drug Design: Models and Methods*. Bentham Science; 2023. pp. 116–35.

21 Kellici TF, Hristozov D, Morao I. AI-based protein structure predictions and their implications in drug discovery. In: *Computational Drug Discovery: Methods and Applications*. John Wiley & Sons; 2024. pp. 1227–53.

22 Nussinov R, Zhang M, Liu Y, Jang H. AlphaFold, artificial intelligence (AI), and allostery. *J. Phys. Chem. B* 2022;126(34):6372–83.

23 Valdés-Tresanco MS, Valdés-Tresanco ME, Jiménez-Gutiérrez DE, Moreno E. Structural modeling of nanobodies: a benchmark of state-of-the-art artificial intelligence programs. *Molecules* 2023;28(10):3991.

24 Choudhury C, Murugan NA, Priyakumar UD. Structure-based drug repurposing: Traditional and advanced AI/ML-aided methods. *Drug Discov. Today* 2022;27(7):1847–61.

25 Kim J, McFee M, Fang Q, et al. Computational and artificial intelligence-based methods for antibody development. *Trends Pharmacol. Sci.* 2023;44(3):175–89.

26 Durham J, Zhang J, Humphreys IR, et al. Recent advances in predicting and modeling protein–protein interactions. *Trends Pharmacol. Sci.* 2023;48(6):527–38.

27 Narayan A, Chanana A, Shanker O, et al. Role of artificial intelligence in pharmaceutical drug development. *Curr. Indian Sci.* 2024;2(1):e2210299X313252.

28 Mosalaganti S, Obarska-Kosinska A, Siggel M, et al. AI-based structure prediction empowers integrative structural analysis of human nuclear pores. *Science* 2022;376(6598):eabm9506.

29 Heinzinger M, Rost B. Artificial intelligence learns protein prediction. *Cold Spring Harb. Perspect. Biol.* 2024;16(9):a041458.

30 Graille M, Sacquin-Mora S, Taly A. Best practices of using AI-based models in crystallography and their impact in structural biology. *J. Chem. Inf. Model.* 2023;63(12):3637–46.

31 Dhakal A, McKay C, Tanner JJ, Cheng J. Artificial intelligence in the prediction of protein–ligand interactions: recent advances and future directions. *Brief. Bioinform.* 2022;23(1):bbab476.

32 Wodak SJ, Vajda S, Lensink MF, et al. Critical assessment of methods for predicting the 3D structure of proteins and protein complexes. *Annu. Rev. Biophys.* 2023;52(1):183–206.

33 Wuyun Q, Chen Y, Shen Y, et al. Recent progress of protein tertiary structure prediction. *Molecules* 2024;29(4):832.
34 Mallik BB, Stanislaw J, Alawathurage TM, Khmelinskaia A. De novo design of polyhedral protein assemblies: before and after the AI revolution. *ChemBioChem* 2023;24(15):e202300117.
35 Chowdhury R. Advances in protein structure, function, and design. *Front. Bioeng. Biotechnol.* 2023;10:1108962.
36 Tang X, Dai H, Knight E, et al. A survey of generative AI for de novo drug design: new frontiers in molecule and protein generation. *Brief. Bioinform.* 2024;25(4):bbae338.
37 Chen Z, Wang R, Guo J, Wang X. The role and future prospects of artificial intelligence algorithms in peptide drug development. *Biomed. Pharmacother.* 2024;175:116709.
38 Cen LP, Ng TK, Ji J, et al. Artificial intelligence-based database for prediction of protein structure and their alterations in ocular diseases. *Database* 2023; 2023:baad083.
39 Szarecka A, Dobson C. A tool to teach evolution of protein sequences and structures: prediction of protein structure by building homology models. *ABT* 2024;86(2):108–15.
40 Agajanian S, Alshahrani M, Bai F, et al. Exploring and learning the universe of protein allostery using artificial intelligence augmented biophysical and computational approaches. *J. Chem. Inf. Model.* 2023;63(5):1413–28.
41 Bergonzo C, Grishaev A. Critical assessment of RNA and DNA structure predictions via artificial intelligence: the imitation game. *J. Chem. Inf. Model.* 2025;65(7):3544–54.
42 Xiao H, Zou Y, Wang J, Wan S. A Review for artificial intelligence based protein subcellular localization. *Biomolecules* 2024;14(4):409.
43 Liu H, Chen Q. Computational protein design with data-driven approaches: Recent developments and perspectives. *Wiley Interdiscip. Rev. Comput. Mol. Sci.* 2023;13(3):e1646.
44 Rosignoli S, Pacelli M, Manganiello F, Paiardini A. An outlook on structural biology after AlphaFold: tools, limits and perspectives. *FEBS Open Bio.* 2025;15(2):202–22.
45 Kim S, Lee W, Kim HI, et al. Recent advances and future challenges in predictive modelling of metalloproteins by artificial intelligence. *Mol. Cells* 2025;48(4):100191.
46 Li S, Wu S, Wang L, et al. Recent advances in predicting protein–protein interactions with the aid of artificial intelligence algorithms. *Curr. Opin. Struct. Biol.* 2022;73:102344.
47 Liu H, Zhuo C, Gao J, et al. AI-integrated network for RNA complex structure and dynamic prediction. *Biophys. Rev.* 2024;5(4):041304.
48 Ziemianowicz DS, Kosinski J. New opportunities in integrative structural modeling. *Curr. Opin. Struct. Biol.* 2022;77:102488.
49 Fatima N, Khan S, Zahid S. A critical address to advancements and challenges in computational strategies for structural prediction of protein in recent past. *Comput. Biol. Chem.* 2025;117:108430.

50 Wang W, Huang X. AI empowers exploration of protein universe. *Sci. China Life Sci.* 2025;68(2):588–9.

51 Guzmán-Vega FJ, González-Álvarez AC, Peña-Guerra KA, et al. Leveraging AI advances and online tools for structure-based variant analysis. *Curr. Protoc.* 2023;3(8):e857.

52 Hetmann M, Parigger L, Sirelkhatim H, et al. Folding the human proteome using BioNeMo: a fused dataset of structural models for machine learning purposes. *Sci. Data* 2024;11(1):591.

53 Mallick B, Dutta A, Mondal P, Dutta M. Proteomic analysis and protein structure prediction of *Shigella* phage Sfk20 based on a comparative study using structure prediction approaches. *Proteins: Struct. Funct. Bioinf.* 2024;92(5):637–48.

54 Dahlström KM, Salminen TA. Apprehensions and emerging solutions in ML-based protein structure prediction. *Curr. Opin. Struct. Biol.* 2024;86:102819.

55 Scardino V, Di Filippo JI, Cavasotto CN. How good are AlphaFold models for docking-based virtual screening? *iScience*. 2023;26(1):105920.

56 Karelina M, Noh JJ, Dror RO. How accurately can one predict drug binding modes using AlphaFold models? *eLife* 2023;12:RP89386.

57 Chang Y. A study on the end-to-end AlphaFold 3D protein structure prediction method. In: *Proceedings of the 2024 International Conference on Electronics and Devices, Computational Science (ICEDCS).* IEEE; 2024. pp. 258–62.

58 Hidayat AA, Nirwantono R, Pardamean B. An exploratory computational approach for elucidating a nonstructural protein interaction in dengue virus 2. *Procedia Comput. Sci.* 2024;245:470–9.

59 Guillon C, Robert X, Gouet P. "It's Only a Model": when protein structure predictions need experimental validation, the case of the HTLV-1 tax protein. *Pathogens* 2024;13(3):241.

60 Rosignoli S, Lustrino E, Di Silverio I, Paiardini A. Making use of averaging methods in modeller for protein structure prediction. *Int. J. Mol. Sci.* 2024;25(3):1731.

61 Wang L, Wen Z, Liu SW, et al. Overview of AlphaFold2 and breakthroughs in overcoming its limitations. *Comput. Biol. Med.* 2024;176:108620.

62 Pan A, Shentu J, Zeng Y, et al. Identification of homologous protein models via 3D comparisons using predicted structures. *STAR Protoc.* 2024;5(1):102814.

63 Scietti L, Forneris F. Modeling of protein complexes. In: *Homology Modeling: Methods and Protocols*. Springer US; 2023. pp. 349–71.

64 Waterhouse AM, Studer G, Robin X, et al. The structure assessment web server: for proteins, complexes and more. *Nucleic Acids Res.* 2024;52(W1):W318–23.

65 Borkakoti N, Thornton JM. AlphaFold2 protein structure prediction: implications for drug discovery. *Curr. Opin. Struct. Biol.* 2023;78:102526.

66 Ceasar SA, Ebeed HT. The present state and impact of AI-driven computational tools for predicting plant protein structures. *Protein Pept. Lett.* 2024;31:749–58.

67 Li C, Luo Y, Xie Y, et al. Structural and functional prediction, evaluation, and validation in the post-sequencing era. *Comput. Struct. Biotechnol. J.* 2024;23:446–51.

68 Abramson J, Adler J, Dunger J, et al. Accurate structure prediction of biomolecular interactions with AlphaFold 3. *Nature* 2024;630(8016):493–500.

69 Weissenow K, Heinzinger M, Rost B. Protein language-model embeddings for fast, accurate, and alignment-free protein structure prediction. *Structure* 2022;30(8):1169–77.

70 Batool M, Ahmad B, Choi S. A structure-based drug discovery paradigm. *Int. J. Mol. Sci.* 2019;20(11):2783.

71 Saha D, Borah NJ, Mamud A. Molecular structure and role of microbial proteins: a delimitation employing bioinformatics techniques. In: *Computational Genomics and Structural Bioinformatics in Microbial Science*. Academic Press; 2025. pp. 273–300.

72 Kim G, Lee S, Levy Karin E, et al. Easy and accurate protein structure prediction using ColabFold. *Nat. Protoc.* 2024;20:620–42.

73 Niazi SK, Mariam Z, Paracha RZ. Limitations of protein structure prediction algorithms in therapeutic protein development. *BioMedInformatics* 2024;4(1):98–112.

74 Herrington NB, Li YC, Stein D, et al. A comprehensive exploration of the druggable conformational space of protein kinases using AI-predicted structures. *PLoS Comput. Biol.* 2024;20(7):e1012302.

75 Camponeschi C, Righino B, Pirolli D, et al. Prediction of CD44 structure by deep learning-based protein modeling. *Biomolecules* 2023;13(7):1047.

76 Pant S, Verma S, Pathak RK, Singh DB. Structure-based drug designing. In: *Bioinformatics*. Academic Press; 2022. pp. 219–31.

77 Feng H, Shen L, Liu J, Wei GW. Mayer-homology learning prediction of protein-ligand binding affinities. arXiv Preprint, 2024, arXiv:2408.13299.

78 Carugo O. pLDDT values in AlphaFold2 protein models are unrelated to globular protein local flexibility. *Crystals* 2023;13(11):1560.

79 Peng Z, Wang W, Han R, et al. Protein structure prediction in the deep learning era. *Curr. Opin. Struct. Biol.* 2022;77:102495.

80 Tu G, Fu T, Zheng G, et al. Computational chemistry in structure-based solute carrier transporter drug design: recent advances and future perspectives. *J. Chem. Inf. Model.* 2024;64(5):1433–55.

81 Herrington NB, Stein D, Li YC, et al. Exploring the druggable conformational space of protein kinases using AI-generated structures. bioRxiv, 2023, 2023.08.31.555779. https://pubmed.ncbi.nlm.nih.gov/37693436/

82 Zhang J, Durham J, Cong Q. Revolutionizing protein–protein interaction prediction with deep learning. *Curr. Opin. Struct. Biol.* 2024;85:102775.

83 Jänes J, Beltrao P. Deep learning for protein structure prediction and design – progress and applications. *Mol. Syst. Biol.* 2024;20(3):162–9.

84 Pucci F, Schwersensky M, Rooman M. Artificial intelligence challenges for predicting the impact of mutations on protein stability. *Curr. Opin. Struct. Biol.* 2022;72:161–8.

8

In Silico Pharmacology: Computational Techniques and Applications in Drug Design

Priya Sharma[1], Ravinesh Mishra[1], Bhartendu Sharma[1], and Chirag Goda[2]*

[1]*School of Pharmacy and Emerging Sciences, Baddi University of Emerging Sciences and Technology, Baddi, Himachal Pradesh, India*
[2]*Department of Pharmaceutical Sciences, Ibn Sina National College for Medical Studies, Jeddah, Saudi Arabia*

*Corresponding author: ravikcp@gmail.com

Introduction

In silico pharmacology represents a paradigm shift in modern drug discovery, leveraging computational techniques to model, analyze, and predict drug interactions within biological systems. This approach complements, and in some cases replaces, traditional in vitro and in vivo methods, offering a faster, cost-effective, and ethically advantageous alternative. The term "in silico," derived from the Latin phrase "in silicum" (performed on silicon chips), distinguishes computational simulations from conventional experimental techniques, such as in vitro (test tube) and in vivo (living organism) studies. The field has evolved significantly since its inception in the mid-20th century, driven by advancements in molecular modeling, artificial intelligence (AI), and high-performance computing. Early computational methods, such as molecular mechanics (MM) and quantitative structure–activity relationship (QSAR) modeling, laid the foundation for contemporary techniques like molecular docking, virtual screening, and molecular dynamics (MD) simulations. These tools enable researchers to identify potential drug targets, optimize lead compounds, and predict pharmacokinetic and toxicological properties with remarkable precision [1]. In silico pharmacology mainly uses two basic strategies: structure-based drug design (SBDD) and ligand-based drug design (LBDD). SBDD makes use of the three-dimensional (3D) structure of proteins or nucleic acids to construct molecules that bind them preferentially. Using the structure of biological molecules in space, the approach helps create molecules that only bind to the chosen target. X-ray crystallography, cryo-electron microscopy, and homology modeling allow for a clear view of structures; meanwhile, molecular docking and

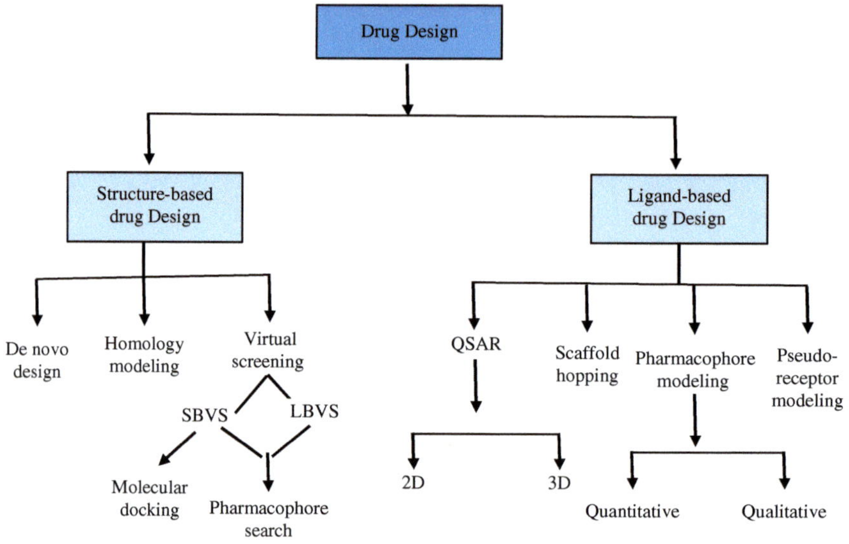

Figure 8.1 Drug design strategies used to produce novel drugs.

simulations are applied to improve drug–target binding. On the other hand, LBDD can form pharmacophore models and QSAR relationships using known bioactive molecules, helping scientists discover drugs without the target structure. These complementary drug design strategies and their associated workflows are illustrated in Figure 8.1. The addition of AI and machine learning (ML) to the field has greatly improved the analysis of large data, permitted the creation of new drugs, and made it possible to foresee the changes in complex biological systems. With tools like AlphaFold, scientists predict the structure of proteins more effectively, which speeds up the process of finding and validating important targets. On top of that, advanced technologies including quantum computing, blockchain, and virtual reality (VR) are expected to make computations more powerful, leading to fresh opportunities in innovation [2].

Scope of In Silico Pharmacology

In silico pharmacology enables the systematic modeling of drug–target interactions across multiple stages of drug discovery, as depicted in Figure 8.2.

Drug Discovery

In silico pharmacology plays a foundational role in early-stage drug discovery by leveraging computational tools to:

- identify potential drug targets through bioinformatics and systems biology;
- screen large compound libraries using virtual screening and molecular docking;
- design novel molecules using de novo drug design and QSAR modeling;
- reduce cost and time by narrowing down candidates before lab synthesis and in vitro testing [3].

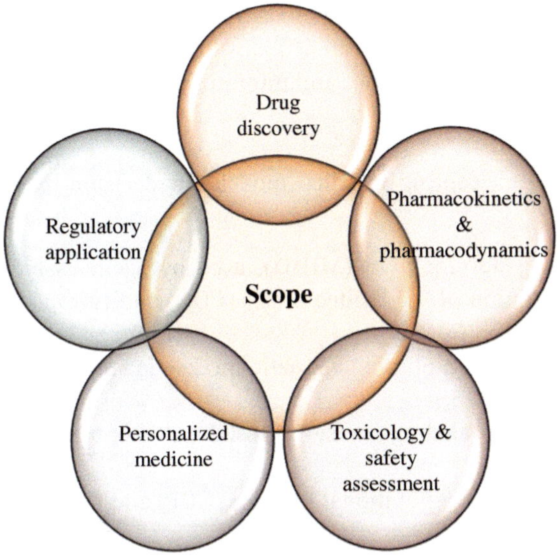

Figure 8.2 Scope of in silico pharmacology.

Pharmacokinetics and Pharmacodynamics

Computational models simulate and predict how drugs behave in the body:

- **Pharmacokinetics (PK):** predict absorption, distribution, metabolism, and excretion (ADME) of compounds using in silico tools like SwissADME or ADMETlab [4].
- **Pharmacodynamics (PD):** simulate their effects on specific cell targets, guess how powerful they will be, and model the connection between the dose given and the outcome.

These simulations help optimize dosage regimens and predict clinical outcomes more effectively.

Toxicology and Safety Assessment

In silico toxicology models assess the safety of new compounds without animal testing. They:

- predict potential toxic effects such as mutagenicity, hepatotoxicity, cardiotoxicity, or carcinogenicity;
- use ML and statistical tools trained on known toxicological databases (e.g., ToxCast, Tox21);
- support regulatory compliance by identifying structural alerts and adverse effects early [5].

Personalized Medicine

In silico pharmacology facilitates precision therapy by:

- analyzing individual genetic profiles (pharmacogenomics) to predict drug response;

- simulating patient-specific models to assess drug efficacy and risks;
- supporting the development of tailored treatment plans based on AI-driven models, leading to better therapeutic outcomes and minimized side effects [6].

Regulatory Application

Regulatory bodies are increasingly accepting in silico models for decision-making. They:

- support model-informed drug development (MIDD), used by agencies like the Food and Drug Administration of the United States (FDA) and European Medicines Agency (EMA);
- validate bioequivalence, toxicity, and pharmacokinetic models for regulatory submissions;
- enable risk assessment and drug labeling decisions based on computational evidence;
- promote standardization via good computational practice (GCP) and transparency in model reporting [7].

Historical Perspective and Evolution of Computational Approaches in Drug Design

Pharmaceutical development in thc present day experienced a transformation because research-based tools developed extensively since the 1950s. During the 1960s and 1970s, scientists commenced molecular modeling until they shifted their focus to advance in silico techniques which combined AI solutions with big data capacities. Researchers at drug design initiated their computational work through MM software implementation for calculating molecular structure interactions based on mathematical equations. Molecular interaction modeling could begin thanks to computers in the mid-20th century because Emil Fischer established the lock-and-key theory for molecular recognition in 1894 [8]. The public discovery of SBDD occurred for the duration of the 1980s when scientists succeeded with X-ray crystallography in determining exact protein 3D structures. Two essential automated docking programs called DOCK and AutoDock emerged from research done at this time for completing virtual screens of protein targets. Scientific research applied SBDD to generate HIV protease inhibitors while making possible a new approach to rational drug creation. Previous knowledge of target-binding molecules was used by scientists during the development of LBDD. The scientists used QSAR and pharmacophore modeling to predict how chemicals may be active, working only with their own chemical structures, regardless of the target. The development of computerized drug discovery started with high-throughput screening (HTS) and cheminformatics after the year 2000 [1]. The discovery of genomics and proteomics enabled bioinformatics to expand its capabilities to design computational drugs for biological and peptide pharmaceuticals [9]. AI joined with deep learning has completely revolutionized drug development through computational design since the recent time period started. The DeepMind-created AlphaFold software represented

Table 8.1 Advantages and limitations of in silico techniques.

Advantages	Limitations
Cost-effective: eliminates the need for lengthy laboratory tests	Without experimental confirmation, accuracy can be lacking
Time-saving: speeds up the processes of screening and prediction	Reliant on algorithms and input data quality
Minimizes the use of animals in studies	
Makes it possible to virtually screen large compound libraries	Oversimplifying biological relationships is common
Makes it easier to identify potential drugs early	Restricted by the modeling methods and processing power available today
Beneficial for formulating hypotheses and directing studies	May overlook infrequent or surprising encounters
Allows macromolecular interactions to be modeled	In vivo metabolism and dynamics might not be taken into consideration by models
Consistent and repeatable outcomes using the same parameters	High-end databases and applications may need pricey licenses
Minimizes the use of animals in studies	Unable to accurately reproduce intricate biological systems

a breakthrough in drug–target interaction modeling abilities [10]. The combination of generative models and reinforcement learning technologies is now used to develop new molecules with specific properties, which enhances the speed of lead optimization work [11].

Advantages and Limitations of In Silico Methods

The benefits and drawbacks of in silico methods, which are frequently employed in toxicology, drug development, molecular modeling, and related domains, are shown in Table 8.1.

Fundamentals of Molecular Modeling

At the beginning of modern molecular modeling chemistry, research students prefer computer utilization instead of mastering original subject material. A lack of scientific understanding among recreational keyboard users leads them to convert meaningful content into visual displays that reduce its worth. MM represents an objective modeling approach for molecular characteristics because it stands alone as the dependable computational method that relies on fundamental theory. The field of molecular modeling establishes systematic procedures to build simulation models which reveal spatial and temporal relation changes between molecules [12].

MM and Force Fields

MM decides to analyze molecular systems through classical physics by treating atoms with point masses while modeling bonds with springs. The predefined force fields within MM determine system potential energy calculations by considering bond-related interactions and electrostatic and van der Waals forces apart from bond lengths and angular deformations and torsional changes,

$$E_{\text{total}} = E_{\text{bond}} + E_{\text{angle}} + E_{\text{dihedral}} + E_{\text{nonbonded}}$$

Widely used force fields include:

- AMBER – ideal for proteins and nucleic acids.
- CHARMM – supports proteins, lipids, and carbohydrates.
- OPLS-AA – optimized for small organic molecules.
- GROMOS – useful in aqueous and biological simulations.

These models are computationally efficient and suitable for large biomolecular systems, although they do not account for electronic properties.

Quantum Mechanics in Drug Design

Using the Schrödinger equation explicitly to handle electrons makes quantum mechanical (QM) methods deliver more accurate fundamental molecular modeling as an approach. Scientists need to use these methods to examine chemical reactivity as well as electronic structures and bond formation/breaking in enzyme active sites and metal-containing systems [13]. Applications in drug design include:

- determination of partial charges;
- calculation of molecular orbitals and HOMO–LUMO gaps;
- modeling catalytic mechanisms and transition states.

Due to their high computational demand, QM techniques are typically applied to small systems or used in combination with MM approaches (QM/MM) for targeted regions of interest [14].

MD Simulations

MD simulations provide researchers observations regarding the temporal behavior of molecular systems through their computational method. The method uses computational methods to solve atomic kinetic equations that model biomolecular positional changes while analyzing both biological structure flexibility and binding and unbinding states and their environment reactions [15]. MD simulations are critical in the following:

- exploring protein–ligand interactions over time;
- evaluating structural stability;
- estimating thermodynamic parameters and free energies.

Advanced MD techniques, such as accelerated MD, replica exchange MD, and coarse-grained MD, enhance sampling efficiency and applicability to complex systems [16].

Energy Minimization Techniques

Molecular structures undergo energy minimization preparation before scientists search for local potential energy surface minima as part of their structure refinement process. The procedure allows the removal of steric contacts while generating feasible geometric structures that are necessary for simulations or docking research [17].

Common algorithms include:

- Steepest Descent: effective for large distortions.
- Conjugate Gradient: better near minima.
- Newton–Raphson: precise but computationally intensive.

Minimization ensures a realistic, low-energy starting point for downstream simulations like MD or docking.

Structure-based Drug Design

SBDD uses the shape of the target to help make molecules that will link it with great specificity. To start, a team identifies and depicts the site at which a drug and a receptor interact, usually by using X-ray crystallography, NMR spectroscopy, or cryo-electron microscopy. Then, computational approaches are used to study the geometry, electrical properties, and hydrophobicity of the binding site. The characteristics are used to modify or pick suitable molecules, and their ability to bind is then determined using docking and simulation software. Through SBDD, researchers are able to maximize interactions that create specificity and increase the success of the drug. Against HIV and certain types of cancers, drugs like HIV protease inhibitors and kinase inhibitors have been developed through this technology. Because of recent advancements in AI and AlphaFold, SBDD is now more accurate and important for speeding up the process of drug discovery.

Principles of SBDD

SBDD relies on coming up with small molecules that will attach themselves specifically and effectively to the active or allosteric site of a biological molecule. The process begins with structural elucidation of the target, followed by analysis of binding site characteristics, including shape, electrostatics, hydrophobicity, hydrogen bonding, and conformational flexibility. The basic steps involved in the SBDD approach are summarized in Figure 8.3. Key objectives of SBDD include:

- identifying druggable binding pockets;
- designing or optimizing compounds for better binding affinity and specificity;
- predicting molecular interactions and potential off-target effects [18].

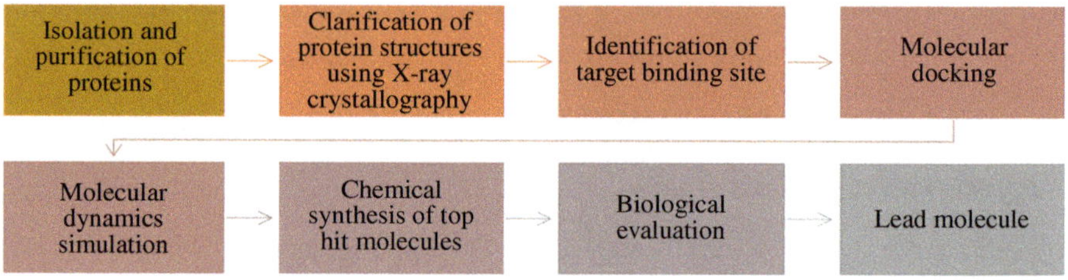

Figure 8.3 Basic steps involved in the SBDD approach.

Table 8.2 Summary of protein structure modeling techniques in in silico drug design.

Method	Description
Homology modeling	Predicts protein structure based on similarity to known templates (e.g., SWISS-MODEL, MODELLER)
Threading/fold recognition	Aligns sequences to known folds, even with low similarity (e.g., Phyre2, I-TASSER)
Ab initio prediction	Predicts structure solely from sequence, without a template (e.g., AlphaFold)

The following categories comprise SBDD:

- molecular docking
- in silico ADME
- MD
- de novo drug design

Protein Structure Prediction and Modeling

Accurate protein structures are critical for SBDD. When experimental structures (e.g., X-ray crystallography or NMR) are unavailable, computational modeling methods are employed. The major protein structure modeling techniques commonly used in in silico drug design are summarized in Table 8.2.

Ligand Docking and Scoring Functions

Molecular docking predicts the most favorable binding location of a ligand on a target protein and estimates the strength of the interaction. It involves pose prediction, which generates possible binding conformations, and scoring functions, which evaluate binding free energies. The key features and capabilities of commonly used ligand docking software in in silico drug design are summarized in Table 8.3.

Structure-based Virtual Screening

Structure-based virtual screening (SBVS) includes computationally screening large libraries of compounds toward a goal structure to become aware of ability hits. The process includes:

Table 8.3 Key features of ligand docking software in in silico drug design.

Docking software	Key features
AutoDock/AutoDock Vina	Widely used; includes free energy scoring functions
Glide (Schrödinger)	High accuracy and flexible docking capabilities
Gold (CCDC)	Genetic algorithm-based pose prediction

- protein preparation and binding site identification;
- ligand library preparation;
- high-throughput docking;
- ranking and selection based on scores.

Fragment-based Drug Design

Fragment-based drug design (FBDD) uses small molecular fragments (MW < 300 Da) to probe the active site of a target protein. These fragments bind weakly but efficiently, providing a starting point for optimization.

Workflow of FBDD

- Fragment screening using biophysical techniques (e.g., X-ray, NMR, and SPR).
- Binding mode analysis via molecular docking or co-crystal structures.
- Fragment growing/linking to improve binding affinity.
- Optimization into lead-like molecules [19].

Ligand-based Drug Design

The approach of drug layout known as ligand-primarily based drug layout additionally operates beneath the call of indirect drug design. A complete know-how of all the molecular interactions that bind to the biological target is essential. The pharmacophore model describes through these molecules the essential structural needs that allow binding to the target. New molecular entities are generated through biological target modeling that uses information about binding molecules to develop such entities for target interaction. A distinct method for this purpose exists through QSAR that establishes connections between tested biological reactions and computationally predicted molecular characteristics. Researchers use the developed correlations to predict how active a new drug will be. The absence of 3D target receptor structure often drives pharmaceutical professionals to conduct LBDD. The collection of active molecules against a target receptor becomes the basis for extracting structural and physicochemical aspects in the LBDD. Researchers use such approach by anchoring their assessments on the fact that biological tasks with matching operation correspond with structural equivalencies [20]. The general workflow of LBDD is illustrated in Figure 8.4.

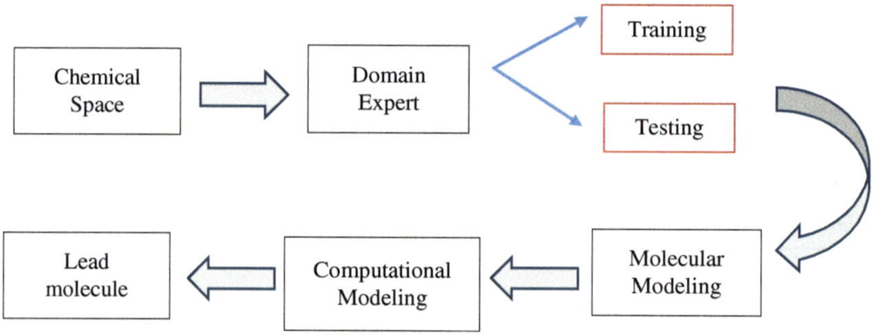

Figure 8.4 The general procedures for designing drugs based on ligands.

Table 8.4 Applications of QSAR.

Application	Uses
Chemical	Boiling point detection Finding the connection between alkenes' carbons The Hammett equation The Taft formula pKa forecast
Biological	Identification of the metabolic route The search for new drugs Partition coefficient prediction Toxicological prediction
Other applications	Control of risks Identification of contaminants that are genotoxic

Quantitative Structure–Activity Relationships

QSAR analysis serves as the main LBDD method that links biological activity changes between drugs. Researchers apply this defined relationship to develop predictive statistical models that predict new substance's biological behavior from various molecular patterns. This procedure helps drug discovery practices to establish leads while it optimizes their effectiveness [21]. The diverse applications of QSAR across chemical, biological, and risk-assessment domains are summarized in Table 8.4.

Pharmacophore Modeling

The pharmacophore model establishes which chemical interaction points between targets and receptors exist along with their spatial relations in ligands. The process of pharmacophore model development entails a combination of hydrophobic areas with aromatic ring systems alongside both positive and negative charge groups and hydrogen bond donors and acceptors. The application of pharmacophore models in

Table 8.5 Pharmacophore modeling techniques.

Tools	Description
Catalyst	An algorithm that reveals 3D configurations of chemical houses shared by manner of a hard and fast of ligands paperwork the idea of the catalyst software. Each configuration is given a score based on its estimated rarity as well as how common it is to the input set
Ligand Scout	Protein–ligand interactions, which include hydrogen bond interactions, price transfers, and lipophilic areas, are detected and categorized by means of Ligand Scout, a totally automatic tool for developing pharmacophore models that function as the muse for excessive-throughput digital screening
DISCO	As an accessory to 3D-QSAR, DISCO is an automatic pharmacophore method that analyzes the records to select out all pharmacophore hypotheses that are in shape
Pharma Gist	A collection of drug-like compounds with a binding affinity to the target protein (up to 32 drug-like molecules) can be used as the input for the free online server Pharma Gist, which creates ligand-based pharmacophore models
Pharm Mapper	A freely accessible online server referred to as Pharm Mapper is regularly used to perceive feasible goal receptors for a positive small chemical by using the pharmacophore mapping approach

virtual screening allows researchers to identify compounds with different frameworks that share identical positions of their fundamental contact points. Pharmacophore model modifications occur through the incorporation of molecular structures, which target molecules adopt within their binding sites. The pharmacophore model serves as a key component in all molecular alignment steps of QSAR research. The commonly used pharmacophore modeling tools and their key characteristics are summarized in Table 8.5 [4].

Drug Discovery in Conjunction with AI

AI functions as a discipline within machine intelligence that teaches computers to analyze data that exists already. AI enables several computational modeling procedures to predict drug compounds' biological actions together with their toxicity levels. The drug discovery manner makes use of AI in diverse methods to allow de novo drug design and virtual screening and to predict QSAR interactions and protein folding whilst comparing drug houses for ADMET. Rational remedy layout relies upon efficient methodologies for layout called system mastering (ML) and deep mastering (DL). The drug improvement system bases its predictions on three fundamental gadget-studying algorithms which consist of support vector machine (SVM) along with random forest (RF) and naive Bayes (NB). Convolutional neural networks (CNNs) function as one of the many methods found within the deep learning techniques [22].

Integrative Approaches in Drug Design

Integrative approaches in drug design aim to replace the usual one-target methods by adopting new strategies that can deal with the complexities found in cancers, diseases of the nervous system, and metabolic ills. These approaches combine multiple disciplines, such as genomics, proteomics, cheminformatics, systems biology, and computational modeling, to create a more holistic view of disease mechanisms and therapeutic interventions. Multitarget drug design and polypharmacology aim to develop compounds that act on several relevant pathways simultaneously, potentially increasing efficacy and reducing drug resistance. Combination therapies leverage synergistic effects of multiple drugs to enhance therapeutic outcomes. Integrative methods can also contribute to personalized medicine by giving treatments that are right for each person's genetics and molecules. Looking at the connection between several biologic systems and patient features, integrative drug design might provide better, safer, and more specific medicines that are needed in modern medicine.

Multitarget Drug Design

Multi-goal drug layout involves agents that work on more than one objective. For illnesses with multiple and redundant processes in the body, the use of multiple and diverse drugs is a helpful move.

Advantages

- **Enhanced efficacy:** modulating multiple pathways can yield synergistic effects.
- **Reduced resistance:** targeting multiple nodes can prevent or delay resistance mechanisms, especially in infectious diseases and oncology.
- **Lower doses:** by using multitarget drugs, it might be possible to achieve the same medical results at lower doses and thus cause fewer side effects.

Design Strategies

- **Hybrid molecules:** bring the active parts of two individual compounds into a common chemical structure.
- **Linker-based approaches:** bring together two active groups by adding a flexible bond between them.
- **De novo design:** make use of computer simulations to develop multitarget molecules from beginning to end.

Polypharmacology Prediction

Polypharmacology refers back to the ability of a drug to have an effect on more than one objective, both beneficially or detrimentally. Expertise in predicting these interactions is vital for both drug repurposing and minimizing adverse results.

Tools and Techniques

- **Ligand-based approaches:** use known ligands to predict off-target interactions through similarity measures.
- **Structure-based methods:** dock compounds across multiple targets to predict binding affinity.
- **ML:** train models using large datasets of drug–target interactions for prediction and analysis.
- **Network pharmacology:** visualize and analyze interaction networks to understand systemic drug effects.

Applications

- Drug repositioning
- Side effect prediction
- Rational design of multitarget agents

Combination Therapy Optimization

Combination therapy involves the usage of extra tablets together to reap better healing effects. This method is regular in treating complex sicknesses, including HIV, most cancers, and tuberculosis.

Goals

- Maximize therapeutic effect while minimizing toxicity.
- Prevent or delay drug resistance.
- Target multiple disease mechanisms simultaneously.

Optimization Methods

- **HTS:** rapidly assess the efficacy of drug combinations in vitro.
- **Dose matrix modeling:** evaluate dose-dependent interactions between drugs.
- **Synergy scoring models:** quantify the interaction effect (e.g., Bliss independence, Loewe additivity).
- **Computational simulations:** predict optimal combinations and dosages using systems biology models.

Precision Medicine and Personalized Drug Design

Precision medicine tailors treatment to man or woman sufferers based on genetic, biomarker, phenotypic, or psychosocial characteristics, personalized drug layout goals to increase therapeutics that account for affected person-specific variations in drug metabolism, ailment mechanisms, and reaction variability.

Key Components

- **Genomic profiling:** identify mutations and polymorphisms that influence drug response.
- **Pharmacogenomics:** study how genes affect an individual's response to drugs.

- **Biomarker discovery:** develop diagnostic tools to guide treatment decisions.
- **Adaptive clinical trials:** flexibly adjust treatments based on real-time patient data.

Impact

- Improved therapeutic efficacy.
- Reduced adverse drug reactions.
- Greater patient adherence and outcomes [23].

Emerging Technologies in In Silico Pharmacology

The rapid advancements in technology are having a big impact on in silico pharmacology and how drugs are developed. A major revolution is quantum computing, where qubits are used to carry out calculations significantly quicker than ordinary computers do. The technology makes it possible to reproduce molecular processes at the quantum level and more accurately model bindings between drugs and targets, protein folding, and intricate electrical structures of biomolecules. Although quantum computing is in its early days due to hardware problems, it could change the field of SBDD by overcoming problems that are difficult to solve with traditional computers, like simulating complicated mechanisms of enzymes. Some of the most exciting new technologies are listed below.

Quantum Computing in Drug Design

Quantum computing, which makes use of quantum bits (qubits) to carry out extraordinarily complicated computations, is a big advancement over classical computing. It makes it possible to simulate molecular interactions in drug development with previously unheard-of accuracy, especially for quantum-level processes like chemical bonding, protein folding, and electrical configuration. Although it has the potential to completely transform SBDD and the prediction of protein–ligand interactions, it is still in the early stages of development because of hardware constraints and scalability issues [24].

Block Chain Technology in Drug Discovery

Data sharing with storage functions becomes achievable through blockchain technology which operates using decentralization and transparency alongside safety measures. The platform designed by blockchain operates for clinical trial data management, drug supply chain authentication, and intellectual property rights tracking in drug discovery. The platform provides transparent and secure access to data that benefits both regulators and pharmaceutical firms as well as researchers. The implementation of blockchain encounters limitations regarding scalability and requires better system interaction capabilities for its wider deployment.

Cloud Computing and Big Data Analytics

The demand for scalable on-demand processing capacity for biomedical data is met by cloud platforms because they offer this capability. With the combination of big data analytics, the machine permits processing of genetic facts together with electronic health facts and real-world evidence and excessive-throughput screening findings. The features enable researchers to conduct predictive modeling studies as well as discover new drugs by integrating AI systems. The primary elements of security together with standardization retain their focus when cloud mechanisms form the basis of infrastructure deployments [25].

Virtual and Augmented Reality in Molecular Visualization

In drug design and molecular modeling procedures, technicians integrate VR and augmented reality (AR) technologies to foster better understanding of molecular structures with their dynamics. Through their 3D interface researchers can observe molecular interactions while performing work with molecules, which delivers authentic understanding of structural changes and docking operations and conformational modification dynamics. The technology functions as both teaching materials for trained professionals as well as students learning about pharmacology and structural biology. Specialized software development along with expensive hardware systems represent the main limitations related to their use [26].

Ethical Considerations and Regulatory Aspects

Ethical considerations and regulatory aspects in in silico drug design have become increasingly significant as computational methods take on a central role in pharmaceutical development. Although these techniques are very efficient and accurate, they create questions about confidentiality, the way algorithms work, and any stereotypes that AI models use. Strict rules on confidentiality and consent should be upheld when using patient-derived data in modeling to respect people's rights. Also, it is becoming more necessary to have clear standards for validating computational predictions. Regulatory authorities like the FDA and EMA are now working on direction for in silico tools in preclinical and clinical research. Trying to address ethical concerns includes making sure almost everyone can use technology and protecting the traditional experimental jobs that may become obsolete. Advancement in the field requires keeping an eye on both new advances, ethics, and adherence to rules.

Ensuring Privacy and Security of Data

Developing in-silico drugs requires using a wealth of patient information from the fields of genomic research and precision medicine.

Genetic and health-related data becomes vulnerable when unauthorized access happens and brings forward ethical problems.

Best Practices

- Data encryption and anonymization adherence to laws such as the Health Insurance Portability and Accountability Act (HIPAA) and the General Data Protection Regulation (GDPR).
- Access controls and safe cloud storage.

Consequences: maintaining integrity and credibility in research, particularly when working with varied populations around the world [27].

Intellectual Property Issues in In Silico Methods

The limits of intellectual property get more complicated as novel drug ideas are produced by AI and ML algorithms.

Principal Issues

- Who is the owner of the medication candidate: the consumer or the algorithm developer?
- How can a chemical "discovered" by an AI system be patented?
- Risks of intellectual property theft via data leaks or model reverse engineering.

Current Discussions

- Acceptance of AI as a "contributor" to innovations.
- AI-generated substances' eligibility for patents in different jurisdictions.
- Laws must be updated to address ownership and safeguard AI-generated outputs [28].

Regulatory Guidelines for Computer-aided Drug Design

The use of computational methods in drug research and discovery is growing, and international regulatory bodies are adjusting.

Authorities That Regulate

FDA, EMA, and Council for International Harmonization (ICH).

Rules and Anticipations

- Computational model validation documentation of the reproducibility and transparency of the algorithm.
- Application of GCP and good machine learning practice (GMLP).

Present Situation

Regulators support early communication and openness in data submission, even though institutionalized mechanisms are still developing.

Ethical Implications of AI-driven Drug Discovery

Although AI has the potential to speed up drug discovery, it also poses serious ethical issues:

- Fairness and bias: AI algorithms that are trained on biased datasets may overlook therapeutic targets for underrepresented populations or produce unfair results.
- Transparency: deep learning models' "black box" nature might obfuscate decision-making, making it challenging to confirm efficacy or safety [29].
- Accountability: whether it's the developer, user, or organization, accountability must be made explicit in the event of AI-led mistakes or unfavorable results.
- Autonomy: relying too much on AI could impair crucial human judgment when assessing potential medications.

To solve these issues, it is becoming more and more advised to use ethical governance frameworks, such as explainable AI (XAI), bias audits, and human oversight.

Future Perspectives

The future of in silico pharmacology is poised for transformative advancements, driven by the convergence of AI, high-performance computing, and integrative biological data. As drug discovery increasingly relies on computational predictions, future directions will emphasize improving the precision, scalability, and interpretability of these models while addressing critical healthcare challenges globally. One major direction is the integration of AI with multi-omics data including genomics, proteomics, metabolomics, and transcriptomics for precision medicine. It is likely that next-generation computational pipelines will make use of patient-specific molecular features to successfully predict the likelihood of drugs working or not. Machine learning algorithms trained on multimodal datasets will facilitate individualized therapy design, enhancing treatment outcomes and reducing adverse reactions. Moreover, AI models such as AlphaFold have demonstrated revolutionary capabilities in protein structure prediction, which can be extended for rapid SBDD [10]. Another promising avenue is the use of quantum computing in molecular simulation. Quantum algorithms offer the potential to model complex biomolecular interactions, electronic transitions, and reaction mechanisms more accurately than classical simulations. As quantum hardware matures, simulations of drug–target complexes and quantum tunneling events in enzymatic reactions will become feasible, significantly boosting rational drug design. The democratization of drug discovery through cloud-based platforms is another anticipated development. Cloud computing allows real-time collaboration and access to advanced molecular modeling tools without the need for expensive infrastructure. This will foster greater inclusion of resource-constrained settings in pharmaceutical research, enabling global partnerships and open-source drug discovery initiatives [2]. Combined with big data analytics, researchers will be able to mine public and proprietary databases for drug repurposing, target identification, and safety profiling with high throughput. Furthermore, VR/AR tools are expected to play a more prominent role in molecular visualization. Interactive 3D environments can aid researchers in comprehending MD, ligand binding modes, and allosteric sites. These immersive technologies enhance decision-making during lead optimization and promote collaborative design discussions across interdisciplinary teams [30]. Another future focus will be the application of systems pharmacology and network-based approaches. As diseases are increasingly understood as disruptions in biological networks, future drug design strategies will aim at multitarget and pathway-level modulation. Computational systems biology models will simulate whole-cell environments, disease progression, and treatment responses under various conditions, allowing for predictive and preventive therapeutic development. Ethical and regulatory advancements will also shape the future landscape. Transparent XAI models will be crucial to gaining regulatory acceptance and public trust. Efforts will be directed toward developing standardized protocols for model validation, data sharing, and digital ethics frameworks to ensure fairness, accountability, and reproducibility in AI-driven drug discovery [31]. Finally, in the field of in silico pharmacology, scientists will be able to positively affect global health plainly. It achieves this by saving money and accelerating the process, which makes it possible to develop useful drugs for neglected tropical diseases, rare genetic diseases, and pandemics. Using technology to study phytochemicals and traditional medicines can discover relevant medicines or drugs that are inclusive of different cultures and regions [32].

Conclusion

In silico pharmacology greatly impacts the field of drug discovery, combining powerfully the practices of experimentation and computer-based simulation. With the assistance of techniques like molecular docking, virtual screening, QSAR modeling, and MD simulations, scientists are now able to predict drug–target relationships, perfect their lead compounds, and assess drug behavior in a body much faster than before. By adopting this new approach, it has become much more efficient to develop drugs, and the costs have lowered, while the need for intense laboratory work has decreased. The chapter highlights that SBDD and LBDD play an important role in the rational development of drugs. SBDD, assisted by new high-resolution protein structures and technologies such as AlphaFold, can point out specific sites to target on proteins for the development of highly targeted and effective therapeutics. Instead, techniques such as pharmacophore modeling and QSAR in LBDD let researchers understand the structure needed for activity in a target, without having detailed information about the target. To handle the complex aspects of disease, scientists also apply multitarget drug design and network pharmacology, which help to modulate many processes and reduce the risk of resistance. Despites, there are several issues in the use of in silico pharmacology. If the data fed into a computer system is inaccurate, the computed answers will also be incorrect. Additionally, since some AI models are hard to interpret, this makes them difficult to use or reproduce. To overcome these weaknesses, we need strong validation processes, teamwork among different experts, and the build-up of frameworks used by all to ensure the reliability of results. Besides, using quantum computing, blockchain, and VR can be expected to make in silico methods more accurate and usable on a larger scale. In the near future, using AI, multi-omics data, and personalized medicine is projected to transform the way drugs are found. AI methods trained with data from genomics, proteomics, and patient records will allow for creating therapies that better suit a person's genes and proteins. There is an increasing acceptance by regulators that computational models should be used in drug approvals. The increase in acceptance, along with advances in technology, guarantees that in silico pharmacology will lead in the field. Further developing computational resources, collaborating between experts, and addressing present issues can help the field greatly improve healthcare globally. As advanced technology becomes a reality, blending computational ideas with hands-on experiments will certainly open doors to important innovations in health care all over the world.

References

1. Hansch C, Fujita T. ρ-σ-π analysis. A method for the correlation of biological activity and chemical structure. *J. Am. Chem. Soc.* 1964;86(8):1616–26.
2. Dean M, Crivella T, Moore T, et al. Cloud computing in biomedical research: Lessons from the NCI cancer genomics cloud pilots. *PLoS One* 2020;15(10):e0238524.
3. Daina A, Michielin O, Zoete V. SwissADME: a free web tool to evaluate pharmacokinetics, drug-likeness and medicinal chemistry friendliness of small molecules. *Sci. Rep.* 2017;7(1):42717.
4. Honorio KM, Moda TL, Andricopulo AD. Pharmacokinetic properties and in silico ADME modeling in drug discovery. *Med. Chem.* 2013;9(2):163–76.
5. Ashley EA. The precision medicine initiative: a new national effort. *JAMA* 2015;313(21):2119–20.
6. Madabushi R, Wang Y, Zineh I. A regulatory perspective on the role of quantitative modeling and simulation in drug development. *CPT Pharmacometrics Syst. Pharmacol.* 2019;8(6):344–6.
7. Fischer E. Einfluss der configuration auf die wirkung der enzyme. *Ber. Dtsch. Chem. Ges.*1894;27(3):2985–93.
8. Kuntz ID, Blaney JM, Oatley SJ, et al. A geometric approach to macromolecule-ligand interactions. *J. Mol. Biol.* 1982;161(2):269–88.
9. Bender A, Cortes-Ciriano I. Artificial intelligence in drug discovery: what is realistic, what are illusions? Part 1: Ways to make an impact, and why we are not there yet. *Drug Discov. Today* 2021;26(2):511–24.
10. Jumper J, Evans R, Pritzel A, et al. Highly accurate protein structure prediction with AlphaFold. *Nature* 2021;596(7873):583–9.
11. Zhavoronkov A, Ivanenkov YA, Aliper A, et al. Deep learning enables rapid identification of potent DDR1 kinase inhibitors. *Nat. Biotechnol.* 2019;37(9):1038–40.

12 Rognan D. The impact of in silico screening in the discovery of novel and safer drug candidates. *Pharmacol. Ther.* 2017;175:47–66.

13 Cramer CJ. *Essentials of Computational Chemistry: Theories and Models.* 2nd ed. Wiley; 2004.

14 Senn HM, Thiel W. QM/MM methods for biomolecular systems. *Angew. Chem. Int. Ed. Engl.* 2009;48(7):1198–229.

15 Karplus M, McCammon JA. Molecular dynamics simulations of biomolecules. *Nat. Struct. Biol.* 2002;9(9):646–52.

16 Dror RO, Dirks RM, Grossman JP, et al. Biomolecular simulation: a computational microscope for molecular biology. *Annu. Rev. Biophys.* 2012;41:429–52.

17 Lemkul JA. From proteins to perturbed Hamiltonians: a suite of tutorials for the GROMACS-2018 molecular simulation package. *J. Chem. Educ.* 2019;96(2):207–18.

18 Noori HR, Spanagel R. In silico pharmacology: drug design and discovery's gate to the future. *In Silico Pharmacol.* 2013;1:1–2.

19 Erlanson DA, Fesik SW, Hubbard RE, et al. Twenty years on: the impact of fragments on drug discovery. *Nat. Rev. Drug Discov.* 2016;15(9):605–19.

20 Andrade EL, Bento AF, Cavalli J, et al. Non-clinical studies required for new drug development: Part I: early in silico and in vitro studies, new target discovery and validation, proof of principles and robustness of animal studies. *Braz. J. Med. Biol. Res.* 2016;49:e5644.

21 Nantsupawat T, Soontrapa S, Klomjit S, Jenkins LA. Rivaroxaban monograph. *Southwest Respir. Crit. Care Chron.* 2015;3(10):27–33.

22 Anderson AC. The process of structure-based drug design. *Chem. Biol.* 2003;10(9):787–97.

23 Klebe G. Recent developments in structure-based drug design. *J. Mol. Med.* 2000;78:269–81.

24 Ferreira LG, dos Santos RN, Oliva G, Andricopulo AD. Molecular docking and structure-based drug design strategies. *Molecules* 2015;20(7):13384–421.

25 Sun H, Scott DO. Structure-based drug metabolism predictions for drug design. *Chem. Biol. Drug Des.* 2010;75(1):3–17.

26 Amzel LM. Structure-based drug design. *Curr. Opin. Biotechnol.* 1998;9(4):366–9.

27 Wang R, Gao Y, Lai L. LigBuilder: a multi-purpose program for structure-based drug design. *J. Mol. Model.* 2000;6:498–516.

28 Sharma V, Wakode S, Kumar H. Structure-and ligand-based drug design: concepts, approaches, and challenges. In: *Chemoinformatics and Bioinformatics in the Pharmaceutical Sciences.* Elsevier Science; 2021. pp. 27–53.

29 Aparoy P, Kumar Reddy K, Reddanna P. Structure and ligand based drug design strategies in the development of novel 5-LOX inhibitors. *Curr. Med. Chem.* 2012;19(22):3763–78.

30 Martínez-Roldán J, Roldán-García C, Gago F. The role of immersive visualization in structure-based drug design. *Br. J. Pharmacol.* 2021;178(21):4175–93.

31 Tonekaboni S, Joshi S, McCradden MD, Goldenberg A. What clinicians want: contextualizing explainable machine learning for clinical end use. *Proc. Mach. Learn. Res.* 2019;106:359–80.

32 Patridge E, Gareiss P, Kinch MS, Hoyer D. An analysis of FDA-approved drugs: natural products and their derivatives. *Drug Discov. Today* 2016;21(2):204–7.

9

Computational Methods for Predicting ADMET Properties: Tools and Strategies to Minimize Late-stage Drug Failures

*Kagan Tolga Cinisli**

Atatürk University Eastern Anatolia High Technology Application and Research Center Directorate, Erzurum, Türkiye

Corresponding author: kagantolgacinisli2525@gmail.com

Introduction

The high attrition rate in drug development is often attributed to poor pharmacokinetic (PK) and toxicity profiles, with approximately 40–50% of failures in clinical phases linked to unfavorable absorption, distribution, metabolism, excretion, and toxicity (ADMET) properties [1]. To address this challenge, computational ADMET prediction tools have become indispensable in early drug discovery. These tools enable researchers to prioritize compounds with optimal bioavailability and safety profiles before costly experimental validation [2]. This chapter explores three key in silico platforms, pkCSM, SwissADME, and Toxicity Estimation Software Tool (TEST), and their role in streamlining drug development by minimizing late-stage failures. The integration of these computational approaches represents a paradigm shift in pharmaceutical research, offering significant advantages over traditional experimental methods [3].

Importance of ADMET Prediction in Drug Discovery

The Challenge of Late-stage Attrition

Drug discovery is a resource-intensive process, with the median cost of bringing a drug to market exceeding $1.14 billion when accounting for failed trials [4]. A significant proportion of these failures stem from inadequate ADMET properties, such as poor solubility, metabolic instability, or toxicity [5, 6]. Computational tools mitigate these risks through several mechanisms: early filtering of compounds with suboptimal PK/toxicity profiles; reducing reliance on in vivo experiments, which are expensive and ethically contentious; and guiding structural optimization to improve drug-likeness [7]. The implementation of computational ADMET screening early in the discovery pipeline can significantly reduce both financial costs and time to market.

Evolution of ADMET Prediction Tools

Traditional methods for ADMET assessment relied on established rules such as Lipinski's Rule of Five and experimental assays conducted in laboratory settings [8]. However, advances in machine learning (ML) and cheminformatics have enabled more accurate in silico models that can process complex chemical data more efficiently [9]. Modern computational tools integrate multiple approaches: quantitative structure–activity relationships (QSARs) for predicting biological activity from chemical structure; molecular dynamics (MD) simulations for understanding molecular interactions; and deep learning techniques, including transformer models, for multitask predictions across various ADMET end points [10, 11]. These technological advancements have dramatically increased the predictive accuracy and scope of computational ADMET tools.

Key Computational Tools for ADMET Prediction

SwissADME: A Comprehensive PK and Drug-likeness Predictor

Developed by the Swiss Institute of Bioinformatics, SwissADME (http://www.swissadme.ch) is a freely accessible web tool that evaluates multiple PK parameters [12]. The platform assesses physicochemical properties including molecular weight, log P (partition coefficient), and topological polar surface area (TPSA). It incorporates consensus models such as iLOGP and XLOGP3 to predict lipophilicity and solubility. SwissADME also features drug-likeness evaluation using the Bioavailability Radar, which visualizes key parameters including size, polarity, and saturation. Additionally, the tool provides medicinal chemistry friendliness assessments through PAINS (pan assay interference compounds) filters and synthetic accessibility scores [13]. A case study demonstrated the application of SwissADME in optimizing artemisinin derivatives, where structural modifications identified through computational analysis improved metabolic stability and oral bioavailability [14]. The user-friendly interface and comprehensive output make SwissADME particularly valuable for researchers without extensive computational backgrounds.

pkCSM: Graph-based ADMET Profiling

The pkCSM platform employs graph-based signatures to predict a wide range of ADMET properties [15]. This approach encodes molecular structure as graphs, with atoms as nodes and bonds as edges, enabling the capture of complex structural patterns. The tool predicts absorption parameters including Caco-2 permeability and human intestinal absorption. Distribution predictions cover volume of distribution and blood–brain barrier penetration. Metabolism predictions focus on cytochrome P450 inhibition, while toxicity predictions include

AMES mutagenicity and hERG inhibition [15]. The primary advantages of pkCSM include high interpretability resulting from transparent descriptor-based modeling and the ability to integrate with other tools such as SwissADME for cross-validation [16]. However, limitations include potentially lower accuracy for novel chemical scaffolds outside its training set and dependence on the quality and breadth of the underlying data.

Toxicity Estimation Software Tool

Developed by the US Environmental Protection Agency, TEST utilizes QSAR and group contribution methods to predict various toxicity end points [17]. The software estimates acute toxicity through parameters such as LD_{50}, evaluates mutagenicity and carcinogenicity potential, and predicts developmental toxicity risks. TEST incorporates multiple QSAR approaches including hierarchical clustering, single-model regression, group contribution, nearest neighbor, and consensus methods to enhance prediction reliability [18]. Key applications of TEST include regulatory compliance assessment for environmental chemicals and prioritizing drug candidates with lower toxicity risks during early discovery phases. The tool's development under EPA guidance ensures alignment with regulatory standards, though it may be less optimized for drug-like molecules specifically designed for human therapeutics [19].

ADMET Property Breakdown

Absorption

Absorption refers to the process by which a drug transitions from its administration form into the systemic circulation. This process significantly influences both the speed and concentration at which a drug reaches its target site of action. Various administration routes exist including oral, intravenous, intramuscular, subcutaneous, and transdermal methods, each with unique absorption characteristics. The process often involves liberation, the mechanism by which the drug is released from its pharmaceutical dosage form, particularly crucial for oral medications [20].

Distribution

Distribution describes how a substance disseminates throughout the body, influenced by both drug characteristics and physiological conditions. The process is primarily affected by diffusion and convection mechanisms, which are further influenced by drug polarity, molecular size, and binding affinity, as well as patient-specific factors including fluid status and body composition. The primary objective of distribution is to achieve effective drug concentration at receptor sites, requiring the drug to reach specific compartmental destinations defined by the volume of distribution while remaining unbound to maintain pharmacological activity [21].

Metabolism

Metabolism involves the body's transformation of drugs into various metabolites, typically aimed at converting compounds into more water-soluble forms for renal elimination. In some cases, such as with prodrugs like codeine, metabolism is essential for generating active metabolites. Metabolic processes occur in multiple tissues, including the gastrointestinal tract, skin, plasma, kidneys, and lungs, with the liver serving as the primary site through phase I (CYP450) and phase II (UGT) reactions. These processes generally lead to drug inactivation, increased hydrophilicity, and facilitated excretion [22].

Excretion

Excretion represents the biological process through which drugs and their metabolites are removed from the body. The primary route is renal elimination, with kidneys performing passive glomerular filtration and active tubular secretion, potentially followed by reabsorption of certain compounds. Alternative excretory pathways include pulmonary, integumentary, and gastrointestinal systems, depending on drug physicochemical properties and metabolic fate [23].

Toxicity

Toxicity assessment is fundamental to drug development, enabling identification and characterization of potential adverse effects. Comprehensive toxicological studies throughout preclinical and clinical phases help determine safe dosage ranges, elucidate toxicity mechanisms, and inform risk–benefit analyses. Toxicity represents a leading cause of drug candidate attrition, often contributing to high development costs when detected late, underscoring the necessity for early rigorous evaluation [24].

Case Studies

In Silico Assessment of the Inhibitory Potential of Bromophenol Against ACE2 Receptor-mediated COVID-19 Viral Entry

SWISSADME/ADMET prediction: the Simplified Molecular Input Line Entry System (SMILES) canonical representation of bromophenol (BP) was submitted to the SwissTargetPrediction server (www.swisstargetprediction.com) to predict potential targets [25]. Subsequent PK, toxicity, and drug-like properties were predicted. Evaluation included adherence to Lipinski's rules, Veber's rule, Egan's rule, polar surface area (TPSA) measurement, consideration of rotatable bonds, ADME properties, and cytochrome P450 metabolism sites (Table 9.1) [13].

Table 9.1 SwissADME/ADMET prediction for β-D-mannose and bromophenol.

		β-D-mannose (control)	**Ligand (synthetic compound)**
	Canonical SMILES	C(C1C(C(C(C(O1)O)O)O)O)O	O=C1CCC(N1Cc1cc(O)c(c(c1Br)Br)O)C(=O)[CH_2^+]
Physicochemical properties	Molecular formula	$C_6H_{12}O_6$	$C_{13}H_{12}Br_2NO_4$
	Molecular weight (g/mol)	180.16	406.05
	Num. rotatable bonds	1	3
	Num. H-bond acceptors	6	4
	Num. H-bond donor	5	2
	Molar refractivity	35.74	84.29
	TPSA (Å^2)	110.38	77.84
Lipophilicity	Consensus log$P$ o/w	−2.26	2
Water solubility	log$S$ (ESOL)	1.15	−3.48
Pharmacokinetics and toxicity	GI absorption	Low	High
	Ames_test	Mutagen	Mutagen
	Caco-2	2.56748	20.1055
	HIA	22.355048	94.470994
	Plasma_Protein_Binding	7.312029	88.587673
	Caecino_Mouse	Negative	Negative
	hERG_inhibition	Low risk	Low risk
	BBB permeant	No	No
	P-gp substrate	Yes	No
	CYP1A2 inhibitor	No	Yes
	CYP2C19 inhibitor	No	No
	CYP2C9 inhibitor	No	No
	CYP2D6 inhibitor	No	No
	CYP3A4 inhibitor	No	No
Drug-likeness	Lipinski	Yes	Yes
	Ghose	No	Yes
	Veber	Yes	Yes
	Egan	Yes	Yes
	Muegge	No	Yes
	Bioavability score	0.55	0.55
Medicinal chemistry	Synthetic accessibility	4.08	2.52

Interpretation of results: Lipinski's Rule of Five (for drug-likeness) predicts oral drug suitability based on specific properties: molecular mass less than 500 daltons (Da), log *P* less than 5, fewer than five hydrogen bond donors, fewer than 10 hydrogen bond acceptors, and molar refractivity between 40 and 130 [8]. The study indicated that BP demonstrated superior properties compared to β-D-mannose control [26]. ADME results showed significant gastrointestinal absorption potential for the ligand [27, 28]. The BP exhibited optimal Caco-2 permeability values and favorable human intestinal absorption results, suggesting strong potential for oral bioavailability [29, 30].

In Silico Study of Synthetic Bromophenol Compounds Against *Staphylococcus aureus* Dihydrofolate Reductase Enzyme

SwissTarget and SwissADME/ADMET prediction: the SMILES canonical representation of synthetic ligand was submitted to SwissTargetPrediction for target identification [31]. PK, toxicity, and drug-likeness properties were predicted for both synthetic ligand and trimethoprim (TMP) control. Evaluation included adherence to Lipinski's rules, Veber's rule, Egan's rule, TPSA, rotatable bonds, and ADME properties (Table 9.2) [26].

Result interpretation: ADME considerations are crucial in drug development, with early studies helping minimize pharmacokinetics-related failures [29]. The synthetic ligand confirmed adherence to Lipinski's rules and related criteria [32–34]. The synthetic accessibility score (2.51) indicated synthetic feasibility [35, 36]. Based on ADME/T results, the synthetic ligand demonstrated strong gastrointestinal absorption capacity [37, 38].

Table 9.2 SwissADME/ADMET prediction for TMP and bromophenol derivatives.

		TMP	Synthetic ligand
	Canonical SMILES	COC1=CC(=CC(=C1OC)OC)C C2=CN=C(N=C2N)N	OC(=O)C1CCC(=O)N1CC1=CC O=C(O) C(Br)=C1Br
Physicochemical properties	Molecular formula	$C_{14}H_{18}N_4O_3$	$C_{12}H_{11}Br_2NO_5$
	Molecular weight (g/mol)	290.32	409.03
	Num. rotatable bonds	5	3
	Num. H-bond acceptors	5	5
	Num. H-bond donor	2	3
	Molar refractivity	79.77	81.56
	TPSA (Å^2)	105.51	98.07

		TMP	Synthetic ligand
Lipophilicity	Consensus log*P* o/w	1.16	1.57
Water solubility	log*S* (ESOL)	−2.31	−3.40
Pharmaco-kinetics and toxicity	GI absorption	High	High
	Ames_test	Mutagen	Mutagen
	Caecino_Mouse	Positive	Negative
	hERG_inhibition	Low risk	Low_risk
	BBB permeant	No	No
	P-gp substrate	Yes	No
	CYP1A2 inhibitor	No	Yes
	CYP2C19 inhibitor	No	No
	CYP2C9 inhibitor	No	No
	CYP2D6 inhibitor	No	No
	CYP3A4 inhibitor	No	No
Drug-likeness	Lipinski	Yes	Yes
	Ghose	Yes	Yes
	Veber	Yes	Yes
	Egan	Yes	Yes
	Muegge	Yes	Yes
	Bioavability score	0.55	0.56
Medicinal chemistry	Synthetic accessibility	2.58	2.51

Target prediction: SwissTargetPrediction anticipates likely protein targets for small compounds [39]. For the synthetic ligand (SMILES: OC(=O)C1CCC(=O)N1CC1=CC(O)=C(O)C(Br)=C1Br), the top 25 predicted targets (proteins) included enzymes promoting drug reactions, protease family members, and kinase targets (Figure 9.1).

Comparative Analysis and Integration of Tools

Strengths and Limitations

A comparative summary of the strengths and limitations of these commonly used ADMET prediction tools is presented in Table 9.3.

Synergistic Use in Drug Development

Early-stage screening benefits from SwissADME filtering for drug-likeness followed by pkCSM refinement of PK/toxicity profiles [40]. Lead optimization incorporates

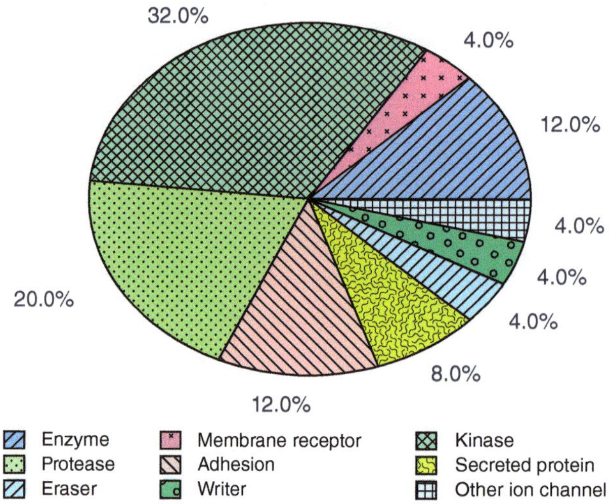

Figure 9.1 Graph displaying the predicted target structure of the synthetic ligand (SMILE canonical: OC(=O)C1CCC(=O)N1CC1=CC(O)=C(O)C(Br)=C1Br).

Table 9.3 Comparative strengths and limitations of commonly used in silico ADMET prediction tools.

Tool	Strengths	Limitations
SwissADME	User-friendly interface, comprehensive PK profiling, real-time visualization, free accessibility	Limited toxicity prediction capabilities, dependence on predefined chemical space
pkCSM	Broad ADMET coverage, interpretable models, graph-based approach	Lower accuracy for novel chemotypes, requires structural input
TEST	Robust toxicity prediction, regulatory acceptance, multiple QSAR approaches	Less optimized for drug-like molecules, primarily environmental focus

TEST to identify toxicity liabilities potentially missed by other tools [41]. A case study involving a transformer-based model demonstrated successful integration of ADMET predictions with IC_{50} screening, reducing attrition in HIV (human immunodeficiency virus) integrase inhibitor development [42].

Future Directions

Emerging Technologies

Deep-PK represents a deep learning platform predicting 73 ADMET end points using graph neural networks [7]. Physiologically based pharmacokinetic (PBPK) modeling enables more accurate translational predictions through simulation of physiological processes [43]. Integration of multi-omics data with ADMET predictions offers potential for more comprehensive profiling [44].

Challenges

Data scarcity for niche targets remains a significant limitation [45]. Model interpretability in AI-driven tools requires further development [46]. Integration of diverse data sources and standardization across platforms present ongoing challenges [47].

Conclusion

Computational ADMET prediction tools have revolutionized early drug discovery by enabling comprehensive evaluation of PK and toxicity properties before costly experimental validation. Platforms such as SwissADME, pkCSM, and TEST offer complementary capabilities that, when integrated, provide robust assessment of drug candidate suitability. These tools address the critical challenge of late-stage attrition by identifying potential ADMET issues early in the development pipeline, thereby conserving resources and accelerating the delivery of safe therapeutics to patients. Despite significant advances, challenges remain including data limitations for novel targets, model interpretability concerns, and the need for improved integration across platforms. Future developments in artificial intelligence, ML, and multi-omics integration promise to enhance predictive accuracy and scope. As computational methods continue to evolve, their role in drug discovery will expand, potentially transforming pharmaceutical development through more efficient identification of viable drug candidates and reduced reliance on animal testing. The ongoing refinement of these tools represents a critical component in addressing the persistent challenge of drug development attrition while advancing toward more personalized and effective therapeutic interventions.

List of Abbreviations

ADMET	Absorption, distribution, metabolism, excretion, and toxicity
AI	Artificial intelligence
BBB	Blood–brain barrier
Caco-2	Human colon adenocarcinoma cell line
CYP	Cytochrome P450
EPA	Environmental Protection Agency
GI	Gastrointestinal
hERG	Human Ether-à-go-go-Related Gene
HIA	Human intestinal absorption
IC_{50}	Half maximal inhibitory concentration

LD_{50}	Lethal dose, 50%
ML	Machine learning
PBPK	Physiologically based pharmacokinetics
PK	Pharmacokinetics
QSAR	Quantitative structure–activity relationship
SA	Synthetic accessibility
SMILES	Simplified Molecular Input Line Entry System
TPSA	Topological polar surface area
TEST	Toxicity Estimation Software Tool
UGT	Uridine 5′-diphospho-glucuronosyltransferase

References

1. Kola I, Landis J. Can the pharmaceutical industry reduce attrition rates? *Nat. Rev. Drug Discov.* 2004;3(8):711–5.
2. Pires DEV, Blundell TL, Ascher DB. pkCSM: Predicting small-molecule pharmacokinetic and toxicity properties using graph-based signatures. *J. Med. Chem.* 2015;58(9):4066–72.
3. DiMasi JA, Grabowski HG, Hansen RW. Innovation in the pharmaceutical industry: new estimates of R&D costs. *J. Health Econ.* 2016;47:20–33.
4. Paul SM, Mytelka DS, Dunwiddie CT, et al. How to improve R&D productivity: the pharmaceutical industry's grand challenge. *Nat. Rev. Drug Discov.* 2010;9(3):203–14.
5. Arrowsmith J, Miller P. Phase II and phase III attrition rates 2011–2012. *Nat. Rev. Drug Discov.* 2013;12(8):569.
6. Hughes JP, Rees S, Kalindjian SB, Philpott KL. Principles of early drug discovery. *Br. J. Pharmacol.* 2011;162(6):1239–49.
7. Dong J, Wang NN, Yao ZJ, et al. ADMETlab: a platform for systematic ADMET evaluation based on a comprehensively collected ADMET database. *J. Cheminform.* 2018;10:29.
8. Lipinski CA, Lombardo F, Dominy BW, Feeney PJ. Experimental and computational approaches to estimate solubility and permeability in drug discovery and development settings. *Adv. Drug Deliv. Rev.* 2001;46(1–3):3–26.
9. Vamathevan J, Clark D, Czodrowski P, et al. Applications of machine learning in drug discovery and development. *Nat. Rev. Drug Discov.* 2019;18(6):463–77.
10. Wu Z, Pan S, Chen F, et al. A comprehensive survey on graph neural networks. *IEEE Trans. Neural Netw. Learn. Syst.* 2021;32(1):4–24.
11. Jiménez-Luna J, Grisoni F, Weskamp N, Schneider G. Artificial intelligence in drug discovery: recent advances and future perspectives. *Expert Opin. Drug Discov.* 2021;16(9):949–59.
12. Daina A, Michielin O, Zoete V. SwissADME: A free web tool to evaluate pharmacokinetics, drug-likeness and medicinal chemistry friendliness of small molecules. *Sci. Rep.* 2017;7:42717.

13 Daina A, Zoete V. A BOILED-Egg to predict gastrointestinal absorption and brain penetration of small molecules. *ChemMedChem* 2016;11(11):1117–21.

14 Ferreira JF, Luthria DL, Sasaki T, Heyerick A. Flavonoids from *Artemisia annua* L. as antioxidants and their potential synergism with artemisinin against malaria and cancer. *Molecules* 2010;15(5):3135–70.

15 Pires DEV, Blundell TL, Ascher DB. pkCSM: predicting small-molecule pharmacokinetic and toxicity properties using graph-based signatures. *J. Med. Chem.* 2015;58(9):4066–72.

16 Tian S, Wang J, Li Y, et al. The application of in silico drug-likeness predictions in pharmaceutical research. *Adv. Drug Deliv. Rev.* 2015;86:2–10.

17 U.S. Environmental Protection Agency. Toxicity Estimation Software Tool (TEST) [software]. Version 5.1. 2021. Available from: https://www.epa.gov/chemical-research/toxicity-estimation-software-tool-test

18 Martin TM, Harten P, Venkatapathy R, et al. A hierarchical clustering methodology for the estimation of toxicity. *Toxicol. Mech. Meth.* 2008;18(2–3):251–66.

19 Yang C, Tarkhov A, Marusczyk J, et al. New publicly available chemical query language, CSRML, to support chemotype representations for application to data mining and modeling. *J. Chem. Inf. Model.* 2015;55(3):510–28.

20 Lennernäs H. Human intestinal permeability. *J. Pharm. Sci.* 1998;87(4):403–10.

21 Smith DA, Di L, Kerns EH. The effect of plasma protein binding on in vivo efficacy: misconceptions in drug discovery. *Nat. Rev. Drug Discov.* 2010;9(12):929–39.

22 Guengerich FP. Cytochrome P450 and chemical toxicology. *Chem. Res. Toxicol.* 2008;21(1):70–83.

23 Rowland M, Tozer TN. *Clinical Pharmacokinetics and Pharmacodynamics: Concepts and Applications*. 4th ed. Lippincott Williams & Wilkins; 2010.

24 Olson H, Betton G, Robinson D, et al. Concordance of the toxicity of pharmaceuticals in humans and in animals. *Regul. Toxicol. Pharmacol.* 2000;32(1):56–67.

25 Weininger D. SMILES, a chemical language and information system. 1. Introduction to methodology and encoding rules. *J. Chem. Inf. Comput. Sci.* 1988;28(1):31–6.

26 Bakchi B, Krishna AD, Sreecharan E, et al. An overview on applications of SwissADME web tool in the design and development of anticancer, antitubercular and antimicrobial agents: A medicinal chemist's perspective. *J. Mol. Struct.* 2022; 1259:132712.

27 Egan WJ, Merz KM, Baldwin JJ. Prediction of drug absorption using multivariate statistics. *J. Med. Chem.* 2000;43(21):3867–77.

28 Veber DF, Johnson SR, Cheng HY, et al. Molecular properties that influence the oral bioavailability of drug candidates. *J. Med. Chem.* 2002;45(12):2615–23.

29 Alqahtani S. In silico ADME-Tox modeling: progress and prospects. *Expert Opin. Drug Metab. Toxicol.* 2017;13(11):1147–58.

30 Zanger UM, Schwab M. Cytochrome P450 enzymes in drug metabolism: regulation of gene expression, enzyme activities, and impact of genetic variation. *Pharmacol. Ther.* 2013;138(1):103–41.

31 Li Z, Wan H, Shi Y, Ouyang P. Personal experience with four kinds of chemical structure drawing software: review on ChemDraw, ChemWindow, ISIS/Draw, and ChemSketch. *J. Chem. Inf. Comput. Sci.* 2004;44(5):1886–90.

32 Protti ÍF, Rodrigues DR, Fonseca SK,. Do drug-likeness rules apply to oral prodrugs? *ChemMedChem* 2021;16(9):1446–56.

33 Ghose AK, Viswanadhan VN, Wendoloski JJ. A knowledge-based approach in designing combinatorial or medicinal chemistry libraries for drug discovery. 1. A qualitative and quantitative characterization of known drug databases. *J. Comb. Chem.* 1999;1(1):55–68.

34 Muegge I, Heald SL, Brittelli D. Simple selection criteria for drug-like chemical matter. *J. Med. Chem.* 2001;44(12):1841–6.

35 Thakkar A, Chadimová V, Bjerrum EJ, et al. Retrosynthetic accessibility score (RAscore) – rapid machine learned synthesizability classification from AI driven retrosynthetic planning. *Chem. Sci.* 2021;12(9):3339–49.

36 Ertl P, Schuffenhauer A. Estimation of synthetic accessibility score of drug-like molecules based on molecular complexity and fragment contributions. *J. Cheminform.* 2009;1:8.

37 Bitew M, Desalegn T, Demissie TB, et al. Pharmacokinetics and drug-likeness of antidiabetic flavonoids: molecular docking and DFT study. *PLoS One* 2021;16(12):e0260853.

38 Zadorozhnii P, Kiselev V, Kharchenko A. In silico prediction of ADME-Tox properties: approaches, challenges and potential applications. *Comput. Biol. Chem.* 2021;92:107481.

39 Gfeller D, Grosdidier A, Wirth M, et al. SwissTargetPrediction: a web server for target prediction of bioactive small molecules. *Nucleic Acids Res.* 2014;42:W32–8.

40 Tian S, Wang J, Li Y, et al. The application of in silico drug-likeness predictions in pharmaceutical research. *Adv. Drug Deliv. Rev.* 2015;86:2–10.

41 Martin TM, Harten P, Venkatapathy R, et al. A hierarchical clustering methodology for the estimation of toxicity. *Toxicol. Mech. Meth.* 2008;18 (2–3):251–66.

42 Goh GB, Hodas NO, Vishnu A. Deep learning for computational chemistry. *J. Comput. Chem.* 2017;38(16):1291–307.

43 Jones HM, Chen Y, Gibson C, et al. Physiologically based pharmacokinetic modeling in drug discovery and development: a pharmaceutical industry perspective. *Clin. Pharmacol. Ther.* 2015;97(3):247–62.

44 Wishart DS, Feunang YD, Guo AC, et al. DrugBank 5.0: a major update to the DrugBank database for 2018. *Nucleic Acids Res.* 2018; 46(D1):D1074–82.

45 Ciallella HL, Russo DP, Aleksunes LM, Zhu H. Predictive modeling of estrogen receptor binding agents using advanced cheminformatics tools and massive public data. *Front. Environ. Sci.* 2020;8:109.

46 Rudin C. Stop explaining black box machine learning models for high stakes decisions and use interpretable models instead. *Nat. Mach. Intell.* 2019;1(5):206–15.

47 Williams AJ, Ekins S, Tkachenko V. Towards a gold standard: regarding quality in public domain chemistry databases and approaches to improving the situation. *Drug Discov. Today* 2012;17(13–14):685–701.

10

In Silico Drug Design and Virtual Screening Techniques

*Dishank Purandare[1], Rahul Jawarkar[2], Bhagwat Nagargoje[1], Yogeshwar Bachhav[3], and Satish Polshettiwar[1]**

[1]*School of Health Sciences and Technology, Dr. Vishwanath Karad MIT World Peace University, Pune, Maharashtra, India*
[2]*Department of Medicinal Chemistry, Dr. Rajendra Gode Institute of Pharmacy, Amravati, Maharashtra, India*
[3]*AiCuris Anti-infective Cures AG, Wuppertal, Germany*

*Corresponding author: satish.polshettiwar@mitwpu.edu.in

Introduction

In silico drug design is a computational approach that has revolutionized the modern drug discovery landscape by significantly enhancing efficiency, reducing costs, and expediting the identification of potential therapeutic agents [1]. It refers to the utilization of advanced computational tools and algorithms to predict, model, and optimize drug-like compounds, facilitating a rational and systematic approach to drug discovery. The integration of in silico methodologies in pharmaceutical research has become indispensable due to their ability to efficiently screen vast chemical libraries, predict molecular interactions, assess pharmacokinetics and toxicology, and optimize lead compounds before experimental validation [2]. This computational approach has significantly contributed to overcoming challenges associated with traditional drug discovery methods, which are often labor-intensive, time-consuming, and costly [3]. Conventional drug discovery heavily relies on high-throughput screening (HTS) and structure–activity relationship (SAR) studies, requiring extensive biochemical assays, cell-based models, and in vivo studies [4]. Although these approaches have led to the development of numerous therapeutics, they suffer from inefficiencies, high attrition rates, and substantial financial burdens, limiting their scalability in modern pharmaceutical research. Additionally, the catastrophe of many drug candidates in later stage clinical trials due to inadequate efficacy, poor bioavailability, or unforeseen toxicity further highlights the necessity for more reliable predictive models [5]. In contrast, computational drug discovery leverages molecular docking, virtual screening, molecular dynamics (MD) simulations, and quantitative structure–activity relationship (QSAR) modeling to predict the behavior of potential drug molecules with high precision [6]. By utilizing in silico methods, researchers can identify promising drug candidates from extensive chemical libraries before conducting experimental validation, significantly plummeting the number of molecules subjected to inflated and time-intensive laboratory testing. One of the crucial importance of in silico drug design is its capability to streamline the drug discovery pipeline by facilitating hit identification, toxicity prediction, and lead optimization. Virtual screening, a fundamental component of in silico drug discovery, employs ligand-based and structure-based methodologies to identify small molecules with high binding affinities to specific drug targets. Ligand-based virtual screening (LBVS) relies on known active compounds to predict novel analogs with similar bioactivity, utilizing machine learning algorithms, pharmacophore modeling, and molecular fingerprints to rank potential candidates [7]. Conversely, structure-based virtual screening (SBVS) employs molecular docking techniques to evaluate ligand–target interactions by simulating binding modes, free energy calculations, and conformational flexibility assessments [8]. These computational techniques provide a high-throughput alternative to traditional screening methods, reducing the reliance on costly wet lab experiments and increasing the probability of identifying viable drug candidates. Furthermore, in silico drug design allows for the primary prediction of ADMET properties, enhancing the selection of compounds with favorable pharmacokinetic and safety profiles [1]. Predictive ADMET modeling incorporates machine learning algorithms, molecular descriptors, and physiologically based pharmacokinetic (PBPK) simulations to evaluate drug-like properties, minimizing the risk of late-stage failures [9]. This approach significantly reduces the number of compounds advancing to clinical trials with suboptimal pharmacological profiles, ultimately increasing the efficiency of the drug development process. Another critical advantage of computational drug discovery is its cost-effectiveness. The traditional drug discovery pipeline often requires billions of dollars and over a decade of research and development before a single therapeutic reaches the market. In contrast, in silico methods substantially lower these costs by prioritizing the most promising candidates before experimental validation. By eliminating the need for extensive trial-and-error experimentation, computational techniques accelerate the identification of lead compounds, optimizing resource allocation and reducing financial risks associated with drug development [10]. Additionally, in silico approaches enable drug repurposing strategies, where existing FDA-approved drugs or investigational compounds are computationally screened against new disease targets, further reducing development timelines and costs [11]. Drug repurposing has been instrumental in addressing urgent medical needs, particularly during pandemic outbreaks, where rapid identification of effective therapeutics is critical [11]. The emergence of AI in computational drug discovery has further augmented the predictive capabilities of in silico models. AI-driven approaches facilitate automated data analysis and molecular property prediction, enhancing the precision of lead identification and optimization. Machine learning algorithms trained on vast datasets can predict molecular docking scores, ADMET properties, and bioactivity with unprecedented precision, allowing the swift identification of novel drug scaffolds. Moreover, generative AI models can design entirely new chemical scaffolds with optimized pharmacological properties, expanding the chemical space available for drug discovery. These improvements emphasize the transformative potential of AI in accelerating drug development and improving hit-to-lead conversion rates. Despite the numerous advantages of in silico drug design, certain limitations must be acknowledged. Computational models rely on the availability of high-quality structural and experimental data to train predictive algorithms effectively. Incomplete or inaccurate data can lead to erroneous predictions, necessitating rigorous validation and integration with experimental approaches. Additionally, molecular docking and virtual screening techniques, while highly efficient, may sometimes fail to consider the dynamic nature of protein–ligand interactions, necessitating MD simulations to assess binding stability over time [12]. The continuous refinement of computational methodologies, coupled with advancements in structural biology and data-driven modeling, is essential to further boost the reliability and applicability of in silico drug discovery. Thus, in silico drug design has become an indispensable tool in modern pharmaceutical research, offering a cost-effective, time-efficient, and highly predictive approach to drug discovery. By integrating virtual screening, molecular docking, QSAR modeling, and AI-driven drug design, computational methodologies have revolutionized the identification and optimization of therapeutic candidates. The ability to predict binding affinities, optimize lead compounds, and assess ADMET properties in silico has significantly streamlined the drug discovery pipeline, minimizing the prices and growing the prospect of success in clinical trials.

Fundamentals of In Silico Drug Design

Theoretical Background of In Silico Drug Design

In silico drug design is a computational approach that enables the discovery and development of probable drug molecules by leveraging advanced algorithms, molecular modeling, and simulation techniques [13]. It has arisen as a crucial component of modern methods due to its capability to predict molecular interactions, optimize drug properties, and streamline the lead identification process. The foundation of in silico drug design lies in understanding molecular interactions, analyzing biological targets, and predicting ligand behavior in a virtual environment. This approach significantly reduces the reliance on trial-and-error experimental methods, thereby expediting the drug discovery procedure and minimizing associated costs. The theoretical framework of in silico drug design is based on two primary methodologies: structure-based drug design (SBDD) and ligand-based drug design (LBDD). While SBDD operates on the three-dimensional (3D) structure of target proteins to design inhibitors, LBDD depends on known active molecules to predict new drug candidates through QSAR and pharmacophore modeling [6–9].

Molecular Modeling Principles

Molecular modeling is a fundamental aspect of in silico drug design that involves computational representation of molecular structures and their interactions with biological targets [14]. It integrates principles from quantum mechanics (QM) and molecular mechanics (MM) to describe the electronic and atomic behavior of molecules. QM-based methods, such as density functional theory (DFT) and ab initio calculations, provide highly accurate electronic structures and energy calculations [15], while MM methods use force fields like CHARMM, AMBER, and OPLS to simulate macromolecular interactions efficiently [16]. Another key component of molecular modeling is molecular docking, which predicts how a ligand fits into the binding pocket of a protein. Docking algorithms assess different orientations and conformations of a ligand, ranking them based on binding affinity using scoring functions. Furthermore, MD simulations analyze the stability and conformational flexibility of protein–ligand complexes over time, allowing for a better understanding of their interactions under physiological conditions [17]. Pharmacophore modeling, on the other hand, identifies essential chemical features required for molecular identification among a ligand and its receptor, providing a framework for designing novel drug candidates with improved binding affinity.

Structure-based Drug Design

This method uses the 3D structure of a target protein to develop new drugs. It relies on structural data gained over X-ray crystallography, nuclear magnetic resonance (NMR) spectroscopy, and cryo-electron microscopy, which offers high-resolution molecular insights into protein–ligand interactions [18]. Once the target structure is determined, binding site analysis is performed to identify key residues that interact with ligands, guiding the design of new inhibitors with enhanced specificity and efficacy. A crucial step in SBDD is molecular docking, which anticipates the preferred binding orientation of a ligand within the active site of a protein. Docking simulations generate multiple ligand conformations, which are then ranked based on scoring functions that estimate binding affinity. Additional optimization techniques, such as induced fit docking (IFD) and free energy perturbation (FEP) calculations, refine binding interactions by accounting for protein flexibility and solvent effects. The lead compounds identified through SBDD undergo iterative cycles of lead optimization, where chemical alterations are introduced to improve potency, selectivity, and pharmacokinetic possessions [19].

Ligand-based Drug Design

LBDD is used when the structure of a target protein is unknown but a set of bioactive compounds is available. This method employs computational techniques to analyze molecular similarities, predict binding affinities, and generate novel compounds with improved therapeutic potential [20]. One of the most widely used LBDD approaches is QSAR modeling, which correlates the structural properties of known molecules with their biological properties. By applying machine learning algorithms and regression models, QSAR studies can predict the pharmacological properties of new compounds based on their molecular descriptors [21]. Other key techniques in LBDD are CoMFA and CoMSIA, which analyze steric and electrostatic interactions to identify molecular features crucial for biological activity. Additionally, molecular fingerprints and similarity search techniques compare new compounds against existing drug libraries to identify potential lead candidates.

LBDD methods are particularly useful in scaffold hopping, where new chemical entities are designed by modifying known drug-like scaffolds to improve their pharmacological profile.

Workflow and Key Components of In Silico Drug Design

The workflow of *in silico* drug design encompasses several stages, from target identification to lead optimization. Every step plays a vital part in guaranteeing the effective identification and refinement of drug candidates.

Target Identification

The initial step is the selection of a biological target associated with a disease. Genomic and proteomic analysis is employed to categorize proteins or enzymes that play a vital role in disease pathology. Once a target is identified, homology modeling is used to predict the 3D structure of unknown proteins based on their sequence similarity to known structures. In cases where structural data is unavailable, computational approaches such as molecular docking and virtual screening are used to evaluate potential binding interactions between small molecules and the target protein [22].

Lead Discovery

Once a target is identified, lead compounds are selected using various computational tools. Virtual screening is one of the most widely used methods, where huge chemical libraries are screened to identify molecules with high binding affinity. Fragment-based drug design (FBDD) is another strategy that involves screening small molecular fragments, which are then optimized to develop more potent inhibitors. Additionally, de novo drug design employs AI and cheminformatics tools to create novel molecular structures with anticipated biological properties [23].

Lead Optimization

After identifying promising lead compounds, further modifications are made to enhance their pharmacological profile. SAR studies play a crucial role in optimizing molecular structures by identifying key functional groups responsible for biological activity. ADMET prediction tools assess the pharmacokinetic properties of drug candidates, confirming their safety and effectiveness. Computational methods such as binding free energy calculations and MD simulations refine ligand interactions to improve drug stability and target selectivity [24].

Software and Tools in In Silico Drug Design

Molecular Docking Software

- **AutoDock and AutoDock Vina:** open-source docking tools used for flexible ligand docking and virtual screening [25].
- **MOE:** a comprehensive platform for docking, pharmacophore modeling, and QSAR studies.

- **Schrödinger Suite (Glide, Maestro):** high-precision docking software with advanced scoring functions and free energy calculations [26].
- **OpenEye (FRED, ROCS):** tools for shape-based virtual screening and molecular similarity analysis [27].

MD Simulation Tools

- **GROMACS, AMBER, and CHARMM:** force-field-based MD simulation platforms for studying biomolecular dynamics.
- **Desmond (Schrödinger):** high-speed MD simulation software with enhanced sampling algorithms.

QSAR and Machine Learning Tools

DeepChem, RDKit, and ChemProp: AI-driven platforms for predicting biological activity and optimizing molecular properties.

ADMET Prediction Tools

SwissADME and pkCSM: web-based tools for evaluating drug-likeness and pharmacokinetics [28]. In silico drug design has revolutionized pharmaceutical research by enabling rapid identification and optimization of drug candidates. Through molecular modeling, virtual screening, and machine learning, computational approaches continue to enhance drug discovery efficiency. The integration of in silico techniques with experimental validation ensures the development of safe, effective, and targeted therapeutics, paving the way for future advancements in precision medicine.

Structure-based Drug Design

This relies on the 3D structure of biological macromolecules to rationally design and improve drug candidates. It provides molecular insights into protein–ligand interactions, allowing for precise modifications that enhance binding affinity, selectivity, and pharmacokinetic properties. SBDD is widely employed in drug discovery due to its ability to streamline lead identification and optimization by integrating computational approaches such as molecular docking, MD simulations, homology modeling, and pharmacophore modeling. The availability of high-resolution structural data, obtained through methods such as X-ray crystallography, NMR spectroscopy, and cryo-EM, has significantly advanced the field. The integration of SBDD with AI and machine learning further enhances its predictive capabilities, making it a crucial approach in modern drug discovery [29].

Importance of Protein Structure in Drug Discovery

Proteins serve as fundamental biological macromolecules that regulate cellular processes through specific molecular interactions. In drug discovery, the structural characterization of target proteins is essential for understanding their functional mechanisms and recognizing potential binding sites for small-molecule therapeutics [30]. The availability of detailed 3D structural information allows researchers to analyze the conformational flexibility of proteins, identify allosteric sites, and design drugs with optimized pharmacological properties. The significance of protein structure in drug discovery is underscored by the concept of SAR, which elucidates how structural modifications in ligands influence their biological activity [31]. Knowledge of protein structure facilitates rational drug design by enabling the prediction of key binding hot spots, optimizing ligand interactions, and minimizing off-target effects [32]. In addition to primary binding pockets, proteins often contain cryptic sites, which are transiently formed and may serve as alternative druggable sites. The identification of such sites has led to the progress of selective inhibitors with improved specificity and reduced toxicity. The impact of protein structure on drug discovery is evident in the success of kinase inhibitors, protease inhibitors, and GPCR modulators, which were developed based on detailed structural insights. Moreover, the emergence of conformational ensemble modeling, which considers the dynamic nature of proteins, has expanded the scope of SBDD beyond static representations, facilitating the design of ligands that stabilize specific conformational states of target proteins.

Protein–Ligand Interactions: Role of Crystallography and NMR

Protein–ligand interactions are overseen by noncovalent forces such as hydrogen bonding, hydrophobic interactions, electrostatic forces, and van der Waals interactions. Understanding these relations at the atomic level is essential for designing ligands with optimal binding affinity and specificity [33]. Structural biology techniques such as X-ray crystallography and NMR spectroscopy provide high-resolution insights into protein–ligand complexes, enabling the rational design of new therapeutic agents.

X-ray Crystallography

This is the most extensively used method for defining protein structures with atomic-level precision. It involves the crystallization of proteins followed by the diffraction of X-rays to generate electron density maps, which are then used to model atomic positions [34]. The key advantages of X-ray crystallography include the following:

- High-resolution structural data (often below 2 Å), allowing for precise mapping of ligand-binding sites.
- Identification of water-mediated interactions that contribute to ligand binding.
- Visualization of conformational changes induced by ligand binding.

Despite its advantages, X-ray crystallography has boundaries, like the necessity for high-quality protein crystals and the inability to capture dynamic conformational states in solution.

NMR Spectroscopy

NMR spectroscopy is a complementary technique that provides structural insights into proteins in solution, capturing their dynamic behavior and transient interactions [35]. Unlike crystallography, NMR does not require protein crystallization and is particularly useful for studying flexible and intrinsically disordered proteins. Applications of NMR in drug discovery include:

- determination of ligand-binding sites through chemical shift perturbation experiments;
- analysis of conformational dynamics using relaxation dispersion techniques;
- screening of fragment libraries to identify weak binders in FBDD.

However, NMR is limited by its lower resolution for large proteins and requires high concentrations of protein samples.

Molecular Docking: Rigid vs. Flexible Docking, Scoring Functions, and Validation

Molecular docking is a computational technique that forecasts the optimal binding pose of a ligand within the active site of a target protein. It is a cornerstone of SBDD and enables the screening of vast chemical libraries to identify lead compounds.

Rigid vs. Flexible Docking

Docking approaches are classified into rigid docking and flexible docking based on the degree of conformational freedom allowed during ligand–protein interactions:

- Rigid docking assumes a fixed conformation of both the protein and the ligand, making it computationally efficient but limited in accuracy when dealing with flexible targets.
- Flexible docking allows for conformational changes in either the ligand, the protein, or both, providing a more realistic representation of molecular interactions but requiring increased computational resources.

Advanced docking algorithms such as IFD incorporate protein flexibility by generating multiple receptor conformations, improving the accuracy of docking predictions [36, 37].

Scoring Functions and Validation

Scoring functions evaluate the binding affinity of docked ligands by estimating free energy changes upon binding. The three major classes of scoring functions include the following:

- Empirical scoring functions (e.g., ChemScore, GlideScore) that incorporate experimental data.
- Knowledge-based scoring functions derived from statistical analysis of known protein–ligand complexes.

- Physics-based scoring purposes that utilize molecular mechanics force fields (e.g., AMBER, CHARMM) for energy calculations.

Docking validation is achieved through cross-docking, re-docking, and retrospective validation using known inhibitors, ensuring the reliability of predictions.

MD Simulations: Force Fields, Thermodynamic Properties, and Free Energy Calculations

MD simulations provide a time-resolved view of protein–ligand interactions, capturing conformational flexibility and binding stability.

Force Fields in MD Simulations

The accuracy of MD simulations depends on the choice of force fields, which define the potential energy functions governing atomic interactions; for example, CHARMM, which is widely used for simulating protein–ligand interactions, and AMBER, which is used for free energy calculations and biomolecular simulations [38].

Free Energy Calculations

Binding free energy calculations, using methods such as FEP and MM/PBSA, provide quantitative estimates of ligand-binding affinity, guiding lead optimization.

Homology Modeling: Sequence Alignment, Model Refinement, and Validation

Homology modeling predicts the 3D structure of proteins when experimental data is unavailable. The process involves the following:

- Sequence alignment to identify homologous templates.
- Model construction using tools such as MODELLER and SWISS-MODEL.
- Refinement through energy minimization and MD.
- Validation using Ramachandran plots and root-mean-square deviation (RMSD) analysis.

Pharmacophore Modeling: Generation and Applications in Lead Discovery

Pharmacophore modeling is a computational approach used in SBDD to identify and represent essential molecular features required for a ligand to bind to a biological target with high affinity and specificity [39]. A pharmacophore is demarcated as the spatial arrangement of functional groups responsible for molecular recognition, including hydrogen bond donors and acceptors, hydrophobic regions, aromatic centers, and charged groups. This concept is fundamental in drug discovery, as it allows for the identification of novel bioactive molecules by matching molecular structures to an idealized pharmacophoric pattern. Pharmacophore modeling can be classified into ligand-based pharmacophore modeling (LBPM) and structure-based pharmacophore modeling (SBPM). In LBPM, common pharmacophoric features are extracted from a set of known active ligands, assuming that structurally similar compounds with similar functional groups will exhibit comparable biological activity. Techniques such as CoMFA and CoMSIA are often employed to refine the generated pharmacophore models and enhance their predictive accuracy [40]. LBPM is particularly useful when structural information of the target protein is unavailable, allowing researchers to screen large compound libraries based on their alignment with known active pharmacophores. In contrast, SBPM derives pharmacophoric features directly from the 3D structure of the target protein and its binding site. This method involves analyzing the geometric and electronic properties of the protein's active site to define the key interaction points required for effective ligand binding. Tools such as LigandScout, PharmMapper, and Phase (Schrödinger Suite) are widely used to generate structure-based pharmacophores [41]. SBPM provides a more rational approach to drug design, as it takes into account the molecular topology and flexibility of the binding site. This method is particularly valuable in identifying allosteric inhibitors and designing drugs that target nontraditional binding pockets. Pharmacophore modeling has numerous applications in lead discovery and drug optimization. It is extensively used in virtual screening, where compound libraries are filtered based on their alignment with an ideal pharmacophore model, significantly reducing the number of candidates that require experimental validation. Scaffold hopping, another key application, involves modifying or replacing the core structure of a known drug while retaining its essential pharmacophoric features, leading to new chemical entities with improved properties such as enhanced selectivity, solubility, or metabolic stability. Additionally, pharmacophore models play a critical role in FBDD by assisting in the rational design of small molecular fragments that can be linked or grown into more potent inhibitors. The integration of pharmacophore modeling with molecular docking and machine learning further enhances its utility, allowing for the rapid identification and optimization of lead compounds in drug discovery. SBDD has transformed drug discovery by integrating structural biology, computational modeling, and molecular simulations. The continued advancement of docking algorithms, MD simulations, and AI-driven predictive models ensures that SBDD remains at the forefront of pharmaceutical research, accelerating the discovery of novel therapeutic agents.

Ligand-based Drug Design

This approach utilizes known active molecules to predict and design new drug candidates. Unlike SBDD, which relies on the 3D structure of the target protein, LBDD focuses on the molecular properties and structural features of bioactive compounds to identify novel ligands with similar pharmacological effects. This method is mainly useful when the structural data of the biological target is unavailable or incomplete [42]. LBDD employs techniques such as QSAR modeling, molecular similarity searching, de novo drug design, and AI-driven virtual screening. These methodologies allow recognition and optimization of lead compounds with improved pharmacokinetic and pharmacodynamic properties, ultimately facilitating the development of effective therapeutic agents.

QSAR Methodologies: 2D- vs. 3D-QSAR, Molecular Descriptors, and Regression Models

QSAR modeling is a central technique in LBDD that establishes mathematical correlations between the structural and physicochemical properties of compounds and their biological activity. The fundamental principle of QSAR is that the activity of a molecule is inherently linked to its structural characteristics, agreeing for the estimate of the pharmacological potential of new compounds.

2D-QSAR vs. 3D-QSAR

QSAR methodologies are broadly classified into 2D-QSAR and 3D-QSAR models based on how molecular structures are represented:

- 2D-QSAR models use simple structural descriptors such as molecular weight, lipophilicity, electronic properties, and topological indices to predict biological activity. These models rely on statistical techniques such as multiple linear regression (MLR), partial least squares (PLS), and support vector machine (SVM) to establish quantitative relationships between molecular features and activity. While 2D-QSAR is computationally efficient and widely applicable, it does not account for conformational flexibility and spatial arrangements of molecules [43].
- 3D-QSAR models incorporate the 3D spatial arrangement of molecular features to improve the predictive accuracy of ligand–receptor interactions. Techniques such as CoMFA and CoMSIA analyze steric and electrostatic field interactions between molecules, offering a more detailed understanding of molecular recognition. These models require molecular alignment based on a common pharmacophore, making them more computationally intensive but also more informative for structure-based optimization [43].

Molecular Descriptors in QSAR

Molecular descriptors are mathematical depictions of chemical structures that encode essential physicochemical and topological properties [44]. These descriptors are classified into several categories:

- **Constitutional descriptors:** simple properties such as molecular weight, number of atoms, and number of functional groups.
- **Topological descriptors:** information derived from the molecular graph, including connectivity indices and Wiener numbers.

- **Electronic descriptors:** parameters such as dipole moment, partial atomic charges, and MEP.
- **Geometrical descriptors:** 3D features such as molecular volume, surface area, and shape indices.

Regression Models in QSAR

Regression models in QSAR analysis are employed to develop predictive equations that correlate molecular descriptors with biological activity. The most commonly used regression models include the following:

- MLR: establishes linear relationships between molecular descriptors and biological activity.
- PLS regression: reduces dimensionality in datasets with highly correlated descriptors.
- Artificial neural networks (ANNs): captures nonlinear relationships by simulating neuron-like processing units.
- Random forest (RF) and SVM: machine learning models that enhance QSAR predictions by managing large and complex datasets.

QSAR models play a crucial role in lead optimization by predicting the potency, selectivity, and ADMET properties of new drug candidates, reducing the need for extensive experimental screening.

Molecular Similarity Searching: Fingerprint-based Methods, Tanimoto Coefficient

Molecular similarity searching is an essential tool in LBDD that identifies structurally analogous molecules based on predefined molecular features. The underlying assumption is that structurally analogous molecules tend to show comparable biological actions.

Fingerprint-based Similarity Searching

Molecular fingerprints are binary representations of chemical structures that encode the presence or absence of specific substructures or functional groups [45]. The most commonly used fingerprinting methods include the following:

- MACCS keys: encodes structural fragments using a predefined set of molecular patterns.
- Extended connectivity fingerprints (ECFPs): circular fingerprints that capture atomic connectivity and functional group patterns.
- Daylight fingerprints: generates hash-based molecular representations for rapid similarity searches.

Tanimoto Coefficient (T_c) in Similarity Analysis

The T_c is a widely used metric for quantifying molecular similarity, where A and B represent the fingerprint sets of two molecules. A Tanimoto coefficient close to 1 indicates high similarity, while a value close to 0 suggests structural dissimilarity. This method is extensively used in virtual screening and lead hopping, enabling the identification of new drug-like molecules with desired pharmacological properties [46].

De Novo Drug Design: Fragment-based Approaches, Combinatorial Library Design

De novo drug design involves the generation of novel molecular structures from scratch based on desired biological properties. This method employs computational techniques to design lead compounds by assembling small molecular fragments into structurally viable drug candidates [47].

Fragment-based Approaches

FBDD identifies small molecular fragments with weak but specific binding affinities and systematically optimizes them into high-affinity ligands. This approach involves the following:

- Fragment screening: using techniques such as NMR, X-ray crystallography, and SPR to identify fragment binders.
- Fragment growing: expanding the core fragment by adding functional groups to enhance binding interactions.
- Fragment linking: combining two weakly binding fragments to achieve improved potency.

Combinatorial Library Design

Combinatorial chemistry enables the rapid generation of diverse molecular libraries through systematic modifications of core scaffolds. Virtual combinatorial libraries are screened using docking and QSAR models to identify promising lead compounds.

AI/ML-driven Ligand-based Approaches

AI and ML have revolutionized LBDD by improving the accuracy of molecular predictions and accelerating virtual screening [48].

Feature Extraction in AI-driven Drug Design

AI-driven drug discovery relies on feature extraction techniques to identify molecular patterns predictive of biological activity. Common feature extraction methods include the following:

- Graph neural networks (GNNs): represent molecules as graphs to capture atomic connectivity.
- Deep autoencoders: reduce high-dimensional molecular data into informative representations.
- Recurrent neural networks (RNNs): predict molecular sequences for de novo drug design.

Deep Learning in Virtual Screening

Deep learning models enhance virtual screening by learning complex structure–activity relationships from large datasets. Some of the widely used deep learning architectures include the following:

- Convolutional neural networks (CNNs): extract spatial features from molecular graphs.
- Generative adversarial networks (GANs): generate novel drug-like molecules with optimized properties.
- Transformer models: improve molecular property prediction using attention mechanisms.

The integration of AI/ML with LBDD enables HTS, efficient lead optimization, and the development of next-generation therapeutics with improved efficacy and safety profiles.

Virtual Screening Techniques

This approach is used to identify potential bioactive compounds from large chemical libraries. It is a cost-effective and time-efficient alternative to HTS, permitting the rapid recognition of lead compounds with desirable pharmacological activities. Virtual screening can be classified into SBVS, LBVS, and hybrid approaches, each leveraging different computational methods to enhance hit identification. SBVS relies on molecular docking techniques to guess the binding affinity of compounds to a known target structure, utilizing algorithms that assess molecular complementarity and interaction energies. This approach is highly dependent on the availability of high-resolution structural data of the target protein, obtained through X-ray crystallography, NMR spectroscopy, or cryo-electron microscopy. SBVS techniques employ flexible docking strategies that consider protein–ligand interactions, conformational adaptability, and solvent effects to improve binding predictions [6, 49]. However, its success is contingent on the accuracy of scoring functions, which estimate the free energy of binding and rank candidate molecules accordingly. In contrast, LBVS is employed when structural information of the target protein is unavailable but a dataset of active ligands exists. This method utilizes molecular similarity searching, QSAR modeling, and pharmacophore-based screening to identify compounds with structural or physicochemical properties similar to known active ligands [7, 49]. Molecular similarity searching employs fingerprint-based methods and T_c to quantify molecular resemblance, while QSAR models establish mathematical correlations between molecular descriptors and biological activity to predict new bioactive compounds. Pharmacophore modeling, a key component of LBVS, identifies the essential features required for ligand–target interactions and applies them as a filter in virtual screening. The accuracy of LBVS depends on the diversity and quality of training datasets, and its predictive power can be enhanced through machine learning algorithms [49]. Hybrid virtual screening approaches integrate elements of both SBVS and LBVS to overcome their respective limitations and improve hit identification rates [50]. These methods combine molecular docking with pharmacophore-based filtering, MD simulations, and AI-driven predictive models to refine candidate selection. Hybrid screening techniques offer improved accuracy by incorporating structural constraints, molecular similarity analysis, and free energy calculations, reducing the likelihood of false positives and enhancing hit prioritization [51]. The efficiency of virtual screening depends on access to large and diverse chemical libraries, and numerous publicly available databases facilitate compound selection. ZINC (Zinc Is Not Commercial), PubChem, ChEMBL, and DrugBank are among the most widely used databases in virtual screening. ZINC is a curated repository of commercially available compounds, providing a vast collection of drug-like molecules that can be readily screened. It includes subsets optimized for lead-like and fragment-based drug discovery, enabling researchers to select compounds based on their physicochemical properties. PubChem, maintained by the National Center for Biotechnology Information (NCBI), is an open-access database containing millions of chemical structures, bioactivity data, and assay results. It supports structure-based and ligand-based screening, offering cheminformatics tools for molecular similarity searching, substructure analysis, and QSAR modeling. ChEMBL is a manually curated database of bioactive molecules with known target interactions, serving as a valuable resource for LBDD. It provides detailed information on compound–target relationships, facilitating the identification of novel ligands with desired pharmacological profiles. DrugBank is a comprehensive database that integrates chemical, pharmacological, and clinical data on approved and investigational drugs, making it an essential tool for drug repurposing and virtual screening campaigns. Hit identification in virtual screening requires robust scoring and ranking strategies to prioritize potential drug candidates. Molecular docking programs utilize scoring functions to estimate binding affinity, incorporating factors such as hydrogen bonding, hydrophobic interactions, electrostatic forces, and entropy contributions. Commonly used scoring functions include empirical, knowledge-based, and physics-based models, each employing different algorithms to evaluate ligand-binding strength. However, scoring functions often generate false positives due to limitations in accurately modeling solvation effects, entropic contributions, and protein flexibility. To mitigate these challenges, consensus scoring is employed, integrating multiple scoring functions to improve ranking reliability. Consensus scoring combines diverse scoring methods to filter out inconsistent predictions, reducing bias and enhancing the robustness of virtual screening results. Additionally, false-positive elimination techniques are essential to refine hit lists and improve screening accuracy. Post-processing steps such as rescoring, MD simulations, and FEP calculations help validate promising candidates by assessing their stability and conformational adaptability. Machine learning algorithms further aid in false-positive elimination by analyzing structure–activity relationships and identifying key molecular features associated with true binders. Advanced computational techniques such as IFD, QM/MM simulations, and AI-driven binding affinity predictions continue to enhance the precision of virtual screening, facilitating the identification of high-quality lead compounds for subsequent experimental validation.

Computational Tools and Databases

Computational tools and databases play a crucial role in in silico drug design by enabling molecular modeling, virtual screening, and data-driven predictions for drug discovery [52]. These tools facilitate processes such as molecular docking, MD simulations, QSAR modeling, and AI-driven drug screening. The integration of publicly available chemical and biological databases further enhances computational drug design by providing access to structural and pharmacokinetic data. Additionally, HTS techniques, including cloud computing, parallel processing, and GPU acceleration, have significantly improved the efficiency and scalability of in silico methodologies, making drug discovery faster and more cost-effective.

Overview of Computational Tools

Computational tools in drug discovery are broadly classified based on their applications, including molecular docking, MD simulations, QSAR modeling, and AI-driven platforms [53]. These tools help predict molecular interactions, optimize lead compounds, and assess pharmacokinetic properties before experimental validation.

Molecular Docking Tools

Molecular docking is an essential technique in SBDD that predicts the preferred binding orientation of a ligand within the active site of a target protein. It is widely used for virtual screening and lead optimization. Some of the most commonly used docking tools, like AutoDock and AutoDock Vina, developed by the Scripps Research Institute, are widely used molecular docking programs. AutoDock utilizes a genetic algorithm to explore ligand conformations and binding affinities, while AutoDock Vina improves docking speed and accuracy by incorporating an advanced scoring function that accounts for hydrophobic, hydrogen bonding, and electrostatic interactions [25]. Similarly, Glide (Schrödinger Suite) is a high-precision docking tool that supports both rigid and flexible docking, integrating IFD to accommodate receptor flexibility, thereby enhancing binding affinity predictions and lead optimization [26].

MD Simulation Tools

MD simulations provide insight into the stability and conformational changes of protein–ligand complexes under physiological conditions. These simulations use force fields to calculate atomic interactions over time. Some widely used MD tools, like GROMACS, are high-performance MD simulation packages renowned for their speed and efficiency in modeling biomolecular systems [54]. GROMACS employs Verlet integrators and leap-frog algorithms to compute atomic motion and energy minimization, making it highly effective for simulating large-scale molecular interactions. Similarly, AMBER is a force field-based MD tool that accurately models biomolecular interactions and performs free energy calculations. It incorporates generalized Born implicit solvation models, making it particularly useful for protein–ligand simulations and drug discovery applications.

QSAR Modeling Tools

QSAR modeling helps predict the biological activity of compounds based on their molecular structure and physicochemical properties. Computational QSAR tools analyze molecular descriptors and statistical correlations to optimize drug candidates:

- OE Chem (OpenEye Scientific Software): OE Chem provides tools for 2D- and 3D-QSAR modeling, molecular similarity searching, and fingerprint-based screening. It employs machine learning algorithms to refine QSAR predictions.
- ChemProp: a deep-learning-based QSAR platform that utilizes GNNs to predict molecular properties. It is widely used for ADMET screening and lead optimization.

Public Chemical and Biological Databases

Public databases provide extensive datasets for computational drug discovery, including structural, pharmacokinetic, and bioactivity data. These repositories are crucial for virtual screening, molecular docking, and machine learning applications.

Protein Data Bank

The Protein Data Bank (PDB) is the largest repository of experimentally determined protein structures obtained through X-ray crystallography, NMR spectroscopy, and cryo-EM [55]. It provides atomic-resolution data essential for molecular docking and SBDD. PDB structures are extensively used in homology modeling, MD simulations, and virtual screening.

SwissADME

SwissADME is an online tool used for predicting the ADMET properties of drug candidates. It evaluates molecular drug-likeness based on Lipinski's Rule of Five, blood–brain barrier (BBB) permeability, and cytochrome P450 metabolism predictions. This tool is widely used for pharmacokinetic screening and lead optimization [28].

BindingDB

BindingDB is a public database containing experimentally measured binding affinities between small molecules and biological targets. It includes kinase inhibitors, enzyme inhibitors, and GPCR modulators, serving as a critical resource for QSAR modeling and virtual screening.

HTS Techniques

HTS techniques have accelerated computational drug discovery by enabling the rapid screening of vast chemical libraries. Advanced computational infrastructure, including cloud computing, parallel processing, and GPU acceleration, enhances the efficiency of in silico methodologies [56].

Cloud Computing in Drug Discovery

Cloud computing provides scalable computational power for virtual screening and molecular simulations. Platforms such as Google Cloud, Amazon Web Services (AWS), and Microsoft Azure offer on-demand access to high-performance computing (HPC) resources [57]. Cloud-based drug discovery facilitates the following:

- distributed molecular docking across multiple virtual machines;
- parallelized QSAR modeling for large-scale compound screening;
- data sharing and collaborative AI-driven drug discovery in pharmaceutical research.

Parallel Processing for Virtual Screening

Parallel processing techniques utilize multiple processing units to simultaneously execute computational tasks, significantly reducing screening time [58]. MPI and OpenMP frameworks enable distributed docking simulations and MD calculations. Applications include:

- massively parallel molecular docking of millions of compounds against target proteins;
- batch processing of pharmacophore-based screening for hit identification;
- simultaneous QSAR modeling across multiple training datasets.

GPU Acceleration in Computational Drug Discovery

GPU acceleration has revolutionized molecular modeling by enabling high-speed computations in docking, MD simulations, and AI-driven screening. CUDA and OpenCL are widely used frameworks for GPU-based molecular modeling. Applications include the following:

- accelerated MD simulations using GROMACS and AMBER to explore protein–ligand dynamics;
- fast ligand docking with GPU-optimized AutoDock-GPU and Glide-XP;
- AI-driven drug discovery using deep learning frameworks such as TensorFlow and PyTorch for SAR predictions.

Challenges and Limitations in In Silico Drug Design

Despite significant advancements in computational drug discovery, in silico techniques face several challenges and limitations that affect their predictive reliability and practical applicability. One of the primary concerns is the accuracy of computational predictions, as virtual screening methods and molecular modeling rely on approximations and empirical scoring functions that may not fully capture the complexity of molecular interactions in a biological environment [59]. The inherent assumptions in force field-based molecular docking and MD simulations often lead to discrepancies between predicted and experimentally observed binding affinities. Force fields such as AMBER, CHARMM, and GROMOS, while widely used, are based on fixed parameter sets that approximate molecular interactions using classical mechanics, neglecting critical quantum mechanical effects and electronic polarization. Consequently, these limitations impact the accuracy of ligand-binding energy calculations, solvation effects, and protein flexibility, reducing the predictive power of SBDD. Furthermore, scoring inaccuracies in molecular docking pose a significant challenge, as most scoring functions prioritize geometric complementarity and enthalpic contributions while failing to account for entropic effects, explicit water-mediated interactions, and induced-fit phenomena. The reliance on simplistic energy functions results in high false-positive and false-negative rates, leading to the selection of nonviable drug candidates and the exclusion of potentially effective molecules [60]. Another major limitation of in silico drug design is its resource-intensive nature, requiring substantial computational power, memory, and storage capacity. High-fidelity molecular simulations, such as FEP calculations, metadynamics, and long-timescale MD simulations, demand HPC clusters, GPUs, and parallel processing architectures. The computational cost associated with large-scale virtual screening, especially when screening millions of compounds against multiple protein conformations, makes these approaches impractical for smaller research institutions with limited access to computational infrastructure. Additionally, QM/MM hybrid models, which improve the accuracy of ligand-binding predictions by integrating electronic structure calculations, are computationally prohibitive due to their intensive memory and processing requirements. These constraints necessitate the use of approximate models, heuristic scoring functions, and data-driven approaches, which introduce additional sources of error and uncertainty in predictions. The issue of false positives and false negatives remains a persistent problem in computational drug discovery. False positives refer to compounds predicted to be active in silico but fail to demonstrate biological activity in vitro or in vivo, whereas false negatives are bioactive compounds that are missed by computational screening due to limitations in virtual screening methodologies. Hit enrichment in virtual screening is influenced by multiple factors, including ligand flexibility, receptor dynamics, and binding site plasticity, which are often oversimplified in docking protocols. Addressing hit enrichment issues requires refinement strategies, such as ensemble docking, consensus scoring, machine-learning-based rescoring, and post-docking MD simulations, to improve prediction accuracy. Consensus scoring, which combines multiple scoring functions, helps mitigate individual scoring biases and enhances the reliability of ranking ligand poses. Additionally, adaptive sampling methods, binding free energy calculations, and AI-driven lead optimization provide more refined predictions by incorporating multiple conformational states and experimental validation data [61]. However, these refinement strategies further increase computational cost and complexity, requiring a balance between accuracy and feasibility. In conclusion, while in silico drug design offers transformative potential in accelerating drug discovery, overcoming these challenges necessitates ongoing advancements in force field development, scoring function optimization, AI-driven predictive modeling, and computational efficiency, ensuring that virtual screening approaches are both scientifically robust and practically viable.

Advances and Future Directions in In Silico Drug Design

The field of in silico drug design is rapidly evolving, driven by advancements in AI, quantum computing, and multi-scale modeling techniques. ML integration has significantly enhanced computational drug discovery by improving the accuracy of molecular docking, virtual screening, and lead optimization. AI-guided drug discovery leverages deep learning models, CNNs, and GNNs to predict protein–ligand interactions, analyze chemical space, and design novel bioactive compounds with optimized pharmacokinetic properties [62]. AI-based approaches such as GANs and reinforcement learning enable the de novo design of drug-like molecules, reducing the need for extensive chemical synthesis and experimental screening. Moreover, the use of active learning strategies allows for iterative model refinement based on experimental feedback, improving the efficiency of hit identification and lead optimization. Quantum computing applications in drug discovery hold immense potential for solving complex molecular interactions with unprecedented speed and accuracy [63]. Classical computational methods are limited by the exponential complexity of quantum mechanical calculations required for accurate molecular simulations. Quantum computing, with its ability to process multiple states simultaneously through quantum superposition and entanglement, offers the potential for ultrafast computations in molecular modeling, particularly in protein folding, ligand-binding free energy calculations, and reaction mechanism elucidation. As quantum hardware continues to develop, its integration with AI and molecular modeling frameworks is expected to revolutionize in silico drug design by providing highly accurate energy landscape predictions and enhanced molecular docking precision. Another critical advancement is multi-scale modeling, which integrates MD simulations with coarse-grained and atomistic models to study biomolecular interactions across different spatial and temporal scales. This approach enables a more comprehensive understanding of protein–ligand dynamics, allosteric regulation, and conformational changes, leading to the development of more effective therapeutic agents. Additionally, the integration of systems biology and personalized medicine into in silico drug discovery is transforming the landscape of pharmacological research. Omics-driven approaches, including genomics, proteomics, and metabolomics, provide patient-specific molecular data that can be used to tailor drug discovery pipelines for precision medicine applications [64]. AI-driven multi-omics analysis facilitates the identification of individualized drug targets, biomarker discovery, and personalized treatment strategies, enhancing the efficacy and safety of therapeutic interventions. In the future, the convergence of AI, quantum computing, and multi-scale modeling will lead to a paradigm shift in drug discovery, enabling real-time molecular simulations, rapid lead optimization, and data-driven precision medicine strategies that accelerate the development of next-generation therapeutics.

Ethical and Regulatory Considerations in Computational Drug Design

As computational drug design continues to evolve, addressing ethical and regulatory considerations is crucial to ensuring the reliability, safety, and fairness of AI-driven drug discovery. Data integrity and reproducibility are fundamental challenges in in silico methodologies, as the accuracy of computational predictions is highly dependent on the quality and reliability of input datasets. Poorly curated data, biases in training datasets, and lack of standardized validation protocols can lead to misleading predictions, increasing the risk of false positives and negatives in virtual screening. To maintain scientific rigor, it is essential to establish transparent data-sharing practices, standardized benchmarking datasets, and rigorous validation frameworks that ensure the reproducibility of computational experiments [65]. Transparency and validation in AI-driven drug discovery are critical for gaining regulatory acceptance and fostering trust in computational models. Black-box machine learning algorithms often lack interpretability, making it challenging to understand decision-making processes in virtual screening and lead optimization. Efforts to develop XAI frameworks aim to enhance the interpretability of deep learning models, allowing researchers and regulatory agencies to evaluate the rationale behind molecular predictions. Furthermore, the compliance of computational drug design with regulatory frameworks set by agencies such as the US FDA and the EMA is essential for ensuring the clinical translation of in silico findings. Regulatory guidelines emphasize the need for computational model validation, robust experimental verification, and adherence to good computational practices (GCPs) in drug discovery [66]. Ethical considerations in AI-driven research include bias mitigation and informed decision-making to prevent algorithmic discrimination in drug development. Machine learning models trained on biased datasets may inadvertently reinforce demographic disparities in drug response predictions, leading to inequitable healthcare outcomes. To address this issue, researchers must implement bias-detection algorithms, diverse training datasets, and fairness-aware AI models that ensure inclusivity and generalizability in computational predictions. Additionally, AI-generated drug candidates raise concerns regarding intellectual property rights, data privacy, and potential misuse in bioterrorism. Establishing ethical guidelines for AI in drug discovery, including transparency in AI-generated molecular designs and stringent cybersecurity measures for proprietary data, is crucial for preventing unethical applications. Moving forward, the integration of ethical AI principles, regulatory compliance, and transparent validation protocols will be essential in shaping a responsible and trustworthy framework for computational drug discovery, ensuring that in silico methodologies contribute to safe, effective, and equitable therapeutic innovations.

Conclusion

In silico drug design and virtual screening techniques have revolutionized modern pharmaceutical research by providing a computationally efficient and cost-effective approach to drug discovery. These methodologies integrate molecular modeling, structure-based and ligand-based approaches, molecular docking, MD simulations, and artificial intelligence-driven tools to accelerate hit identification, lead optimization, and toxicity prediction. The ability to screen vast chemical libraries, predict molecular interactions, and assess pharmacokinetics in silico has significantly reduced the reliance on traditional trial-and-error experimental approaches, leading to faster and more precise drug development pipelines. SBDD leverages high-resolution protein structural data to elucidate protein–ligand interactions, facilitating rational drug design through docking and molecular simulations. LBDD, on the other hand, employs QSAR models, molecular similarity searching, and de novo drug design strategies to identify bioactive compounds in the absence of target protein structures. The integration of hybrid screening methodologies has further enhanced predictive accuracy by combining the strengths of both approaches. Despite the numerous advantages of in silico drug design, several challenges remain. The accuracy of computational predictions is often limited by force-field approximations, scoring function inaccuracies, and the inability to fully capture protein flexibility and solvent effects. The computational cost of large-scale virtual screening and MD simulations can be prohibitively high, requiring high-performance computing infrastructure, cloud-based resources, and GPU acceleration to manage complex datasets. Furthermore, false positives and false negatives in virtual screening continue to pose significant challenges, necessitating the use of consensus scoring, free energy calculations, and experimental validation to refine hit selection. Addressing these challenges will require continuous advancements in AI-driven predictive models, improved scoring functions, and the integration of multi-scale modeling techniques to enhance the reliability of computational drug discovery. Emerging technologies like quantum computing, deep learning, and personalized medicine are set to transform in silico drug design. AI-driven discovery enhances de novo molecule generation, pharmacophore modeling, and ADMET prediction, while quantum computing enables ultrafast simulations of protein folding and molecular interactions. Integrating systems biology and omics-driven analysis will advance precision medicine. However, stringent ethical and regulatory frameworks, including FDA and EMA compliance, bias mitigation, and validation protocols, are essential to ensure transparency, data integrity, and equitable drug discovery. Thus, in silico drug design and virtual screening techniques have transformed the landscape of pharmaceutical research, significantly reducing the time and cost associated with drug development. While challenges persist, ongoing advancements in computational modeling, AI-driven drug discovery, and HTS techniques will continue to enhance the efficiency and accuracy of computational drug design. As these methodologies evolve, they hold immense potential for accelerating the discovery of next-generation therapeutics, improving drug repurposing strategies, and advancing precision medicine to address global healthcare challenges.

Consent for Publications

All authors consent to the publication of the above chapter.

Conflict of Interest

Authors declare no conflict of interest.

Acknowledgments

All authors express their gratitude toward MIT-WPU for providing all the support required.

References

1 Chang Y, Hawkins BA, Du JJ, et al. A Guide to in silico drug design. *Pharmaceutics* 2022;15(1):49–101.
2 Shaker B, Ahmad S, Lee J, et al. In silico methods and tools for drug discovery. *Comput. Biol. Med.* 2021;137:104851
3 Ezike TC, Okpala US, Onoja UL, et al. Advances in drug delivery systems, challenges and future directions. *Heliyon* 2023;9(6):e17488.
4 Wei F, Wang S, Gou X. A review for cell-based screening methods in drug discovery. *Biophys. Rep.* 2021;7(6):504–16.
5 Jain R, Subramanian J, Rathore AS. A review of therapeutic failures in late-stage clinical trials. *Expert Opin. Pharmacother.* 2023;24(3):389–99.
6 Shah M, Patel M, Shah M, et al. Computational transformation in drug discovery: a comprehensive study on molecular docking and quantitative structure activity relationship (QSAR). *Intell. Pharm.* 2024;2(5):589–98.
7 Hamza A, Wei NN, Zhan CG. Ligand-based virtual screening approach using a new scoring function. *J. Chem. Inf. Model.* 2012;52(4):963–74.
8 Maia EHB, Assis LC, de Oliveira TA, et al. Structure-based virtual screening: from classical to artificial intelligence. *Front. Chem.* 2020;8:343–61.
9 Jia L, Gao H. Machine learning for in silico ADMET prediction. *Methods Mol. Biol.* 2022;2390:447–60.
10 Agamah FE, Mazandu GK, Hassan R, et al. Computational/in silico methods in drug target and lead prediction. *Brief. Bioinform.* 2020;21(5):1663–75.
11 Cavalcante BRR, Freitas RD, Siquara da Rocha LO, et al. In silico approaches for drug repurposing in oncology: a scoping review. *Front. Pharmacol.* 2024;15:1400029.
12 Berry M, Fielding B, Gamieldien J. Practical considerations in virtual screening and molecular docking. In: *Emerging Trends in Computational Biology, Bioinformatics, and Systems Biology*. Morgan Kaufmann; 2015. pp. 487–502.
13 Roney M, Aluwi MFFM. The importance of in-silico studies in drug discovery. *Intell. Pharm.* 2024;2(4):578–9.
14 Tripathi MK, Ahmad S, Tyagi R, et al. Fundamentals of molecular modeling in drug design. In: *Computer Aided Drug Design (CADD): From Ligand-Based Methods to Structure-Based Approaches*. Elsevier Science; 2022. pp. 125–55.

15 Zhou T, Huang D, Caflisch A. Quantum mechanical methods for drug design. *Curr. Top. Med. Chem.* 2010;10(1):33–45.

16 Martin MG. Comparison of the AMBER, CHARMM, COMPASS, GROMOS, OPLS, TraPPE and UFF force fields for prediction of vapor–liquid coexistence curves and liquid densities. *Fluid Ph. Equilib.* 2006;248(1):50–5.

17 Sasidharan S, Gosu V, Tripathi T, Saudagar P. Molecular dynamics simulation to study protein conformation and ligand interaction. In: *Protein Folding Dynamics and Stability*. Springer Nature; 2023. pp. 107–27.

18 Batool M, Ahmad B, Choi S. A structure-based drug discovery paradigm. *Int. J. Mol. Sci.* 2019;20(11):2783.

19 Jorgensen WL. Efficient drug lead discovery and optimization. *Acc. Chem. Res.* 2009;42(6):724–33.

20 Yadav V, Reang J, Vinita, Tonk RK. Ligand-based Drug Design (LBDD). In: *Computer Aided Drug Design (CADD): From Ligand-Based Methods to Structure-Based Approaches*; Rudrapal M, Egbuna C, eds. Elsevier; 2022. pp. 57–99.

21 Niazi SK, Mariam Z. Recent advances in machine-learning-based chemoinformatics: a comprehensive review. *Int. J. Mol. Sci.* 2023;24(14):11488.

22 Agu PC, Afiukwa CA, Orji OU, et al. Molecular docking as a tool for the discovery of molecular targets of nutraceuticals in diseases management. *Sci. Rep.* 2023;13(1):13398.

23 Torres PHM, Sodero ACR, Jofily P, Silva-Jr FP. Key topics in molecular docking for drug design. *Int. J. Mol. Sci.* 2019;20(18):4574.

24 Wang S, Dong G, Sheng C. Structural simplification: an efficient strategy in lead optimization. *Acta. Pharm. Sin. B* 2019;9(5):880–901.

25 Trott O, Olson AJ. AutoDock Vina: improving the speed and accuracy of docking with a new scoring function, efficient optimization, and multithreading. *J. Comput. Chem.* 2010;31(2):455–61.

26 Bhachoo J, Beuming T. Investigating protein-peptide interactions using the Schrödinger computational suite. *Methods Mol. Biol.* 2017;1561:235–54.

27 Kearnes S, Pande V. ROCS-derived features for virtual screening. *J. Comput. Aided Mol. Des.* 2016;30(8):609–17.

28 Daina A, Michielin O, Zoete V. SwissADME: a free web tool to evaluate pharmacokinetics, drug-likeness and medicinal chemistry friendliness of small molecules. *Sci. Rep.* 2017;7:42717.

29 Anderson AC. Structure-based functional design of drugs: from target to lead compound. *Methods Mol. Biol.* 2012;823:359–66.

30 Heppner DE. Ascertaining a structural basis in drug discovery and development. *J. Med. Chem.* 2025. https://doi.org/10.1021/acs.jmedchem.5c00326

31 Nero TL, Parker MW, Morton CJ. Protein structure and computational drug discovery. *Biochem. Soc. Trans.* 2018;46(5):1367–79.

32 Śledź P, Caflisch A. Protein structure-based drug design: from docking to molecular dynamics. *Curr. Opin. Struct. Biol.* 2018;48:93–102.

33 Du X, Li Y, Xia YL, et al. Insights into protein-ligand interactions: mechanisms, models, and methods. *Int. J. Mol. Sci.* 2016;17(2):144–78.

34 Narasimhan S. Determining protein structures using X-ray crystallography. *Methods Mol. Biol.* 2024;2787:333–53.

35 Sugiki T, Kobayashi N, Fujiwara T. Modern technologies of solution nuclear magnetic resonance spectroscopy for three-dimensional structure determination of proteins open avenues for life scientists. *Comput. Struct. Biotechnol. J.* 2017;15:328–39.

36 Kuder KJ. Docking foundations: from rigid to flexible docking. *Methods Mol. Biol.* 2024;2780:3–14

37 Raval K, Ganatra T. Basics, types and applications of molecular docking: a review. *IP Int. J. Compr. Adv. Pharmacol.* 2022;7:12–6.

38 Zhang H, Kim S, Giese TJ, et al. CHARMM-GUI free energy calculator for practical ligand binding free energy simulations with AMBER. *J. Chem. Inf. Model.* 2021;61(9):4145–51.

39 Lu X, Yang H, Chen Y, et al. The development of pharmacophore modeling: generation and recent applications in drug discovery. *Curr. Pharm. Des.* 2018;24(29):3424–39.

40 Tyagi R, Singh A, Chaudhary KK, Yadav MK. Pharmacophore modeling and its applications. In: *Bioinformatics*. Academic Press; 2022. pp. 269–89.

41 Choudhury C, Narahari GS. Pharmacophore modelling and screening: concepts, recent developments and applications in rational drug design. *Chall. Adv. Comput. Chem. Phys.* 2019;27:25–53.

42 Aparoy P, Reddy KK, Reddanna P. Structure and ligand based drug design strategies in the development of novel 5-LOX inhibitors. *Curr. Med. Chem.* 2012;19(22):3763–78.

43 Bahia MS, Kaspi O, Touitou M, et al. A comparison between 2D and 3D descriptors in QSAR modeling based on bio-active conformations. *Mol. Inform.* 2023;42(4):e2200186.

44 Danishuddin, Khan AU. Descriptors and their selection methods in QSAR analysis: paradigm for drug design. *Drug Discov. Today* 2016;21(8):1291–302.

45 Muegge I, Mukherjee P. An overview of molecular fingerprint similarity search in virtual screening. *Expert Opin. Drug Discov.* 2016;11(2):137–48.

46 Huang N, Shoichet BK, Irwin JJ. Benchmarking sets for molecular docking. *J. Med. Chem.* 2006;49(23):6789–801.

47 Mouchlis VD, Afantitis A, Serra A, et al. Advances in de novo drug design: from conventional to machine learning methods. *Int. J. Mol. Sci.* 2021;22(4):1676.

48 Selvaraj C, Chandra I, Singh SK. Artificial intelligence and machine learning approaches for drug design: challenges and opportunities for the pharmaceutical industries. *Mol. Divers.* 2022;26(3):1893–913.

49 Lavecchia A, Di Giovanni C. Virtual screening strategies in drug discovery: a critical review. *Curr. Med. Chem.* 2013;20(23):2839–60.

50 Giordano D, Biancaniello C, Argenio MA, Facchiano A. Drug design by pharmacophore and virtual screening approach. *Pharmaceuticals (Basel)* 2022;15(5):646–62.

51 Curcio A, Rocca R, Chiera F, et al. Hit identification and functional validation of novel dual inhibitors of HDAC8 and tubulin identified by combining docking and molecular dynamics simulations. *Antioxidants (Basel)* 2024;13(11):1427–48.

52 Ekins S, Mestres J, Testa B. In silico pharmacology for drug discovery: methods for virtual ligand screening and profiling. *Br. J. Pharmacol.* 2007;152(1):9–20.

53 Iwata H. Transforming drug discovery: the impact of AI and molecular simulation on R&D efficiency. *Bioanalysis* 2024;16(23–24):1211–7.

54 Abraham MJ, Murtola T, Schulz R, et al. GROMACS: high performance molecular simulations through multi-level parallelism from laptops to supercomputers. *SoftwareX* 2015;1–2:19–25.

55 Burley SK, Berman HM, Kleywegt GJ, et al. Protein Data Bank (PDB): The single global macromolecular structure archive. *Methods Mol. Biol.* 2017;1607:627–41.

56 Blay V, Tolani B, Ho SP, Arkin MR. High-throughput screening: today's biochemical and cell-based approaches. *Drug Discov. Today* 2020;25(10):1807–21.

57 D'Agostino D, Clematis A, Quarati A, et al. Cloud infrastructures for in silico drug discovery: economic and practical aspects. *Biomed. Res. Int.* 2013; 2013:138012.

58 Murugan NA, Podobas A, Gadioli D, et al. A review on parallel virtual screening softwares for high-performance computers. *Pharmaceuticals (Basel)* 2022;15(1):63–85.

59 Niazi SK, Mariam Z. Computer-aided drug design and drug discovery: a prospective analysis. *Pharmaceuticals (Basel)* 2023;17(1):22–44.

60 Park BS, Mori M. Balancing false discovery and false negative rates in selection of differentially expressed genes in microarrays. *Open Access Bioinform.* 2010;2010(2):1–9.

61 Genheden S, Ryde U. The MM/PBSA and MM/GBSA methods to estimate ligand-binding affinities. *Expert Opin. Drug Discov.* 2015;10(5):449–61.

62 Paul D, Sanap G, Shenoy S, et al. Artificial intelligence in drug discovery and development. *Drug Discov. Today* 2021;26(1):80–93.

63 Chow JCL. Quantum computing in medicine. *Med. Sci. (Basel)* 2024;12(4):67–83.

64 Mathé E, Hays JL, Stover DG, Chen JL. The omics revolution continues: the maturation of high-throughput biological data sources. *Yearb. Med. Inform.* 2018;27(1):211–22.

65 Alves VM, Auerbach SS, Kleinstreuer N, et al. Curated data in – Trustworthy *in silico* models out: the impact of data quality on the reliability of artificial intelligence models as alternatives to animal testing. *Altern. Lab. Anim.* 2021;49(3):73–82.

66 Ajmal CS, Yerram S, Abishek V, et al. Innovative approaches in regulatory affairs: leveraging artificial intelligence and machine learning for efficient compliance and decision-making. *AAPS J.* 2025;27(1):22.

11

De Novo Drug Design: From Fragment-based to AI-driven Approaches

Taufik Mulla[1], *Drashti Dave*[2], *Vetriselvan Subramaniyan*[3*], *and Ambika Nand Jha*[4]

[1]*Faculty of Pharmacy, Institute of Pharmaceutical Sciences, Parul University, Vadodara, Gujarat, India*
[2]*Department of Pharmaceutics, Sigma Institute of Pharmacy, Sigma University, Vadodara, Gujarat, India*
[3]*Department of Medical Sciences, School of Medical and Life Sciences, Sunway University, Petaling Jaya, Malaysia*
[4]*School of Pharmacy, Sharda University, Greater Noida, Uttar Pradesh, India*

*Corresponding author: vetris@sunway.edu.my

Introduction

De novo drug design signifies a new direction for pharmaceutical drug discovery using computational ability to create completely novel bioactive compounds de novo as opposed to the modification or repurposing of existing compounds (drug candidates). In traditional drug discovery, we begin with existing ligands and scaffolds. De novo drug design allows the discovery and creation of completely new chemical entities that are customized to target biological systems or targets of interest [1]. In an age of precision medicine, high-throughput computational modeling is making the de novo design of drugs increasingly applicable and important as drug discovery timelines can be reduced, costs lowered, and the chemical space limited for therapeutic applications expanded [2].

History and Development

The idea of designing drugs de novo began in the late 1980s along with the start of computational chemistry and molecular modeling. Early implementations of de novo, however, were constrained by primitive computational power and simplistic scoring protocols. However, as more advanced computer processing became available to the drug design community and as more structural data on protein targets emerged, de novo increasingly began to gain traction [3]. The incorporation of fragment-based drug design methodologies and now more prominently artificial intelligence (AI) has altered the landscape of drug discovery, with de novo methodologies now establishing the foundation of modern medicinal chemistry [4]. In the traditional drug discovery pipeline, the primary method for finding hits was through high-throughput screening (HTS) that evaluated large libraries of different compounds. HTS is one-dimensional and resource-intensive; as a method, it is effective but will present leads that may be functionally less effective. Conversely, de novo design can be applied in a manner that minimizes inefficiencies by allowing meaningful, target-guided generation of compounds with knowledge of the molecular architecture of the binding space for the target [5].

Concept and Scope

At its core, de novo drug design involves the construction of candidate molecules that fit a predefined pharmacophoric profile of a biological target. These molecules are designed to bind with high affinity and specificity, minimizing off-target interactions and potential toxicity. The approach encompasses both ligand-based and structure-based methods, although the latter is more common due to the availability of crystallographic and cryo-electron microscopy (cryo-EM) data [6, 7]. The de novo process can be divided into three broad stages:

Target identification and characterization: a clear understanding of the biological target, usually a protein or nucleic acid, is crucial. This includes the 3D structure of the active site, the nature of its binding interactions (hydrophobic, hydrogen bonding, ionic, etc.), and the biological function of the target in disease pathways [8].

Molecular generation and assembly: novel molecules are constructed using either atom-by-atom growth or fragment-based assembly techniques. The latter, often integrated with knowledge-based libraries, allows for more efficient exploration of chemical space while preserving drug-like properties [9].

Scoring and optimization: generated molecules are evaluated based on binding affinity, synthetic accessibility, and pharmacokinetic properties [10]. Multi-objective optimization methods, including machine learning (ML) models, are increasingly used to balance potency, selectivity, and drug-likeness.

Rationale and Importance

De novo design offers distinct advantages over traditional drug discovery methodologies. These include:

Expanded chemical space exploration: the traditional chemical libraries are only a minuscule fraction of the theoretically possible chemical space. De novo methods can explore unexplored areas, potentially leading to first-in-class therapeutics [11].

Customization and precision: molecules can be tailored specifically to the 3D conformation and dynamics of the target site, enhancing binding specificity.

Time and cost efficiency: rational design, aided by in silico tools, accelerates the early stages of drug development, often finding high-quality leads faster than empirical screening [12].

Integration with AI and ML: AI can generate novel scaffolds, predict absorption, distribution, metabolism, excretion, and toxicity (ADMET) profiles, and improve synthetic routes, significantly enhancing the power of de novo design [13].

Key Strategies in De Novo Design

Several methodologies are employed in de novo drug design, each with distinct strengths:

Atom-based design: this approach builds molecules atom by atom, guided by geometric and chemical rules [14]. While highly customizable, it is computationally intensive and prone to generating synthetically inaccessible structures [15].

Fragment-based design: fragments, small chemical building blocks, are combined or grown within the binding site of the target protein. This strategy benefits from higher synthetic feasibility and lower molecular complexity [16]. Tools like LUDI, SPROUT, and GROW have been developed for fragment-based de novo design [17].

Reaction-based design: this approach ensures synthetic accessibility by constructing molecules using known chemical reactions. The Retrosynthetic Combinatorial Analysis Procedure (RECAP) and Breaking of Retrosynthetically Interesting Chemical Substructures (BRICS) methods are notable examples [18].

AI-assisted generation: ML models, particularly generative adversarial networks (GANs), recurrent neural networks (RNNs), and variational autoencoders (VAEs), are trained on molecular datasets to generate novel compounds that adhere to target-specific constraints [19].

Table 11.1 summarizes these approaches, comparing key parameters such as customization potential, synthetic feasibility, and computational demands.

Table 11.1 Comparison of key strategies in de novo drug design [20, 21].

Strategy	Customization potential	Synthetic feasibility	Computational demand	Example tools/ technologies
Atom-based	High	Low	High	LUDI, SPROUT
Fragment-based	Moderate	High	Moderate	GROW, MOE, CAVEAT
Reaction-based	Moderate	Very high	Moderate	RECAP, BRICS
AI-assisted generation	Very high	Variable	High	GANs, VAEs, DeepChem, Reinvent

Integration with Target Structures

The power of de novo drug design is most effectively harnessed when structural data of the target is available. X-ray crystallography, nuclear magnetic resonance (NMR) spectroscopy, and cryo-EM have revolutionized structural biology, allowing precise modeling of ligand–receptor interactions [22]. Binding site information, such as shape, electrostatic profile, and flexibility, guides the placement and orientation of molecular fragments or atoms during the design process [23]. In cases where no experimental structure is available, homology modeling and ab initio predictions can be employed to construct approximate models, although these come with increased uncertainty [24].

Assessment and Scoring Functions

A crucial step in de novo design is evaluating the quality of the generated molecules. Several scoring functions are used:

Energy-based scoring: estimates the binding affinity based on molecular mechanics force fields (e.g., AMBER, CHARMM) [25].
Knowledge-based scoring: derives statistical potentials from known protein–ligand complexes.
ML-based scoring: uses trained models to predict binding affinity and drug-likeness metrics.

Other properties assessed include Lipinski's Rule of Five compliance, polar surface area, and rotatable bond count, all important for oral bioavailability.

Limitations and Challenges

Despite its promise, de novo drug design faces several limitations:

Synthetic inaccessibility: molecules generated purely in silico may be difficult or impossible to synthesize in practice [26].
Computational cost: high-resolution simulations, especially those incorporating solvent and protein flexibility, are resource-intensive.
False positives: poorly calibrated scoring functions may rank nonbinders highly.
Validation bottlenecks: experimental validation of hits stays time-consuming and expensive [27].

However, advances in cheminformatics, cloud computing, and synthetic biology are progressively addressing these issues.

Fragment-based Drug Discovery

Fragment-based drug discovery (FBDD) has appeared as a revolutionary approach in modern drug discovery, offering a smarter and often more efficient route to identify potential drug candidates. Unlike traditional HTS that uses large, complex molecules, FBDD starts small, quite literally, with low molecular-weight chemical fragments. These fragments may bind weakly to biological targets but serve as crucial starting points for designing potent therapeutic compounds through strategic optimization [3].

Principles of FBDD

At its core, FBDD works by screening libraries of small chemical fragments, typically under 300 daltons, for their ability to bind to biological targets such as enzymes or receptors. Even though these fragments exhibit only modest binding affinity, they often possess high ligand efficiency, meaning they interact efficiently with the target despite their size [28]. The idea is to use these small pieces as "molecular seeds" and then build upon them, either by growing, linking, or merging fragments, to enhance their binding affinity and specificity. The charm of FBDD lies in its minimalistic nature. Since the chemical space of smaller molecules is significantly smaller than that of larger ones, researchers can explore more diverse interactions using fewer compounds. This not only saves time and resources but also increases the chance of discovering novel chemical scaffolds [29].

Key Techniques in FBDD

Identifying fragments that bind to a target is no small feat. Because these fragments bind weakly, extremely sensitive biophysical techniques are employed to detect and validate their interactions. Three cornerstone techniques dominate the FBDD landscape:

X-ray Crystallography

X-ray crystallography remains the gold standard in FBDD. It provides a high-resolution, 3D view of how a fragment interacts with the target protein. Once a fragment is found to bind, researchers can visualize the exact binding mode, offering valuable insights for subsequent chemical modifications [30]. This structural guidance greatly increases the potential for the development of powerful inhibitors or modulators.

NMR Screening

Another common method that is particularly advantageous for the identification of weak interactions is NMR spectroscopy. It provides a means of observing binding events of molecules directly or indirectly, without having to crystallize the protein or buffer conditions being studied. Ligand-detected NMR and protein-detected NMR provide different aspects of valuable information on fragment binding and dynamics [4].

Molecular Docking

Some computational methods, such as molecular docking, function as supplemental methods to predict how a fragment may fit into the active site of a target protein of interest. This must be considered less definitive than experimental methods. However, it can be a means to decrease the "scope" of fragment libraries to examine and help guide experimental activities, especially when combined with structure-based methods [31].

Success Stories in FBDD

The implications of FBDD are best illustrated by their real-world success stories. There are multiple drugs that have entered the market or clinical space that originated, at least in part, through fragment-based methods. Vemurafenib, a BRAF inhibitor used in treating metastatic melanoma, is a classic example. The discovery process began with fragment screening that identified a low-affinity binder to BRAF kinase. Through iterative optimization guided by structural biology, a potent and selective inhibitor was developed [32].

Another notable example is Venetoclax, a breakthrough therapy for chronic lymphocytic leukemia (CLL). It targets the BCL-2 protein, which plays a role in cell survival. FBDD enabled the identification and refinement of fragments that disrupted BCL-2 interactions, eventually leading to a high-affinity, orally available drug [33]. AZD9291 (Osimertinib), used in non-small cell lung cancer, also stemmed from fragment-based origins. Fragments targeting mutant EGFR were structurally optimized to produce a third-generation tyrosine kinase inhibitor with remarkable selectivity and efficacy [34]. These instances demonstrate that FBDD is flexible and can be applied across various therapeutic areas, such as oncology, infectious diseases, and neurological disorders [35].

Computational Advancements in De Novo Drug Design

The digital age has potentialized more than just technological innovation; drug discovery has changed. The days of drug discovery relying on a purely trial-and-error paradigm are over. Researchers are using powerful computational tools that support the de novo design of rationally designed new molecules. The term "de novo drug design" is used to describe a whole suitcase of new drugs that are designed, rationalized, and created through chemistry, biology, and computer science to speed up the identification of new drug candidates with higher accuracy and lower expense.

Molecular Docking and Structure-based Drug Design

In the world of de novo drug design, at the center of the process lies structure-based drug design (SBDD). SBDD aims to design new molecules that will bind to a target protein based on the chains of three-dimensional (3D) structures. For this to happen, a relatively good representation of a typical biological target or at least its active site must be inferred to develop new drug molecules accordingly or to obtain new 3D representations of the binding site of a biologically relevant molecule. Aid to drug discovery in this manner has been aided through many technology providers who use similar molecular docking routines of computerized, high-performance chemical libraries to infer predictive qualities. Molecular docking is probably the most used in SBDD [36]. Molecular docking acts like a dating service for small molecules and proteins; it predicts how well a small compound fits into its target protein's binding pocket (the ligand). A helpful analogy for molecular docking is a jigsaw puzzle, as the more accurately the shape match is congruous through docking, the greater the formation of a drug-like molecule that can inhibit or modulate the behavior of the protein. The use of scoring algorithms for binding affinity and molecular geometric orientation allows docking to rank-order compounds for ordering and testing. Docking has been an important part of speeding up early-phase drug discovery. If there are crystal structures or homology models available, by using docking, researchers can virtually test thousands (if not millions) of compound–protein interactions in a timescale far shorter than performing the experiments in the lab [37]. The advancement in AI combined with the development of flexible docking methods will be valuable for increasing accuracy and predictability regarding the simulations.

Pharmacophore Modeling and Ligand-based Methods

When structural information on the targeted protein is limited, or in some cases unavailable, scientists turn to ligand-based drug design (LBDD). This method, LBDD, starts with known active compounds and identifies the common characteristics required for biological activity. One of the most powerful components of LBDD is pharmacophore modeling [38]. A pharmacophore acts like a chemical fingerprint; it describes the required molecular characteristics (examples would include hydrogen bond donors, hydrogen bond acceptors, aromatic rings, and hydrophobic areas) needed for a compound to bind to its specific target protein. Once a pharmacophore model is built from a set of active ligands, it can be used as a model to screen compound libraries for additional new molecules in the context of their common characteristics [39]. Pharmacophore modeling is especially useful when dealing with G-protein coupled receptors (GPCRs), enzymes, or ion channels, where detailed structural data might be incomplete. When combined with ML and data mining, pharmacophore-based screening becomes even more efficient and predictive [40]. Another ligand-based strategy is quantitative structure–activity relationship (QSAR) modeling. Here, mathematical models are developed to relate a molecule's chemical structure to its biological activity. These models can predict how modifications to a molecule might impact its effectiveness, streamlining the optimization process [41].

Multi-objective Optimization and Virtual Screening

Drug discovery isn't just about finding a compound that binds tightly to a target; it's a multifaceted challenge. Ideal drug candidates must also exhibit good pharmacokinetic properties (ADME), minimal toxicity, and selectivity over related targets. This is where multi-objective optimization steps in. Multi-objective optimization involves balancing several goals at once, like maximizing potency while minimizing side effects or optimizing solubility alongside metabolic stability. Computational algorithms, including genetic algorithms and Pareto-based optimization methods, help researchers navigate these trade-offs effectively [42]. Another pillar of computational drug design is virtual screening. This technique allows researchers to sift through millions of compounds, either from chemical libraries or generated de novo, in silico (on the computer) to identify candidates with the highest potential for success. Virtual screening can be structure-based, using molecular docking, or ligand-based, relying on pharmacophore or QSAR models [43]. Recent advances in deep learning (DL) and AI have enhanced virtual screening capabilities dramatically. These tools not only predict binding affinities more accurately but can also design completely novel molecules with desirable drug-like properties, giving rise to AI-driven de novo design platforms [15].

AI-driven Drug Discovery and Quantum Computing: Shaping the Future of Pharmaceuticals

The world of drug discovery is being reshaped by disruptive digital technologies. AI and quantum computing are no longer just buzzwords; they're powerful allies in the race to develop safer, more effective drugs faster than ever before [6]. Together, they hold the potential to revolutionize everything from target identification to molecular design and clinical predictions [11]. Let's explore how AI and quantum computing are transforming this landscape, unlocking new levels of efficiency, creativity, and precision in pharmaceutical research.

AI-driven Drug Discovery

AI and ML in Molecular Generation

AI, especially ML, has brought about a major change in basic assumptions in drug design. Sifting through thousands of molecules the traditional way is time-consuming and can be expensive [44]. Thankfully, AI is stepping in and revolutionizing things. With access to vast amounts of data that range from chemical forms to how these molecules act biologically, AI is now able to generate new molecules with certain desired traits. One of the ways AI accomplishes this is through generative models such as VAEs and RNNs. These models are able to design new compounds from scratch and tend to perform better than humans in terms of novelty, how well the molecules might function as drugs, and how precisely they attack diseases. Not only does this technology accelerate the idea-generating process but also opens up for more individuals to be involved [45]. GANs are another interesting tool. They operate by involving two neural networks, the generator, which generates new chemical structures, and the discriminator, which evaluates those structures for quality. In drug discovery, this competition drives the improvement of the overall quality of the molecules generated [42]. Reinforcement learning is taking it one step further by employing trial and error to train models what actions produce the most desirable outcomes, like increased binding to target sites or reduced toxicity. When paired with generative models, this process can be used to fine-tune molecules for other desired characteristics like how well they dissolve or how they process in the body [46].

AI Is Not All About Discovering New Molecules

It's also enhancing the ones we currently have. Molecular optimization algorithms can propose minor modifications to improve the traits that make them useful as drugs. These models analyze patterns in past successes and failures to propose promising modifications in real time [47]. Moreover, predictive modeling is one of AI's greatest strengths. Tools like deep neural networks (DNNs) and graph neural networks (GNNs) are increasingly used to predict pharmacokinetic parameters (ADMET), off-target effects, and clinical outcomes. By integrating omics data, electronic health records, and real-world evidence, AI helps forecast a molecule's journey from lab bench to bedside [48].

Integration of Quantum Computing in Drug Discovery

Quantum Mechanics-based Simulations

Quantum computing, still in its early stages, offers a completely different approach to solving drug discovery challenges by simulating molecular interactions at the quantum level. Classical computers struggle to accurately model electron behaviors in complex molecules, but quantum mechanics-based simulations can manage this with greater precision. For example, quantum algorithms like the Variational Quantum Eigensolver (VQE) and Quantum Approximate Optimization Algorithm (QAOA) help predict molecular properties such as energy levels, binding affinities, and reaction pathways with unprecedented accuracy [49]. This allows researchers to fine-tune drug candidates much earlier in the pipeline.

Synergy Between AI and Quantum Computing

What's even more promising is the synergistic relationship between AI and quantum computing. Quantum computers can enhance AI by accelerating the training of ML models, particularly in optimization-heavy tasks like molecule generation and protein folding. Conversely, AI can help correct quantum errors and make quantum algorithms more efficient [50]. One exciting development is quantum-enhanced machine learning (QML), which combines the best of both worlds. These hybrid models are being evaluated for drug target identification, molecular fingerprinting, and multi-objective optimization, offering a glimpse into the next era of computational drug design [51].

Future Potential and Challenges

While quantum computing holds immense potential, there are still significant challenges to overcome. Current quantum hardware (known as NISQ: Noisy Intermediate-Scale Quantum devices) has limitations in qubit stability, error rates, and computational scalability. This makes real-world applications in drug discovery limited for now [51]. However, the future looks promising. Tech giants like IBM, Google, and startups like D-Wave and Rigetti are working to improve quantum hardware. With continued advancements, quantum computers could soon perform simulations that are impossible on classical machines, potentially cutting years off the drug development process [52].

AI in Personalized Drug Design and Precision Medicine

The intersection of AI with personalized medicine is revolutionizing how we understand and treat diseases. Rather than adopting a one-size-fits-all model, AI enables highly tailored approaches that consider an individual's genetic makeup, lifestyle, and disease profile. This revolution promises not only more effective treatment outcomes but also a deeper understanding of the underlying biology.

AI-assisted Biomarker Identification

At the heart of precision medicine is the identification of biomarkers, biological signatures that help diagnose disease, predict outcomes, or determine therapeutic responses. Traditional biomarker discovery methods are labor-intensive and often yield inconsistent results (Table 11.2). AI, particularly ML and DL, can sift through complex genomic, proteomic, and transcriptomic datasets to identify potential biomarkers with remarkable speed and accuracy [53]. AI models trained on vast datasets can detect patterns that would be invisible to the human eye. These insights help predict which patients are likely to respond to a specific treatment, reducing trial-and-error approaches and enhancing therapeutic efficacy.

Patient-specific Drug Development

AI is also being leveraged to create drugs specifically designed for individual patients. For example, DL algorithms can simulate how a drug interacts with a patient's unique genomic or proteomic profile, predicting side effects or resistance before the drug is even synthesized [54]. This approach ensures the right drug is given to the right patient at the right time. Some platforms use AI to integrate data from wearable sensors, medical history, and even lifestyle factors to develop patient-specific treatment regimens. This kind of personalized formulation has the potential to dramatically reduce adverse effects and boost compliance (as shown in Table 11.3) [55].

Table 11.2 AI vs. traditional biomarker discovery.

Parameter	Traditional methods	AI-based methods
Speed	Slow and manual	High-speed analysis
Scalability	Limited	Handles large omics data
Accuracy	Variable	High, with better pattern recognition
Personalization	Limited	Highly tailored predictions
Cost	Expensive	Potentially cost-effective

Table 11.3 Applications of AI in personalized medicine.

AI application	Benefit
Genomic profiling	Identifies patient-specific mutations
Virtual screening	Narrows down personalized drug options
Predictive modeling	Forecasts treatment success/failure
Dosage optimization	Tailors dose for maximum safety
Real-time monitoring	Adjusts therapy based on live data

AI-driven Drug Repurposing Strategies

Drug repurposing, finding new uses for existing drugs, has gained renewed interest, especially during health crises like the COVID-19 pandemic. AI excels here by mining existing literature, clinical data, and molecular databases to predict alternative uses for known compounds [56]. This approach saves time and money, and reduces the risk of failure since repurposed drugs already have known safety profiles. Platforms like BenevolentAI and Atomwise use natural language processing and DL to identify potential candidates for diseases that lack effective therapies.

Challenges and Ethical Considerations in AI-driven Drug Discovery

Despite the groundbreaking advancements AI has brought to drug discovery, there are significant hurdles to overcome. These include technological, ethical, and regulatory concerns that could influence the future trajectory of AI in healthcare.

Data Biases and Interpretability Issues

A major challenge is bias in training datasets, often stemming from underrepresentation of certain ethnicities, age groups, or genders (Table 11.4). This can lead to skewed predictions and unfair clinical outcomes [57]. Additionally, many AI models, especially DL architectures, are "black boxes" that lack transparency. This means they can make accurate predictions without offering an explanation. In drug development, where safety and trust are paramount, interpretability is essential [58].

Table 11.4 Common biases in biomedical datasets.

Bias type	Impact on AI model
Ethnic bias	Inequitable treatment recommendations
Gender bias	Over- or under-prediction of responses
Sample size bias	Poor generalization to broader populations
Disease-specific bias	Skewed prioritization of certain conditions

Table 11.5 Regulatory hurdles in AI-driven drug discovery.

Challenge	Description
Lack of guidelines	No clear path for approval of AI-designed drugs
Algorithm validation	Difficulty in proving reliability
Update policies	AI models evolve, how to regulate changes?
Global standards	Need for international harmonization

Source: [62]/American Medical Association.

Ethical Implications of AI in Drug Discovery

AI poses several ethical dilemmas. For example, who is accountable if an AI-suggested drug causes harm? Is it the algorithm developer, the pharmaceutical company, or the prescriber? Furthermore, there's concern over the privacy of patient data used to train AI models. Secure data handling and informed consent must be prioritized. Ethical frameworks are needed to ensure that AI technologies are developed and deployed responsibly, equitably, and transparently [59].

Regulatory Challenges and Future Perspectives

Current regulatory frameworks are not fully equipped to handle the complexities of AI (major regulatory hurdles are shown in Table 11.5). Regulatory bodies like the FDA and EMA are exploring ways to evaluate AI-driven drug design tools, but there's a lack of standardized protocols for validation and approval [60]. Emerging efforts aim to establish good machine learning practices (GMLPs) to guide AI development in healthcare. Collaboration among technologists, regulators, and healthcare professionals is essential for developing systems that are innovative and reliable [61].

Case Studies of Successful Drug Candidates: From Fragment-based Roots to AI-led Innovations

The drug discovery journey has traveled far, transitioning from manually tweaking small chemical pieces to fully harnessing the power of AI. Let's have a glance at some of the actual success stories that reflect this journey in progress.

Vemurafenib: A Success Story for FBDD

One of the most notable tales in FBDD is Vemurafenib, for the treatment of late-stage melanoma. Researchers at Plexxikon and Roche set out searching with small chemical pieces that would bind to the BRAF V600E mutation, which is a known melanoma driver. Utilizing such tools as X-ray crystallography, they slowly mutated the pieces for better fit to the target. The result? A drug that lengthened lives, FDA-approved in 2011. It was a big proof that beginning small, literally, could result in huge successes [63].

Venetoclax: Conquering the "Undruggable"

Venetoclax is another shining example of how FBDD can overcome difficult targets. Created by AbbVie and Genentech, it was developed to block BCL-2, a protein that aids cancer cells in survival. The journey started with simple fragments that were weakly bound to the target. But through the aid of NMR methods and detailed analysis of structure, the team optimized those hits into a potent, selective oral medicine. Approved for CLL, Venetoclax has transformed the prognosis for many patients [32].

DDR1 Inhibitors: AI Designs in Just 21 Days

In a dramatic illustration of how much has been accomplished, scientists at in silico medicine utilized AI to develop an entirely novel drug candidate, quickly. Their algorithms, based on DL and generative models, designed a DDR1 kinase inhibitor in under a month. They didn't stop at predictions on the computer. The drug was actually prepared and found to function in laboratory tests. This experiment demonstrated that with the appropriate AI tools, it was possible to transform drug discovery from weeks to years [42].

COVID-19 Crisis: AI Joins the Frontlines

During the height of the COVID-19 pandemic, companies like BenevolentAI and DeepMind jumped in with their ML models to hunt for potential treatments. They screened existing drugs and designed new candidates to target the virus's main protease. Although none of these candidates made it to the clinic yet, these rapid efforts showed AI's promise during public health emergencies [64].

AlphaFold: Cracking Protein Structures for Everyone

Although not a drug in itself, DeepMind's AlphaFold must be mentioned. Its AI-driven protein structure predictions, usually more precise than those of conventional lab processes, have been revolutionary. By cracking one of biology's greatest enigmas, AlphaFold is enabling scientists across the world to design better drugs in shorter periods [65].

Conclusion

The field of drug discovery is being revolutionized by the collaboration between AI and contemporary computational methods. From structure-guided design and fragment-based strategies to the convergence of quantum computing and precision medicine, each innovation is a step toward the accelerated, cleverer, and more targeted process of developing therapies. AI has not merely optimized intricate workflows, such as molecular docking, biomarker discovery, and compound screening, but opened up new possibilities through tools such as reinforcement learning and

GANs. The future possibilities are thrilling and revolutionary. AI-aided de novo drug design has the promise of designing completely novel molecules with ideal properties, refined not only for disease targets but also for individual patient profiles. The combination of quantum computing should accelerate simulations, so that real-time predictions become more exact and complete. A fair strategy will be essential in the future, one that promotes innovation while balancing against ethics, transparency of data, and regulatory preparedness. Through ongoing collaboration among clinicians, technologists, researchers, and policymakers, AI is well positioned not only to augment drug discovery but to transform it into a more patient-focused, streamlined, and vibrant process.

References

1 Erlanson DA, McDowell RS, O'Brien T. Fragment-based drug discovery. *J. Med. Chem.* 2004;47(14):3463–82. https://doi.org/10.1021/jm040031v

2 Hajduk PJ, Greer J. A decade of fragment-based drug design: strategic advances and lessons learned. *Nat. Rev. Drug Discov.* 2007;6(3):211–9. https://doi.org/10.1038/nrd2220

3 Murray CW, Rees DC. The rise of fragment-based drug discovery. *Nat. Chem.* 2009;1(3):187–92. https://doi.org/10.1038/nchem.217

4 Congreve M, Carr R, Murray C, Jhoti H. A 'rule of three' for fragment-based lead discovery? *Drug Discov. Today* 2003;8(19):876–7. https://doi.org/10.1016/S1359-6446(03)02831-9

5 Schulz MN, Hubbard RE. Recent advances in fragment-based lead discovery. *Curr. Opin. Pharmacol.* 2009;9(5):615–21. https://doi.org/10.1016/j.coph.2009.04.009

6 Chen H, Engkvist O, Wang Y, et al. The rise of deep learning in drug discovery. *Drug Discov. Today* 2018;23(6):1241–50. https://doi.org/10.1016/j.drudis.2018.01.039

7 Schneider G. Automating drug discovery. *Nat. Rev. Drug Discov.* 2018;17(2):97–113. https://doi.org/10.1038/nrd.2017.232

8 Walters WP, Barzilay R. Applications of deep learning in molecule generation and molecular property prediction. *Acc. Chem. Res.* 2021;54(2):263–70. https://doi.org/10.1021/acs.accounts.0c00699

9 Zhavoronkov A. Artificial intelligence for drug discovery, biomarker development, and generation of novel chemistry. *Mol. Pharm.* 2018;15(10):4311–3. https://doi.org/10.1021/acs.molpharmaceut.8b00930

10 Vamathevan J, Clark D, Czodrowski P, et al. Applications of machine learning in drug discovery and development. *Nat. Rev. Drug Discov.* 2019;18(6):463–77. https://doi.org/10.1038/s41573-019-0024-5

11 Segler MHS, Preuss M, Waller MP. Planning chemical syntheses with deep neural networks and symbolic AI. *Nature* 2018;555(7698):604–10. https://doi.org/10.1038/nature25978

12 Gómez-Bombarelli R, Wei JN, Duvenaud D, et al. Automatic chemical design using a data-driven continuous representation of molecules. *ACS Cent. Sci.* 2018;4(2):268–76. https://doi.org/10.1021/acscentsci.7b00572

13 Jiménez-Luna J, Grisoni F, Weskamp N, Schneider G. Artificial intelligence in drug discovery: Recent advances and future perspectives. *Expert Opin. Drug. Discov.* 2021;16(9):949–59. https://doi.org/10.1080/17460441.2021.1922622

14 Sliwoski G, Kothiwale S, Meiler J, Lowe Jr. EW. Computational methods in drug discovery. *Pharmacol. Rev.* 2014;66(1):334–95. https://doi.org/10.1124/pr.112.007336

15 Lounkine E, Keiser MJ, Whitebread S, et al. Large-scale prediction and testing of drug activity on side-effect targets. *Nature.* 2012;486(7403):361–7. https://doi.org/10.1038/nature11159

16 Li H, Gao Z, Kang L, et al. TarFisDock: a web server for identifying drug targets with docking approach. *Nucleic Acids Res.* 2006;34:W219–24. https://doi.org/10.1093/nar/gkl114

17 Grebner C, Matter H, Plowright AT. The state of the art in the in silico prediction of physicochemical and ADMET properties. *Drug Discov. Today* 2020;25(3):556–67. https://doi.org/10.1016/j.drudis.2020.01.017

18 Yang X, Wang Y, Byrne R, et al. Concepts of artificial intelligence for computer-assisted drug discovery. *Chem. Rev.* 2019;119(18):10520–94. https://doi.org/10.1021/acs.chemrev.8b00728

19 Merk D, Friedrich L, Grisoni F, Schneider G. De novo design of bioactive small molecules by artificial intelligence. *Mol. Inform.* 2018;37 (1–2):1700153. https://doi.org/10.1002/minf.201700153

20 Schneider G, Fechner U. Computer-based de novo design of drug-like molecules. *Nat. Rev. Drug Discov.* 2005;4(8):649–63. https://doi.org/10.1038/nrd1799

21 Lyu J, Wang S, Balius TE, et al. Ultra-large library docking for discovering new chemotypes. *Nature* 2019;566(7743):224–9. https://doi.org/10.1038/s41586

22 Medina-Franco JL, Martínez-Mayorga K. Scaffold-based analysis of drug-like molecules: concepts and applications. *J. Chem. Inf. Model.* 2009;49(4):731–41. https://doi.org/10.1021/ci8004377

23 Bajorath J. Integration of virtual and high-throughput screening. *Nat. Rev. Drug Discov.* 2002;1(11):882–94. https://doi.org/10.1038/nrd961

24 Elton DC, Boukouvalas Z, Fuge MD, Chung PW. Deep learning for molecular design – a review of the state of the art. *Mol. Syst. Des. Eng.* 2019;4(4):828–49. https://doi.org/10.1039/C9ME00039A

25 Schneider P, Walters WP, Plowright AT, et al. Rethinking drug design in the artificial intelligence era. *Nat. Rev. Drug Discov.* 2020;19(5):353–64. https://doi.org/10.1038/s41573-020-0070-1

26 Gao W, Coley CW. The synthesizability of molecules proposed by generative models. ("Artificial intelligence and machine learning approaches for drug design ..."). *J. Chem. Inf. Model.* 2020;60(12):5714–23. https://doi.org/10.1021/acs.jcim.0c00174

27 Stokes JM, Yang K, Swanson K, et al. A deep learning approach to antibiotic discovery. *Cell* 2020;180(4):688–702.e13. https://doi.org/10.1016/j.cell.2020.01.021

28 Erlanson DA, Wells JA, Braisted AC. Site-directed ligand discovery. *Nat. Rev. Drug Discov.* 2004;3(8):660–72. https://doi.org/10.1038/nrd1460

29 Blundell TL, Jhoti H, Abell C. High-throughput crystallography for lead discovery in drug design. *Nat. Rev. Drug Discov.* 2002;1(1):45–54. https://doi.org/10.1038/nrd705

30 Hajduk PJ, Huth JR, Fesik SW. NMR-based screening in drug discovery. *Nat. Rev. Drug Discov.* 2005;4(3):211–9. https://doi.org/10.1038/nrd1650

31 Tsai J, Lee JT, Wang W, et al. Discovery of a selective inhibitor of oncogenic B-Raf kinase with potent antimelanoma activity. *Proc. Natl. Acad. Sci. USA* 2008;105(8):3041–6. https://doi.org/10.1073/pnas.0711741105

32 Souers AJ, Leverson JD, Boghaert ER, et al. ABT-199, a potent and selective BCL-2 inhibitor, achieves antitumor activity while sparing platelets. ("Acute lymphoblastic leukemia as a clonally unrelated second primary ...") *Nat. Med.* 2013;19(2):202–8. https://doi.org/10.1038/nm.3048

33 Cross DAE, Ashton SE, Ghiorghiu S, et al. AZD9291, an irreversible EGFR TKI, overcomes T790M-mediated resistance to EGFR inhibitors. *Cancer Discov.* 2014;4(9):1046–61. https://doi.org/10.1158/2159-8290.CD-14-0337

34 Kawatkar SP, Patel KR, Vavia PR. Fragment-based drug discovery: the hit fragment optimisation and the role of structural biology. *Eur. J. Med. Chem.* 2020;199:112390. https://doi.org/10.1016/j.ejmech.2020.112390

35 Murray CW, Roughley SD, Matassova N, Rees DC. Fragment-based drug discovery applied to Hsp90. *Drug Discov. Today Technol.* 2012;9(1):e3–11. https://doi.org/10.1016/j.ddtec.2011.07.001

36 Meng XY, Zhang HX, Mezei M, Cui M. Molecular docking: a powerful approach for structure-based drug discovery. *Curr. Comput. Aided Drug Des.* 2011;7(2):146–57. https://doi.org/10.2174/157340911795677602

37 Kitchen DB, Decornez H, Furr JR, Bajorath J. Docking and scoring in virtual screening for drug discovery: methods and applications. *Nat. Rev. Drug Discov.* 2004;3(11):935–49. https://doi.org/10.1038/nrd1549

38 Schuster D, Laggner C, Langer T. Pharmacophore modeling and virtual screening: concepts and applications. *Curr. Comput. Aided Drug Des.* 2006;2(1):31–49. https://doi.org/10.2174/157340906775252960

39 Cherkasov A, Muratov EN, Fourches D, et al. QSAR modeling: where have you been? Where are you going to? *J. Med. Chem.* 2014;57(12):4977–5010. https://doi.org/10.1021/jm4004285

40 Bajusz D, Rácz A, Héberger K. Multi-objective optimization in drug discovery: current strategies and future directions. *Drug Discov. Today* 2015;20(10):1183–93. https://doi.org/10.1016/j.drudis.2015.05.009

41 Lionta E, Spyrou G, Vassilatis DK, Cournia Z. Structure-based virtual screening for drug discovery: principles, applications and recent advances. *Curr. Top Med. Chem.* 2014;14(16):1923–38. https://doi.org/10.2174/1568026614666140929124445

42 Zhavoronkov A, Ivanenkov YA, Aliper A, et al. Deep learning enables rapid identification of potent DDR1 kinase inhibitors. ("The role of nanomedicine and artificial intelligence in cancer health ..."). *Nat. Biotechnol.* 2019;37(9):1038–40. https://doi.org/10.1038/s41587-019-0224-x

43 Schneider G. Virtual screening: an endless staircase? *Nat. Rev. Drug Discov.* 2010;9(4):273–6. https://doi.org/10.1038/nrd3139

44 Kadurin A, Nikolenko S, Khrabrov K, et al. druGAN: an advanced generative adversarial autoencoder model for de novo generation of new molecules. *Mol. Pharm.* 2017;14(9):3098–104. https://doi.org/10.1021/acs.molpharmaceut.7b00346
45 Popova M, Isayev O, Tropsha A. Deep reinforcement learning for de novo drug design. *Sci. Adv.* 2018;4(7):eaap7885. https://doi.org/10.1126/sciadv.aap7885
46 Altae-Tran H, Ramsundar B, Pappu AS, Pande V. Low data drug discovery with one-shot learning. *ACS Cent. Sci.* 2017;3(4):283–93. https://doi.org/10.1021/acscentsci.6b00367
47 Cao Y, Romero J, Olson JP, et al. Quantum chemistry in the age of quantum computing. *Chem. Rev.* 2019;119(19):10856–915. https://doi.org/10.1021/acs.chemrev.8b00803
48 Benedetti M, Lloyd E, Sack S, Fiorentini M. Parameterized quantum circuits as machine learning models. *Quantum Sci. Technol.* 2019;4(4):043001. https://doi.org/10.1088/2058-9565/ab4eb5
49 Schuld M, Killoran N. Quantum machine learning in feature Hilbert spaces. *Phys. Rev. Lett.* 2019;122(4):040504. https://doi.org/10.1103/PhysRevLett.122.040504
50 Preskill J. Quantum computing in the NISQ era and beyond. *Quantum* 2018;2:79. https://doi.org/10.22331/q-2018-08-06-79
51 McArdle S, Endo S, Aspuru-Guzik A, et al. Quantum computational chemistry. *Rev. Mod. Phys.* 2020;92(1):015003.
52 Ragoza M, Hochuli J, Idrobo E, et al. Protein–ligand scoring with convolutional neural networks. *J. Chem. Inf. Model.* 2017;57(4):942–57. https://doi.org/10.1021/acs.jcim.6b00635
53 Sanchez-Lengeling B, Aspuru-Guzik A. Inverse molecular design using machine learning: generative models for matter engineering. *Science* 2018;361(6400):360–5. https://doi.org/10.1126/science.aap7885
54 Lavecchia A. Deep learning in drug discovery: opportunities, challenges and future prospects. *Drug Discov. Today* 2019;24(10):2017–32. https://doi.org/10.1016/j.drudis.2019.07.003
55 Bian Z, Li J, Li Z, et al. Solving chemistry problems using quantum computers. *Commun. Chem.* 2019;2(1):1–9. https://doi.org/10.1038/s42004-019-0141-9
56 Libbrecht MW, Noble WS. Machine learning applications in genetics and genomics. *Nat. Rev. Genet.* 2015;16(6):321–32. https://doi.org/10.1038/nrg3920
57 Esteva A, Robicquet A, Ramsundar B, et al. A guide to deep learning in healthcare. *Nat. Med.* 2019;25(1):24–9. https://doi.org/10.1038/s41591-018-0316-5
58 Zhou Y, Wang F, Tang J, et al. Artificial intelligence in COVID-19 drug repurposing. *Brief. Bioinform.* 2020;22(2):726–41. https://doi.org/10.1093/bib/bbz178
59 Char DS, Shah NH, Magnus D. Implementing machine learning in health care – addressing ethical challenges. *N. Engl. J. Med.* 2018;378(11):981–3. https://doi.org/10.1056/NEJMp1714229
60 Floridi L, Cowls J, King T, Taddeo M. How to design AI for social good. *Science* 2018;361(6404):447–8. https://doi.org/10.1126/science.aat7306
61 Topol EJ. High-performance medicine: the convergence of human and artificial intelligence. *Nat. Med.* 2019;25(1):44–56. https://doi.org/10.1038/s41591-018-0300-7

62 Beam AL, Kohane IS. Big data and machine learning in health care. *JAMA* 2018;319(13):1317–8. https://doi.org/10.1001/jama.2018.0265

63 Bollag G, Hirth P, Tsai J, et al. Clinical efficacy of a RAF inhibitor needs broad target blockade in BRAF-mutant melanoma. *Nature* 2010;467(7315):596–9. https://doi.org/10.1038/nature09454

64 Zhou Y, Wang F, Tang J, et al. Artificial intelligence in COVID-19 drug repurposing. *Brief. Bioinform.* 2021;22(2):726–41. https://doi.org/10.1093/bib/bbaa320

65 Jumper J, Evans R, Pritzel A, et al. Highly accurate protein structure prediction with AlphaFold. *Nature* 2021;596(7873):583–9. https://doi.org/10.1038/s41586-021-03819-2

12

Artificial Intelligence in Computer-aided Drug Design

*Teena Garg[1], Shikha Baghel Chauhan[1], Indu Singh[1], and Anshuman Sinha[2,3]**

[1]*Amity Institute of Pharmacy, Amity University, Noida, Uttar Pradesh, India*
[2]*Department of Pediatrics, Ann & Robert H. Lurie Children's Hospital, Chicago, IL, USA*
[3]*Feinberg School of Medicine, Northwestern University, Chicago, IL, USA*

*Corresponding author: anshuman.pharma@gmail.com

Introduction

Artificial intelligence (AI) comprises a vast chemical space with a large number of compounds in excess of 1,060, which also fosters a lot of analytical therapies. It is a reality that the development of medicines is a time-consuming and very costly endeavor that is hindered by a lack of creative technology [1]. It is a fact that AI is able to accomplish the development of medicines that is hindered by a lack of creative technology. Nevertheless, AI faces many challenges regarding data quality, including its size, growth rate, complexity, and unpredictability. Though quantitative structure–activity relationship (QSAR)-based computational models can swiftly forecast thousands of chemicals or basic physical chemical properties like log$P$ or log$D$, these models lack effectiveness when it comes to more complicated biologic properties like chemical efficacy and side effects. Moreover, QSAR based models face many difficulties like limited training data, experimental data inaccuracies, and a lack of experimental validations [2]. The datasets of chemical medicine research of pharmaceutical corporations can hold millions of molecules that conventional machine learning (ML) tools can't process. Implementation of newly developing methods using AI, e.g., deep learning (DL) methods, and model research using these methods, can be beneficial for the assessment of pharmacological agents' safety and efficacy using huge data models. A QSAR ML competition for demonstrating the ability of DL to support drug development in the pharmaceutical sector was organized by Merck in 2012. When compared to the usual ML technique on 15 drug candidate absorption, distribution, metabolism, excretion, and toxicity (ADMET) attributes, the predictive efficacy of the DL models increased greatly [3]. A large virtual universe offers a geographical representation of molecules using a visualization technique indicating distributions and attributes of molecules. The aim of chemical space representation in searching bioactive agents is to derive positional knowledge about molecules in chemical space. Virtual screening (VS) helps in identifying, which molecules to test by choosing the best ones. The general public can browse several chemical databases like PubChem, ChemBank, DrugBank, and ChemDB. Besides structure-based and ligand-based strategies, some other strategies concerning in silico techniques for virtual screening of virtual chemical spaces offer improved analysis of profiles, faster removal of nonlead molecules, and even more cost-efficient selection of therapeutic molecules [4]. Coulomb matrices and fingerprint recognition of molecules represent two drug development strategies, which take physical and chemical attributes as well as toxicological attributes in choosing a lead molecule. The chemical structure of a compound can be forecasted using various parameters such as predictive models, molecular similarities, process of molecule formation, and in silico techniques [5, 6]. With 95,000 decoys, 40 receptors, and 2,950 ligands, when tested, DeepVS method proved to be excellent [7]. For optimizing the efficacy of a cyclin-independent kinase2 inhibitor, a multi-objective automated replacement method was also used. QSAR model builder has been employed for designing potential medicinally active candidates and has advanced to AI QSAR model builder techniques like linear discriminant analysis (LDA), support vector machine (SVM), random forest (RF), or decision trees, which can be employed to accelerate QSAR analysis [8]. With a comparison of six AI techniques to conventional procedures for anonymous substances for their biologic activity, King et al. observed little statistical difference [9].

Emergence of AI in Drug Design

Recently, AI has come up as a revolutionary technology in the area of drug design, overturning conventional approaches in this area and facilitating faster development of new medications. Conventionally, developing a drug is an arduous task involving significant costs and taking over a decade for development and launch of a molecule in the market. However, AI provides an interim solution in making every step of this complex procedure smoother and efficient. Through this technology, it is possible to analyze large amounts of information and make predictions about molecule behavior using ML and DL concepts, and analytics in this area now enable researchers to make use of large amounts of information and "discover potential drug candidates at an unmatched speed and accuracy" and even perform tasks like analyzing millions of molecule combinations and finding patterns and links not easily recognizable by humans in this area [10]. Furthermore, AI is reinventing the concepts of target identification, lead optimization, and toxicity prediction in a manner where safer and more efficacious compounds reach the stage of clinical trials. Generative models, for example, are being employed to develop compounds from scratch with the sole intention of targeting a particular biological target. Notably, it is important to point out that AI is not a replacement for human intelligence but a tool complementary to the human creative intellect [11]. This partnership of human research experts and AI gives rise to an innovative culture where ideas can be verified and made real at a pace that has never been achieved before.

While this has so much potential, there are challenges involved in the use of AI in drug development. Issues such as the authenticity of the data used in AI and the acceptability of the AI models by the relevant authorities are some that remain to be considered. However, the trend is evident. The pharmaceutical industry, learning institutions, as well as startups, are getting deeply invested in AI technology. This marks a paradigm shift in how we tackle the development of drugs [12]. As developments in the sector continue to take place, findings through AI offer a great promise in helping patients heal faster, heal safely, and heal according to their needs.

Fundamentals of AI

AI is the name of the process of simulation of human intelligence in machines, and particularly in computer systems, to execute tasks such as language understanding, learning, problem-solving, and perception. AI is an innovative method in discovering drugs in the biomedical world, in the analysis of vast information, in the detection of complex patterns, and in the prediction of the corresponding responses of the biological systems, which cannot be explained in the normal computational method. With the incorporation of learning algorithms in AI, the possibilities of computer-aided drug design have been widened.

ML Basics

AI technology named ML applies data to teach algorithms in an unwarranted task specific manner. For the prediction of drug target interaction, toxicology, efficacy, and absorption, ML models are designed and developed, employing chemical structure, biologic data, and pharmacologic knowledge. For the classification problem (active vs. inactive compound) and regression analysis (binding affinity prediction), supervised learning is employed, in which models are trained on labeled data [13]. Unsupervised learning strategies like clustering analysis and dimensional reduction help in the comprehension of chemical space and the alignment of aligned molecules. Based on the desired application and the availability of the labeled dataset, semi-supervised learning and reinforcement learning (RL) strategies are sometimes employed.

DL and Neural Networks

DL, being the advanced version of ML, employs multilayer neural networks, also called deep neural networks (DNNs), and is based on the structure and function of the human brain. CNN and RNN techniques have found tremendous applications in the synthesis of medications while handling high-dimensional data, such as protein sequences, three-dimensional (3D) molecular representations, and omics data. There is no need for human feature engineering, as the models don't require the identification of features from the data by human engineers. This is accomplished by understanding complex and nonlinear relations between the data of biological samples and successfully applies to tasks such as virtual screening, protein structure prediction (e.g., AlphaFold), and de novo synthesis of molecules [14].

Other AI Techniques: RL, Natural Language Processing

Other DL technologies, including RL and natural language processing, are gradually gaining prominence in drug development, not relying on ML and DL algorithms. RL is based on encouraging good behavior and thus teaches an agent to take a series of steps to reach a goal. This is very helpful in developing a chemical entity or optimizing molecular properties through a procedure and error trial process. Another area, natural language processing, is used by computers to understand and generate human language [15]. This is applied in medical article extraction, analysis of clinical reports, and even development of medications by leveraging very large language models that have been trained on patents and chemical literature. These are very cutting-edge technologies that have given rise to new approaches to dealing with the complexity that is associated with biological systems and new therapies, thus increasing application areas within the realm of AI and computer-aided drug design (CADD) [16].

Integration of AI in CADD

The whole process of conventional drug discovery has been radically changed and supplemented with the integration of AI in CADD in order to hasten the process and obtain accurate forecasts, along with the ability to design new compounds. AI, including ML and DL algorithms, provides the capability to analyze large amounts of biological and chemical data to detect hidden patterns and relations, which are difficult to obtain manually. These findings enable the identification of new potential candidates, the estimation of their pharmacological properties, an enhancement in their attributes, and even designing a new compound with desirable properties [17]. The entire procedure of drug development becomes much more efficient and economical through the use of AI.

Virtual Screening and Molecular Docking

Virtual screening is actually one of the first applications of AI in drug development, meeting the requirement to find potential candidates from gigantic libraries based on their predicted biological activity. The AI models that use experimental data to select candidates with improved binding affinity and selectivity are advancing the more traditional virtual screening techniques, such as ligand and structure-based virtual screening. These include improving scoring functions and binding pose prediction in molecule docking, also predicting the best binding pose of the small ligand in the binding site of the target. DL models are more accurate compared to other conventional scoring techniques for predicting binding affinities, thus improving the choice of lead compounds for further investigation [18].

Predictive Modeling for ADMET Properties

ADMET is a vital part of drug development, since a poor ADMET profile is often the culprit responsible for the failure of a great number of candidate compounds in clinical trials [19]. AI-driven prediction algorithms take a closer look at chemical structures and biological data with the aim of predicting ADMET features as early as possible in design. Examples of ML models are SVMs, RFs, and DNNs trained with historical datasets on ADMET for very accurate predictions regarding pharmacokinetic behavior and toxicity hazards. The technique finally makes medicinal chemists perform logical, less expensive drug research when they apply proactive prediction to changing the molecule for better bioavailability, adverse effects, and safety assurance [20].

De Novo Drug Design Using AI

Contrary to the evaluation of existing molecules, the design of new drugs involves the design of novel compound molecules from scratch. Major breakthroughs in this area have been achieved through the application of AI, in the form of generative models including RL approaches, generative adversarial networks (GANs), and variational autoencoders (VAEs) [21]. The above AI approaches design new molecules on the background of achieved biological needs, desired potencies, and desired safety profiles based on learning the rules of chemical space. Multi-objective optimization is one of the abilities of some of these AI approaches to seek balances between chemical synthetics ability, drug-likeness, and efficacy. This ability of AI to "design" molecules creatively expands the boundaries of molecular design in designing new drugs through the generation of an ever-increasing number of molecules to be studied instead of adopting the trial-and-error approaches [22].

Key AI Algorithms and Models in Drug Design

AI has transformed the domain of drug development through the application of innovative computational tools and techniques, which can efficiently deal with large amounts of data present in the biochemical fields. A number of prime algorithms of AI have earned popularity in the domain of CADD [23].

Support Vector Machines

In drug development, vector machines are well-known supervised learning algorithms that can be applied to issues such as regression and classification. The process involves identifying the hyperplane that separates data points into distinct categories; for example, active and inactive compounds. The fact that SVMs are not sensitive to high-dimensional data makes them perfectly suitable for target identification and virtual screening. They are also very good predictors of biological activity, toxicity, and pharmacokinetics [24].

Random Forests

A number of decision trees are created through training with the RF algorithm, which is a method of ensemble learning where the class (in classification) and/or mean value (in regression) from each decision tree is predicted. The RF models are applied in drug discovery applications such as predicting ADMET properties, binding affinities, and molecular properties. They are useful models for HTS and lead optimization due to their ability to process large datasets and prevent overfitting [25].

Deep Neural Networks

DNNs are composed of multiple layers of interlinked neurons that possess the capability of detecting complex patterns and correlations in the data. DNNs have also found applications in drug design tasks such as simulating biologic reactions, optimizing chemical structures, and predicting drug target interactions. The reason behind the enhanced capability of DNNs in new target identification and de novo drug development lies in its ability to learn from basic molecular forms such as molecular graphs and SMILES strings [26].

Generative Adversarial Networks

Two parallel neural networks, known as a generator and a discriminator, compete with each other to develop more realistic data. This is because GANs are highly creative in the process of creating drugs due to their ability to develop new chemical substances possessing the required pharmacological properties [27]. GANs can highly accelerate the early stages of developing a drug by developing new chemicals that are synthetically possible and physiologically active due to their ability to learn the distribution of chemical information.

Transformer Models and AI-driven Language Models

Originally designed for tackling challenges based on natural language processing tasks, transformer architectures have applications galore in designing medications. The representation of molecules as sequences can be interpreted and generated using models such as BERT and GPT and their variants derived from chemistry (e.g., ChemBERTa and MolBERT). Using chemical information as a language has produced remarkable results for molecule generation, retrosynthesis prediction, and results of a reaction using these models. Enhanced accuracy and originality of AI-driven drug design are derived from their capability to interpret relationships between chemical information and context [28].

Databases and Tools Supporting AI-based CADD

A big boost to the incorporation of AI within CADD comes from the availability of large biological and chemical dataset resources combined with sophisticated AI resources/tools available for use. AI-driven CADD is based on a foundation comprising publicly available databases that provide a judiciously chosen set of biological and chemical data required for activities such as model validation and prediction [29].

Public Databases (e.g., ChEMBL, PubChem)

Two of the commonly used public databases for applying AI in drug development include ChEMBL and PubChem databases. The ChEMBL database is a well curated database that encompasses pharmacological profiles, functional tests, binding data, and bioactivity studies of small molecules similar to drugs. The quantitative indices necessary for building prediction models in drug development best described by ChEMBL databases include IC_{50}, EC_{50}, and K_i values. The extensive set of chemical substances as well as biological activities in PubChem databases, a database administered by National Center for Biotechnology Information (NCBI), supplies a wide array of comprehensive details about chemical substances and biological activities necessary for training in AI algorithms in relation to structure–activity relationships (SARs) [30].

AI Software and Platforms

In order to make the application of ML and DL in CADD easier, several specific platforms and tools of AI have emerged. An opensource software kit named DeepChem was developed to specifically apply the capabilities of DL to tackle difficult tasks in the fields of quantum chemistry, materials science, and pharmaceutical development [31]. In the context of molecular science, it provides easy access to predefined models, processing of data, and customized learning. Target discovery and structure-based drug design have been transformed because of the development of DeepMind's AlphaFold, a milestone in protein structure prediction, in which AI was successfully applied to predict the 3D structure of a protein from the sequence of amino acids with atomic accuracy. AutoQSAR and LiveDesign, two of Schrödinger's AI-assisted tools, feature both ML models that help speed up lead optimization, prediction, and the design of a molecule [32]. Various platforms together could help scientists extend the boundaries of traditional drug discovery to carry out increasingly complex simulations, screens, and predictions.

Applications of AI In Different Stages of Drug Discovery

AI has become an integral component across multiple stages of the drug discovery pipeline, ranging from early target identification to preclinical evaluation. AI-driven approaches enable the analysis of large-scale biological datasets, accelerate virtual screening, optimize lead compounds, and improve the prediction of pharmacokinetic and toxicity profiles. The specific roles, methodologies, and representative AI techniques applied at each stage of drug discovery are summarized in Table 12.1 [33, 34].

Table 12.1 Applications of AI at different stages of the drug discovery pipeline.

Stage	Description	Role of AI	Key examples/ techniques
Target identification and validation	Finding biological molecules like proteins or genes that contribute to disease, and confirming their involvement	Analyzing large-scale omics datasets (genomics, proteomics, metabolomics) Predicting new targets through advanced pattern recognition Exploring disease pathways using AI-driven network analysis	ML for gene disease link prediction DL to understand biological network behavior NLP for extracting information from scientific publications
Lead compound discovery	Searching for chemical compounds that can interact effectively with the validated target	High-throughput virtual screening of chemical libraries Rapid prediction of binding strengths Designing new molecules using generative AI models	DeepDock platform for virtual screening RL for novel molecule creation GANs for designing innovative structures
Lead optimization	Enhancing early-stage compounds to improve their therapeutic potential and minimize side effects	Building QSAR models Forecasting ADMET properties (absorption, distribution, metabolism, excretion, toxicity) Simulating chemical modifications through AI methods	SVM for ADMET prediction RF algorithms for toxicity risk evaluation AI-powered MD simulations to assess stability
Preclinical prediction models	Estimating drug behavior in biological systems before starting human trials	Modeling pharmacokinetics and pharmacodynamics Predicting outcomes in animal models Early identification of risks and toxicity profiles using AI	DL frameworks like DeepADMET for property prediction Bayesian networks for assessing safety and toxicity AI models forecasting blood brain barrier permeability

Challenges and Limitations

Despite the potential offered by AI in transforming the field of CADD, several problems and constraints hamper its use. One of the biggest concerns is the use of a large, diverse, and high-quality dataset. This is because the predictions made by the

AI models are not always reliable when they are trained using improper, biased, or inconsistent biological, chemical, and pharmacokinetic data. In addition to that, the complexity associated with biological systems, especially in dynamic physiological systems, makes it difficult for AI models to comprehensively understand intricate interactions. Another important area of concern is referred to as the "black box problem," which refers to the inability to interpret complex or advanced AI models. Scientists may feel hesitant about accepting and adhering to the predictive results of an AI model when they fail to understand how such models arrive at their results, particularly when such results are as crucial as those in drug development. Further challenges are posed by the potential issues surrounding ethics and data privacy and consent on the one hand and potential bias when designing algorithms on the other hand. All these challenges indicate that for successful integration of AI into CADD systems, there is a need for careful optimism and better standards and collaborations.

Data Quality and Availability Issues

The quality, amount, and complexity of input data affect AI model performance for CADD. Unfortunately, many "dark areas" have not been yet discovered within drug development because many sets of data remain either privately owned, inadequate, or biased toward targets and chemical spaces for which many studies have been conducted. The quality of data integrity can be further compromised because of experimental methodological inconsistencies, errors of reporting, and lab variation. Moreover, many cases of data insufficiencies impose limitations regarding innovative targets, rare diseases, and chemical substances, making it difficult for AI to generalize [35]. Methods of synthetic data augmentation have been occasionally employed; however, these might introduce additional artifacts. Obviously, poor quality and insufficiency of data can seriously jeopardize model training, promoting overfitting, poor external validation, and ultimately, failure to predict really innovative and effective medicaments.

Interpretability of AI Models

However, one of the major hurdles in the application of AI models in CADD is still interpretability. This is because many of the state-of-the-art models, including DL networks like GANs or CNNs, act as "black boxes," where it is difficult to interpret how a particular result is derived from the input variables. This lack of transparency is where regulatory authorities, medicinal chemists, or other scientists place their lack of trust. Lack of transparency is always difficult in those domains like drug design where the safety, efficacy, and mode of action of the drug candidates should be well justified. Based on the aim to provide explanations to these models, new explainable AI (XAI) strategies like feature attribution, decision trees, and model reduction are being introduced [35]. Nevertheless, it is still an essential aim to achieve the right balance between interpretability and model complexity.

Regulatory and Ethical Concerns

Significant ethical and regulatory issues are also raised by the application of AI to CADD. When interpretability is difficult in AI models or when the use of opaque training data makes it hard to validate or reproduce studies, then strict validation and reproducibility required by agencies like the Food and Drug Administration (FDA) and European Medicines Agency (EMA) can be challenging. There is also much debate around the classification and regulation of molecules designed by AI. Some questions include: whether the AI system must be considered as a coinventor and what liability issues could arise when predictions of AI lead to undesirable outcomes. Ethical considerations on data privacy, especially when using genetic or clinical data obtained from patients, are of great importance to prevent breaches and their exploitation. Bias in AI algorithms about historical injustices or scientific shortcomings will result in less successful treatments of underprivileged populations. Therefore, responsible usage of AI in CADD incorporates concepts of ethical AI, transparency in algorithm development, and constant consultation with regulatory agencies [35].

Future Perspectives

The future for AI in CADD looks very promising, as it has the ability to revolutionize pharmaceutical discovery and facilitate the creation of new medications. It is expected that as AI software develops, there will be more accurate predictions for a complicated biological system, hence paving the way for the creation of new drugs that have improved efficacy and safeness levels. To improve medicinal designs, future systems will rely on multi-omics data, evidence, and patient provided information. Additionally, the need for XAI will emerge, as it will allow scientists to understand the reasoning behind the predictions generated by AI, hence promoting confidence and easy approval for regulatory authorities. An entire automated process for drug development will emerge when AI is combined with robots, lab automation, and cloud data management systems. This will greatly reduce costs and time required for the launching of a new drug [36].

AI in Personalized Medicine and Precision Drug Design

AI is also anticipated to have a significant role in personalized medicines in which a patient gets a treatment tailored to his or her needs based on genetic background, surroundings, and lifestyle. Regarding precisely designing drugs through AI technology, these AI algorithms search through a huge amount of information like EHRs, proteomics, metabolomics, and genomes to predict patient responses to potential drugs. These ML algorithms identify biomarkers to define patient groups so that a person can precisely chalk out a pharmaceutical formulation for maximum efficiency with least side effects. AI aids in making a treatment plan perfectly consistent with a person's metabolic profile so that in a way, AI literally raises a treatment's efficiency [36].

Emerging Trends: Quantum Computing and AI

With processing strengths that are thus much more superior than conventional computers, quantum computing thus initiates a paradigm shift regarding the incorporation of AI-assisted drug discovery. With respect to conventional computing, quantum computing offers the possibility of speeding up diverse operations of computer-assisted drug discovery, including virtual screening of compounds, simulation of protein folding, as well as molecular docking. Drug–receptor interactions would be unlocked at speeds that would be unthinkable in the current domain owing to the superiority of quantum ML, which would be able to scan 10 times larger search spaces. In an attempt to simulate complex biological events, which would be inexplicable at present levels of computation, there is a growing effort at present regarding the investigation of quantum AI hybrids. With the development of quantum technology, there would emerge fresh opportunities regarding molecular design, including targets that would be "undruggable" as identified by conventional solutions [36].

Collaboration Between AI Experts and Medicinal Chemists

The collaborative efforts of AI researchers and medicinal chemists play a pivotal role in unlocking the true potential offered by AI-based drug design modeling. Although AI systems are capable of data-efficient optimization and ML-based pattern recognition, it is necessary that the significant understanding of biomolecular systems and chemistry expertise possessed by medicinal chemists is leveraged to analyze and validate AI predictions, and predictions can also serve as an interface in giving a significant direction to AI-based predictions since they possess significant expertise in chemistry and biomolecular systems and can also play a pivotal role in demarking a strategic plan to move ahead in dealing with issues created by data biases and data overfitting during AI-based drug design modeling by bringing a multidisciplinary twist to combine efforts of AI researchers and life science professionals as future success depends upon collaborative training and research programs that support a two-way learning process among AI and life sciences professionals [36].

Conclusion

AI and CADD are technologies that have been greatly influenced by the developments associated with the application of AI. The main advantage associated with the application of AI is that it offers a faster, more accurate, and more economically viable alternative for drug design compared with the current drug design approach and other previous drug design methods that were less accurate and more time-consuming and costly. With the current limitations associated with the accuracy and interpretability of the associated models that pose challenges associated with the adoption of AI in drug development, potential future improvements may result in more accurate, personalized, and efficient drug development associated with the application of quantum computation and explainable AI. Future intersections associated with computational scientists and medicinal chemists and AI researchers will be critical for the complete exploitation of the potential associated with the development of drugs and AI.

List of Abbreviations

ADMET	Absorption, distribution, metabolism, excretion, and toxicity
AI	Artificial intelligence
ANN	Artificial neural network
BBB	Blood–brain barrier
CADD	Computer-aided drug design
CNN	Convolutional neural network
DL	Deep learning
DNN	Deep neural network
EC_{50}	Half maximal effective concentration
EMA	European Medicines Agency
FDA	Food and Drug Administration
GAN	Generative adversarial network
IC_{50}	Half maximal inhibitory concentration
K_i	Inhibition constant
LDA	Linear discriminant analysis
ML	Machine learning
NCBI	National Center for Biotechnology Information
NLP	Natural language processing
Omics	Genomics, Proteomics, and Metabolomics
PD	Pharmacodynamics
PK	Pharmacokinetics
QSAR	Quantitative structure–activity relationship
RF	Random forest
RL	Reinforcement learning
RNN	Recurrent neural network
SaaS	Software as a Service
SAR	Structure–activity relationship
SMILES	Simplified Molecular Input Line Entry System
SVM	Support vector machine
VAE	Variational Autoencoder
VS	Virtual screening
XAI	Explainable artificial intelligence

References

1 Vyas M, Thakur S, Riyaz B, et al. Artificial intelligence: the beginning of a new era in pharmacy profession. *Asian J. Pharm.* 2018;12(2):72–6.
2 Mak KK, Pichika MR. Artificial intelligence in drug development: present status and future prospects. *Drug Discov. Today* 2019;24(3):773–80.
3 Sellwood MA, Ahmed M, Segler MHS, Brown N. Artificial intelligence in drug discovery. *Future Sci. OA* 2018;4(8):FSO20252028.
4 Zhu H. Big data and artificial intelligence modeling for drug discovery. *Annu. Rev. Pharmacol. Toxicol.* 2020;60:573–89.
5 Ciallella HL, Zhu H. Advancing computational toxicology in the big data era by artificial intelligence: data-driven and mechanism-driven modeling for chemical toxicity. *Chem. Res. Toxicol.* 2019;32(4):536–47.
6 Chan HS, Shan H, Dahoun T, et al. Advancing drug discovery via artificial intelligence. *Trends Pharmacol. Sci.* 2019;40(8):592–604.
7 Brown N. *In Silico Medicinal Chemistry: Computational Methods to Support Drug Design*. Royal Society of Chemistry; 2015.
8 Pereira JC, Caffarena ER, dos Santos CN. Boosting docking-based virtual screening with deep learning. *J. Chem. Inf. Model.* 2016;56(12):2495–506.
9 Firth NC, Brown N, Blagg J. MOARF, an integrated workflow for multiobjective optimization: implementation, synthesis, and biological evaluation. *J. Chem. Inf. Model.* 2015;55(6):1169–80.
10 Vamathevan J, Clark D, Czodrowski P, et al. Applications of machine learning in drug discovery and development. *Nat. Rev. Drug Discov.* 2019;18(6):463–77.
11 Zhavoronkov A, Ivanenkov YA, Aliper A, et al. Deep learning enables rapid identification of potent DDR1 kinase inhibitors. *Nat. Biotechnol.* 2019;37(9):1038–40.
12 Lee JW, Maria-Solano MA, Vu TN, et al. Big data and artificial intelligence (AI) methodologies for computer-aided drug design (CADD). *Biochem. Soc. Trans.* 2022;50(1):241–52.
13 Siddiqui B, Yadav CS, Akil M, et al. Artificial intelligence in computer-aided drug design tools for the finding of potent biologically active small molecules: traditional to modern approach. *Comb. Chem. High Throughput Screen.* 2025;28(1):115.
14 Satpathy R. Artificial intelligence techniques in the classification and screening of compounds in computer-aided drug design. In: *Artificial Intelligence and Machine Learning in Drug Design and Development*. Elsevier; 2024. pp. 473–97.
15 Seo S, Lee JW. Applications of big data and AI-driven technologies in computer-aided drug design. In: *Computational Drug Discovery and Design*. Springer; 2023. pp. 295–305.
16 Wang K, Huang Y, Wang Y, et al. Recent advances from computer-aided drug design to artificial intelligence drug design. *RSC Med. Chem.* 2024;15(12):3978–4000.

17 Chen K, Xu R, Hu X, et al. Recent advances in the development of DprE1 inhibitors using AI and CADD approaches. *Drug Discov. Today* 2024;29:103987.

18 Rusinko A, Rezaei M, Friedrich L, et al. AIDDISON: empowering drug discovery with AI/ML and CADD tools in a secure, web-based SaaS platform. *J. Chem. Inf. Model.* 2024;64(1):38.

19 Sharma VK, Bharatam PV. Artificial intelligence in drug discovery. *Curr. Res. Inf. Pharm. Sci.* 2022;16(1):37.

20 Javed MR. CADD and molecular dynamic simulations: potential impacts to conventional medicines. *Comb. Chem. High Throughput Screen.* 2022;25(4):658–9.

21 Elkholy N, Hassan R, Bedda L, et al. Exploration of SAMI riboswitch inhibitors: in silico discovery of ligands to a new target employing multistage CADD approaches. *Artif. Intell. Chem.* 2024;2(1):100044.

22 Kumari N, Hasija Y. CADD: exploring the digital frontier in drug designing. In: *Proceedings of the 3rd Int Conf Computational Modelling, Simulation and Optimization (ICCMSO)*. IEEE; 2024. pp. 272–7.

23 del Carmen Quintal Bojórquez N, Campos MR. Traditional and novel computer-aided drug design approaches in the anticancer drug discovery process. *Curr. Cancer Drug Targets* 2023;23(5):333–45.

24 Ye F, Lin M, Jin J, Broussy S. Computer-aided drug design: drug discovery, computational modelling, and artificial intelligence. *Front. Chem.* 2022;10:968687.

25 Nascimento IJ, de Aquino TM, da Silva-Júnior EF. The new era of drug discovery: the power of computer-aided drug design. *Lett. Drug Des. Discov.* 2022;19(11):951–5.

26 Narayan A, Chanana A, Shanker O, et al. Role of artificial intelligence in pharmaceutical drug development. *Curr. Indian Sci.* 2024;2(1):e2210299X313252.

27 Medina-Franco JL, López-López E. What is the plausibility that all drugs will be designed by computers by the end of the decade? *Expert Opin. Drug. Discov.* 2024;19(5):507–10.

28 Chakravarthi CH, Mulpuru V, Mishra N. Artificial intelligence and bioinformatics: a powerful synergy for drug design and discovery. In: *Artificial intelligence: A Multidisciplinary Approach Towards Teaching and Learning*. Bentham Science; 2024. pp. 26–79.

29 Gupta P, Chanana A, Kulkarni YR, et al. Computer-aided drug design: innovation and its application in reshaping modern medicine. *Curr. Artif. Intell.* 2024;2:e29503752321279.

30 Niazi SK, Mariam Z. Computer-aided drug design and drug discovery: a prospective analysis. *Pharmaceuticals (Basel)* 2024;17(1):22.

31 Kohli RK, Santoshi S, Yadav SS, Chauhan V. Applications of AI in computer-aided drug discovery. In: *Applying AI-based IoT Systems to Simulation-based Information Retrieval*. IGI Global; 2023. pp. 77–89.

32 Hussain W, Rasool N, Khan YD. Insights into machine learning-based approaches for virtual screening in drug discovery. *Curr. Drug. Discov. Technol.* 2021;18(4):463–72.

33 Rudrapal M, Khan J (eds.). *CADD and Informatics in Drug Discovery*. Springer; 2023.

34 Wu Z, Chen S, Wang Y, et al. Current perspectives and trend of computer-aided drug design: a review and bibliometric analysis. *Int. J. Surg.* 2024;110(6):3848–78.

35 Pargaien AV, Rana AJ, Chilwal B et al. The role of artificial intelligence in computer-aided drug design. In: *Challenges in Information, Communication and Computing Technology*. CRC Press; 2025. pp. 524–8.

36 Yuemin B. From computer-aided drug design to artificial intelligence-driven drug design. *Chin. J. Nat.* 2025;47(1):18.

13

Computational Chemistry and Drug Discovery: Intersection of AI, QSAR, and Molecular Modeling

*Purna Nagasree Kurre[1], Murali Krishna Kumar Muthyala[2], and Mallikarjuna Rao Pichika[3,4]**

[1]*Department of Pharmaceutical Chemistry, Avanthi Institute of Pharmaceutical Sciences, Vizianagaram, Andhra Pradesh, India*
[2]*AU College of Pharmaceutical Sciences, Andhra University, Visakhapatnam, Andhra Pradesh, India*
[3]*Faculty of Medical Sciences, Newcastle University, Newcastle upon Tyne, UK*
[4]*Newcastle University Medicine Malaysia, Iskandar Puteri, Malaysia*

*Corresponding author: Mallikarjuna-Rao.Pichika@newcastle.edu.my, pmallirao2000@gmail.com

Introduction

The human body functions through a complex interplay of heterogeneous systems, orchestrated by intricate chemical signaling mechanisms. Health is sustained only when all biochemical pathways operate optimally and in concert. Drugs, typically small organic molecules, are designed to selectively modulate specific biochemical pathways that have become dysregulated, giving rise to pathological conditions [1]. The discovery of a safe and efficacious therapeutic agent is, therefore, a formidable challenge, as it must precisely target a specific biological entity without adversely affecting other physiological processes [2]. Drug discovery is universally acknowledged as a labor-intensive, time-consuming, and economically demanding endeavor. This is primarily due to its high attrition rate and low probability of success. It is estimated that the development of a single clinically approved drug can span 10–15 years and require an investment exceeding USD 1–2 billion [3]. Among the various stages in this process, the identification of a novel lead compound is considered the costliest and failure-prone. Consequently, researchers have long sought to develop rational, knowledge-driven strategies and decision-support systems to enhance the efficiency and cost-effectiveness of drug discovery [4]. Modern drug design methodologies are fundamentally rooted in the principles of pharmacodynamics and molecular recognition. Most therapeutic agents exert their biological effects by binding to specific receptor macromolecules, primarily proteins, and modulating their function [5]. The efficacy of this interaction is governed not only by the drug's concentration at the receptor site but also by its binding affinity.

The three-dimensional architecture of the molecule, including conformational flexibility, dictates its ability to complement the binding pocket. Moreover, the spatial arrangement of functional groups and electronic distribution determines the strength and specificity of noncovalent drug–receptor interactions [6]. From a pharmacokinetic perspective, a drug molecule must successfully traverse multiple biological barriers en route from the site of administration to the site of action. This necessitates the optimization of key physicochemical properties that enable partitioning of drug between diverse biological environments [7]. The ability to achieve and maintain adequate concentrations at the target site is largely determined by the drug's pharmacokinetic profile comprising absorption, distribution, metabolism, and excretion (ADME). Hence, molecular attributes such as lipophilicity, solubility, molecular weight, and ionization state are pivotal in determining both pharmacokinetic behavior and therapeutic efficacy [8]. During its passage from the site of ingestion to the site of action, a drug encounters a multitude of heterogeneous chemical environments. Navigating this journey requires the molecule to possess an optimal combination of physicochemical properties that facilitate smooth transitions between compartments. For instance, drugs targeting the central nervous system must exhibit adequate lipophilicity to cross the highly lipophilic blood–brain barrier [9]. Thus, bioactivity may be conceptualized as a cumulative outcome of various molecular properties or descriptors:

$$\text{Bioactivity of a molecule} = f(\text{Molecular descriptors})$$

This equation highlights the fundamental assumption underlying quantitative structure–activity relationship (QSAR) modeling, where molecular descriptors serve as quantitative proxies for chemical structure and reactivity.

Fundamentals and Challenges in QSAR Model Development

The fundamental principle of QSAR studies is that the biological activity of a molecule is a function of its chemical structure and physicochemical properties. The origins of QSAR can be traced back to Crum-Brown and Fraser (1868), who first proposed a link between molecular structure and biological effect. The field matured significantly in the 1960s with the work of Corwin Hansch, who introduced the *Hansch equation*, correlating biological activity with lipophilicity, electronic properties, and steric effects [10]. Around the same time, Free and Wilson (1964) developed an additive model to quantify the contributions of molecular substituents. Between the 1970s and 1990s, QSAR expanded to include multivariate statistical techniques such as *multiple linear regression* (MLR), *principal component analysis* (PCA), and *partial least squares* (PLS), alongside a growing library of molecular descriptors [11]. The 2000s witnessed the adoption of machine learning (ML) methods, enabling the modeling of nonlinear structure–activity relationships and handling of large, multidimensional datasets [12]. More recently, the advent of *deep learning* (DL), *graph neural networks* (GNNs), and *transfer learning* has revolutionized QSAR by facilitating end-to-end learning from representations like SMILES strings or molecular graphs [13].

QSAR modeling serves two main purposes:

1. **Descriptive QSAR**, which helps elucidate the structural or physicochemical features responsible for observed biological activity.
2. **Predictive QSAR**, which aims to forecast biological activity for novel or untested compounds, even if structurally unrelated to the training dataset.

QSAR Model Development Workflow

QSAR model development follows a structured, multistep workflow combining cheminformatics, statistical modeling, and rigorous validation. The process begins with data collection and curation, where experimental activity data (e.g., IC_{50}, EC_{50}, MIC) are retrieved from databases such as ChEMBL, PubChem, or peer-reviewed literature. Careful data preprocessing is vital to eliminate duplicates, correct structural errors, resolve stereochemistry, normalize units, and ensure consistent molecular representations [14]. The next step is descriptor generation, involving the calculation of numerical features that encode molecular information. These include physicochemical (e.g., log P, molecular weight), topological, geometrical, and quantum-chemical descriptors (e.g., HOMO–LUMO gap). Tools such as PaDEL-Descriptor, Dragon, Mordred, and RDKit are commonly used [11]. Additionally, molecular fingerprints and graph-based embeddings are widely applied in ML- and DL-based QSAR approaches. Feature selection or dimensionality reduction follows, to remove redundant or irrelevant descriptors, which could otherwise lead to overfitting. Techniques such as PCA, *recursive feature elimination* (RFE), genetic algorithms, or mutual information-based filtering are commonly applied [12]. The selected features are then used to train statistical or ML models. Algorithms commonly employed include MLR, support vector machines (SVMs), random forests (RFs), *k*-nearest neighbors (k-NNs), and neural networks. Model performance is evaluated through internal validation (e.g., *k*-fold or leave-one-out cross-validation) to assess robustness and prevent overfitting, and external validation using independent test sets to assess applicability [15, 16]. A critical step in QSAR modeling is the definition of the applicability domain (AD); the chemical space in which the model makes reliable predictions. AD can be defined using leverage-based, distance-based, or probability-density methods [17]. Finally, model transparency, interpretability, and documentation, especially in regulatory contexts, must comply with the *OECD Principles for the Validation of QSAR Models*, which require a defined end point, an unambiguous algorithm, a defined applicability domain, appropriate measures of goodness-of-fit and predictivity, and a mechanistic interpretation when possible [18]. When rigorously applied, this QSAR workflow supports not only the prediction of biological activities but also assists in lead optimization, virtual screening, toxicity prediction, and regulatory submissions.

Challenges in QSAR Model Development

Despite decades of progress, QSAR model development faces several persistent and interrelated challenges:

- **Data quality and heterogeneity:** one of the foremost issues is variability in experimental activity data due to differences in assay protocols, conditions, and reporting standards. These inconsistencies introduce noise and reduce the reliability of the dataset [15].
- **Structural representation errors:** QSAR performance is highly sensitive to accurate molecular representation. Common problems include incorrect stereochemistry, tautomerism, and salt forms, all of which can lead to erroneous descriptors [14].
- **Curse of dimensionality:** modern descriptor sets often contain thousands of features. Without proper feature reduction, this can result in multicollinearity, model overfitting, and poor generalization, especially for small datasets [12].
- **Applicability domain definition:** many QSAR models are deployed without clear AD definitions, leading to unreliable extrapolations. This compromises the trustworthiness of predictions, particularly in safety-critical domains [17].
- **Model interpretability:** the increasing use of "black-box" models such as deep neural networks complicates mechanistic interpretation. While these models may offer superior predictive performance, their lack of transparency poses challenges for regulatory acceptance and scientific insight [18].
- **Data imbalance:** especially in classification tasks (e.g., active vs. inactive compounds), imbalanced datasets can bias models toward the majority class, leading to poor sensitivity for minority classes.
- **Validation and reproducibility:** the absence of standardized validation protocols such as external validation sets, Y-randomization, and consensus modeling continues to undermine reproducibility and scientific rigor [16].
- **Regulatory acceptability:** to gain regulatory approval, QSAR models must satisfy stringent requirements, including model transparency, validation against well-defined end points, and clear applicability domains. Meeting these criteria is still a challenge in many cases [18].

The development of robust, interpretable, and predictive QSAR models demands meticulous attention to data quality, descriptor selection, algorithmic rigor, and model validation. When these elements are well executed, QSAR offers immense potential to accelerate drug discovery, optimize chemical space exploration, and enhance regulatory risk assessment in a cost-effective manner. The increasing complexity of chemical space and biological systems necessitates more sophisticated approaches in QSAR modeling, and this is where artificial intelligence (AI) tools become indispensable. Traditional QSAR techniques, though valuable, are often constrained by linear assumptions, limited descriptor handling capacity, and reduced performance in modeling nonlinear or high-dimensional relationships. AI methods, particularly ML and DL, have revolutionized QSAR by enabling models to uncover intricate, nonlinear patterns between molecular structures and their biological activities across large and diverse datasets [12, 19]. AI tools excel in handling high-dimensional feature spaces, automating feature selection, and integrating diverse data types such as molecular descriptors, fingerprints, three-dimensional (3D) conformations, and even images or text (e.g., SMILES strings). They also enhance the predictive accuracy and scope of models through ensemble learning, neural networks, and transfer learning, especially in low-data scenarios. Moreover, advances in explainable AI (XAI) are addressing the historical challenge of interpretability, enabling researchers to understand *why* a model makes certain predictions critical for regulatory acceptance and mechanistic insight. Given the scale, diversity, and ambiguity of modern chemical and pharmacological data, AI is not merely an enhancement but a necessity for the future of QSAR modeling. It empowers researchers to build faster, more reliable, and more interpretable models, facilitating efficient lead optimization, virtual screening, toxicity prediction, and personalized drug design.

Current ML Trends in QSAR Modeling

DL and Neural Networks

DL architectures, particularly convolutional neural networks (CNNs), GNNs, and recurrent neural networks (RNNs), are increasingly used in QSAR due to their ability to learn complex, nonlinear relationships and eliminate the need for handcrafted descriptors:

- **Graph convolutional networks (GCNs):** directly operate on molecular graphs, capturing atom-level and bond-level features [20].
- **Transformer-based models:** (e.g., ChemBERTa, MolBERT) use SMILES strings as input and have shown high accuracy in molecular property prediction.

End-to-end Modeling with Raw Data

Traditional QSAR models rely on molecular descriptors, but current ML trends favor end-to-end pipelines that accept SMILES, molecular graphs, or images of chemical structures as input, using DL to extract relevant features automatically [21].

AutoML for QSAR

Automated machine learning (AutoML) tools like TPOT, Auto-sklearn, and Google's AutoML are increasingly being adopted for QSAR workflows. These platforms optimize feature selection, model choice, and hyperparameters without manual tuning, democratizing QSAR model building [22].

Multitask and Transfer Learning

QSAR models now increasingly employ multitask learning, where a single model is trained on multiple related end points (e.g., activity against different targets), improving generalization. Transfer learning enables models pretrained on large public datasets (e.g., ZINC, ChEMBL) to be fine-tuned on smaller, task-specific datasets, enhancing performance on data-scarce problems [23].

Explainable AI (XAI) in QSAR

A major focus in current QSAR research is interpretability. Tools such as SHAP (SHapley Additive exPlanations), LIME, and Integrated Gradients are used to explain predictions from black-box models, making them more acceptable in regulatory settings [24].

Uncertainty Quantification and Applicability Domain

New ML-based QSAR models incorporate Bayesian neural networks or Monte Carlo drop-out methods to estimate prediction uncertainty, a key requirement for regulatory and clinical translation [25].

Active Learning and Few-shot QSAR

QSAR is increasingly adopting active learning strategies where models identify which compounds should be tested next to most improve the model. Few-shot learning enables generalization from very small datasets, which is critical for rare diseases or novel targets [23].

Integration with Molecular Docking and Generative Models

Modern QSAR models are now often integrated with docking scores, molecular dynamics, and generative AI models (e.g., VAEs, GANs) to guide de novo compound generation with predicted properties, bridging the gap between virtual screening and lead optimization [26].

AI Tools for Extracting Chemical Structures from Literature

DECIMER.ai (deep learning for chemical image recognition) is an open-source, AI-driven software platform designed to automate the extraction of chemical structure information from scientific publications. This capability accelerates data-driven research in cheminformatics and QSAR modeling. The core principle of DECIMER is the transformation of graphical chemical depictions, commonly found in scanned articles, patents, or legacy documents, into machine-readable formats such as SMILES and InChI. This is achieved through advanced CNNs and transformer-based architectures trained on millions of molecular depictions derived from public databases like PubChem [27, 28]. DECIMER's modular architecture consists of: (1) DECIMER Image Transformer, which translates structure images into SMILES strings using DL; (2) DECIMER Segmentation, which isolates individual chemical depictions from multi-panel figures; and (3) DECIMER.ai Web App, a user-friendly interface that allows users to upload and process documents efficiently [29]. A distinguishing advantage of DECIMER is its integration within optical chemical structure recognition (OCSR) workflows, allowing for the seamless transition from unstructured graphical data to structured cheminformatics pipelines. In the context of QSAR modeling, DECIMER serves a critical upstream role by enabling high-throughput digitization of chemical structures from older publications that lack machine-readable formats. This dramatically increases dataset diversity and size, essential for building robust QSAR models [14]. When combined with cheminformatics tools such as RDKit and PaDEL-Descriptor, DECIMER supports downstream molecular descriptor generation directly from visual sources, thereby expanding the accessible chemical space for model training and validation. Furthermore, DECIMER aligns with FAIR (findable, accessible, interoperable, and reusable) data principles and integrates with platforms such as ChemDataExtractor and Kekule.js, promoting its use in fully automated QSAR and AI-enabled drug discovery workflows. With ongoing improvements in recognition accuracy, multilingual text parsing, and integration with large language models (LLMs), DECIMER.ai is becoming a cornerstone tool for digitizing chemical knowledge and enriching QSAR datasets with otherwise inaccessible structure–activity information.

BioChemInsight is a cutting-edge open-source toolkit that automates the extraction of both chemical structures and associated bioactivity data from scientific literature and patents, significantly accelerating structure–activity relationship (SAR) dataset construction for QSAR modeling. Its architecture combines several key components: OCSR tools like DECIMER Segmentation and MolVec for identifying structures; a vision-language model (Qwen2.5 VL-32B) for linking molecules with identifiers; and natural language processing (NLP) modules such as PaddleOCR with Gemini 2.0 Flash for extracting bioactivity parameters (e.g., IC_{50}, K_i) and normalizing units [30]. In benchmarking studies on 25 patent documents and 17 research articles, BioChemInsight achieved over 95% accuracy in extracting structure–identifier pairs from tabulated formats, around 80% for unstructured formats, and over 92% accuracy in bioactivity extraction. From a QSAR perspective, BioChemInsight is a powerful upstream enabler, rapidly converting unstructured chemical and biological information into structured, high-quality SAR datasets. This automation transforms weeks of manual curation into a matter of hours.

MolMiner is an innovative, open-source OCSR system that leverages advanced DL architectures to extract molecular structures from scientific images and scanned documents. Unlike traditional rule-based OCSR tools, MolMiner utilizes a "You Only Look Once" (YOLO)-based object detection pipeline, in combination with MobileNetV2 and EasyOCR, to identify and classify atoms, bonds, and textual labels directly from visual inputs. After detecting visual elements, a graph assembly algorithm reconstructs them into connected molecular graphs, which are then converted into canonical SMILES or InChI strings. MolMiner also features an auto-generated dataset module that fabricates millions of synthetic chemical structure images, complete with annotations and segmentation masks to train its deep neural networks. This ensures robust performance across diverse publication formats and image qualities. Technically, MolMiner operates through a three-stage pipeline: (1) ImgDet for atom and bond detection, (2) ImgRec for regional recognition, and (3) TextOCR for extracting element labels and supergroup names. It has demonstrated state-of-the-art accuracy and processing speed compared to existing OCSR tools such as MolVec and OSRA, with competitive performance on benchmark datasets like USPTO, UOB, JPO, and CLEF2012 [31]. MolMiner supports batch PDF processing, image snapshots, and real-time editing through Ketcher integration, enabling correction of ambiguous cases [32]. In QSAR model development, MolMiner plays a pivotal role in upstream data assembly. It enables high-throughput digitization of legacy literature, scanned patents, and image-based datasets, transforming previously inaccessible two-dimensional (2D) graphical chemical data into structured representations such as SMILES or InChI. With ongoing enhancements addressing challenges such as crossing bonds, crowded layouts, and Markush structures, along with its user-friendly correction tools, MolMiner is emerging as a transformative asset in cheminformatics. It bridges the gap between unstructured image data and machine-readable chemical information, thereby fostering scalable, reproducible, and high-quality QSAR model development.

MolScribe is a state-of-the-art, open-source OCSR system designed to convert molecular depictions into structured chemical graphs by explicitly predicting atoms, bonds, and their 2D geometries. It employs an image-to-graph DL architecture comprising convolutional and transformer layers, which identify atom nodes and bond edges directly from pixel-based molecular images. A symbolic-chemistry post-processing module ensures chemical plausibility by enforcing valence, stereochemistry, and shorthand expansion rules [33]. To enhance resilience across diverse chemical drawing styles and printed formats, MolScribe utilizes extensive data augmentation during training. It achieves benchmark accuracies ranging between 76 and 93% on both synthetic and real-world datasets, including USPTO, CLEF, and UOB [29, 34]. The system also provides atom-level confidence scores to support chemist-guided validation and error correction. MolScribe's architecture includes segmentation of molecular depictions, followed by processing through a Swin Transformer encoder and a six-layer transformer decoder, which outputs a graph representation of atoms, coordinates, and bond orders. The final symbolic-chemistry module ensures valence and stereochemical validity [35]. Compared to earlier tools such as OSRA or MolVec, MolScribe's graph-based generation pipeline with confidence scoring substantially reduces manual curation workload. In QSAR model development, MolScribe serves as a key tool for automating the extraction of SMILES and InChI from legacy documents and patents. This broadens and diversifies data inputs for descriptor calculation via tools like RDKit and PaDEL, enabling more comprehensive feature space exploration and model robustness [12].

OpenChemIE is an open-source, multimodal chemical information extraction framework that converts embedded content from academic literature, including figures, tables, and text, into structured, machine-readable data formats. Its architecture is organized into two primary phases: (1) extraction of chemical entities at the modality level and (2) integration into unified molecular and reaction representations. Key modules include MolDetect for locating molecular diagrams in images, ChemNER for identifying chemical entities in text, MolScribe (or equivalent OCSR tools) for translating cropped diagrams into SMILES strings, and MolCoref for resolving coreference between textual and graphical elements [36]. These domain-specific modules are orchestrated using chemistry-aware algorithms that reconcile overlapping information and assign contextual roles, e.g., associating reaction schemes with conditions and yields, yielding structured outputs including reactants, products, and experimental metadata. OpenChemIE achieves strong performance metrics: individual modality components deliver state-of-the-art accuracy, while the end-to-end system yields an F_1 score of ~69.5% when benchmarked against annotated reaction schemes and ~64% agreement with the Reaxys database [37, 38]. From a QSAR modeling perspective, OpenChemIE offers substantial upstream benefits. By semi-automatically harvesting structural and activity-relevant information from legacy PDFs, including patented molecules, reaction yields, and assay conditions, it significantly reduces manual effort and expands the diversity of chemical training data. This enables conversion into descriptor-ready formats (e.g., SMILES → RDKit/PaDEL pipelines), enhances dataset coverage, and supports robust feature selection and model validation.

MolMole, introduced by LG AI Research in 2025, is a state-of-the-art vision-based DL framework for automating molecule detection, reaction diagram parsing, and OCSR directly from scholarly document pages [39]. It employs a unified neural architecture capable of processing full scientific pages, locating molecular structures, segmenting reaction schemes, and translating them into structured molecular data (e.g., MOL files) without manual intervention. A key innovation lies in its training dataset: a 550-page benchmark corpus featuring annotated bounding boxes and reaction labels, coupled with a novel evaluation metric tailored to page-level extraction accuracy. MolMole significantly outperforms existing OCSR systems across this benchmark and other public datasets. Technically, MolMole integrates modules for page layout analysis, object detection (molecules, arrows), image segmentation, and structure-to-graph conversion, all trained jointly to ensure cohesive extraction from complex documents. Its output pipeline produces canonical structure files that seamlessly integrate with cheminformatics platforms such as RDKit or PaDEL-Descriptor for downstream QSAR workflows. Within QSAR pipelines, MolMole plays a pivotal upstream role, enabling high-quality structure extraction from image-rich legacy literature and patents. This automated extraction expands chemical diversity, minimizes manual errors, and enhances efficiency in descriptor generation, feature selection, and model validation, thereby supporting more reproducible and OECD-compliant QSAR models.

CurateGPT is a cutting-edge AI-assisted biocuration platform that combines instruction-tuned LLMs with retrieval-augmented generation (RAG) and structured knowledge bases to convert unstructured scientific literature into curated, machine-readable data [40]. The platform employs an agent-based architecture where each agent specializes in tasks such as ontology mapping, entity recognition, and citation linking. These agents interact with the LLM and external databases to accurately extract chemical structures, bioactivity metrics (e.g., IC_{50}, K_i), assay conditions, and other metadata, while providing traceable citations to support data provenance. In QSAR modeling, CurateGPT acts as a powerful data-engineering tool by curating large-scale compound–activity datasets from heterogeneous sources. It automates normalization of chemical identifiers (e.g., SMILES, InChI), standardizes units, and links assay protocols. Ontology-aware agents ensure FAIR-compliant data output [41], which can feed directly into descriptor calculation pipelines (e.g., RDKit, PaDEL-Descriptor). This improves reproducibility, reduces data curation timelines, and enhances QSAR model applicability.

ChemEx is an open-source chemical information extraction system that integrates OCSR and NLP techniques to convert unstructured documents into structured chemical datasets [42]. It uses OSRA to identify chemical structures from images and convert them into SMILES or MDL Molfile formats. Associated compound labels are detected using GOCR and linked via a graph-matching algorithm. Meanwhile, ChemEx's Text Annotator, built on the UIMA framework and powered by the OSCAR4 tokenizer, identifies textual chemical entities, organisms, and bioassay terms. In QSAR modeling, ChemEx streamlines the transformation of legacy literature into descriptor-ready datasets. Its outputs, including curated structures and annotated assay information feed, directly into molecular descriptor calculators like RDKit or PaDEL. This boosts dataset scale, increases chemical and assay diversity, and enhances consistency, crucial for building robust QSAR models.

AutoTemplate is a specialized AI tool that improves the accuracy and consistency of chemical reaction datasets by detecting and correcting common errors that hinder downstream ML tasks. It operates in two stages: (1) deriving generalized reaction templates from large, high-quality corpora using SMARTS representations, and (2) applying these templates to identify and correct erroneous entries (e.g., missing reactants, incorrect atom mappings). SMARTS matching facilitates automatic correction or exclusion of flawed reactions, significantly enhancing dataset integrity. Though originally designed for reaction datasets, AutoTemplate's applications extend to reaction-aware QSAR modeling, where accurate mapping of reactants and products is critical. Its data correction pipeline ensures the fidelity of molecular descriptors, thereby preventing spurious correlations and improving model reproducibility. AutoTemplate contributes meaningfully to feature engineering in reactivity- or metabolism-focused QSAR workflows.

ChemAxon offers a comprehensive suite of cheminformatics and bioinformatics solutions widely used in academic, pharmaceutical, and agrochemical industries. Key components include Marvin for molecular editing, JChem for database and property management, and Chemicalize for high-throughput property prediction [43]. For QSAR modeling, ChemAxon provides descriptor calculation tools such as cxcalc and the Calculator Plugins, which support a wide array of features including log$P$, log$D$, pKa, solubility, polar surface area, and molecular charge distribution. These tools are easily integrated with scripting environments (e.g., Python, Java) and ML workflows, facilitating seamless feature generation and model building. ChemAxon's capacity to manage large datasets and its support for structural variations (e.g., tautomers, Markush structures) are especially valuable in QSAR-based virtual screening and lead optimization. Although ChemAxon is proprietary, potentially limiting access for some academic users, it remains a gold standard in cheminformatics due to its precision, regulatory acceptance, and widespread industrial adoption [44, 45].

AI Tools for Chemical Data Curation

RDKit (Greg Landrum) is a widely adopted open-source cheminformatics toolkit developed in C++ with Python bindings, designed for the analysis, transformation, and representation of chemical structures and related molecular data. It offers a robust suite of functionalities, including substructure searching, molecular fingerprinting, descriptor calculation, structure standardization, and reaction handling [46]. RDKit supports a variety of chemical file formats such as SMILES, SDF, and MOL, and integrates seamlessly with popular scientific Python libraries such as

NumPy, Pandas, and scikit-learn. This interoperability makes RDKit particularly well suited for data science and ML workflows in cheminformatics and drug discovery. In practical applications, RDKit is extensively used for virtual screening, scaffold hopping, QSAR modeling, molecular clustering, and reaction informatics [47]. Its modular architecture and compatibility with Jupyter Notebooks and workflow platforms like KNIME enhance accessibility for both academic researchers and industrial practitioners. The active developer community and comprehensive documentation further support rapid prototyping and reproducible research. As a flexible, extensible, and high-performance framework, RDKit plays a central role in modern computational chemistry and AI-driven molecular design.

DeepChem is an open-source Python library that bridges DL and molecular science by providing a standardized platform for building AI models in chemistry, biology, and materials science [48]. Built atop TensorFlow and PyTorch, DeepChem offers a high-level API tailored for molecular property prediction, protein–ligand binding affinity estimation, virtual screening, and materials discovery. It supports diverse molecular representations, including SMILES strings, molecular graphs, and 3D conformations. DeepChem includes utilities for data preprocessing, molecular featurization, and model evaluation, facilitating rapid development and deployment of ML models. It is particularly notable for its native support of advanced architectures such as GNNs, CNNs for voxel-based 3D representations, and transformer-based sequence models [49]. The platform is widely used in academia and is increasingly adopted in industry for drug discovery, ADMET modeling, and materials optimization. DeepChem's open and extensible design makes it a foundational toolkit in the emerging discipline of molecular machine learning, contributing substantially to the acceleration of AI-enabled drug discovery pipelines and computational chemistry innovation.

ChemDataExtractor is a widely adopted open-source NLP toolkit designed to automate the extraction of chemical information from unstructured scientific literature, patents, and technical documents [50]. Implemented in Python, it combines rule-based grammars with ML algorithms to identify and extract diverse chemical entities and associated properties, such as melting points, log$P$, solubility, and spectral data. In the context of QSAR modeling, ChemDataExtractor serves as a foundational tool for constructing structured datasets from textual sources, enabling the rapid generation of training data for ML models. This capability is especially important in drug discovery and predictive toxicology, where relevant experimental data are often dispersed across publications rather than centralized in databases.

By interfacing with cheminformatics libraries such as RDKit, ChemDataExtractor allows extracted textual data to be converted into molecular descriptors and fingerprints suitable for building robust predictive models. This integration facilitates seamless incorporation of extracted data into cheminformatics workflows. Despite its strengths, ChemDataExtractor does have limitations. Its performance is influenced by the clarity and uniformity of the source texts, and it often struggles with ambiguous syntax, inconsistent units, and complex or nested information, particularly within tables and figures. Moreover, it lacks native functionality for recognizing chemical structures from images, limiting its usefulness for documents rich in graphical content.

To address this gap, tools such as DECIMER [28] and MolScribe [51] have been developed to perform OCSR. These tools leverage deep learning and computer vision techniques to convert chemical diagrams from scanned documents and PDFs into machine-readable formats such as SMILES and InChI. They are particularly effective for digitizing chemical information from historical or non-digitally native sources. In contrast, ChemScraper [37] focuses on extracting structured chemical data from web pages and online databases using web crawling and metadata parsing, offering a complementary strategy for chemical data mining beyond traditional literature.

AlvaDesc is a powerful and widely used software tool for computing molecular descriptors, which are fundamental to the development of QSAR and quantitative structure–property relationship (QSPR) models. Developed by Kode Chemoinformatics, AlvaDesc can calculate over 5,000 descriptors across various categories, including constitutional, topological, geometrical, WHIM, GETAWAY, 3D-MoRSE, and molecular quantum descriptors. Its support for both 2D and 3D molecular structures allows for comprehensive molecular characterization, an essential requirement for building statistically robust and chemically meaningful QSAR models. A key advantage of AlvaDesc is its compatibility with diverse chemical data formats (e.g., SDF, MOL, SMILES), batch processing capabilities for high-throughput descriptor generation, and seamless integration with statistical and ML platforms such as Python, R, and KNIME. These features facilitate efficient and reproducible development of predictive models for various cheminformatics applications. In QSAR modeling, AlvaDesc is particularly useful during descriptor selection and dimensionality reduction. It offers tools for correlation analysis, feature ranking, and redundancy elimination, which are crucial for mitigating overfitting and improving the interpretability and generalizability of models. The software has been extensively applied in QSAR research for drug discovery, toxicity prediction, environmental risk assessment, and materials design [52].

Other Web-based Platforms for Molecular Descriptor Calculation

ChemDes is a free, integrated web-based platform designed to compute molecular descriptors and fingerprints for cheminformatics and QSAR modeling. It integrates multiple well-established cheminformatics toolkits, including RDKit, the Chemistry Development Kit (CDK), PaDEL-Descriptor, and Chemopy [53]. This integration enables users to calculate a comprehensive set of 3,679 molecular descriptors and 59 types of molecular fingerprints via a user-friendly web interface, thereby eliminating the need for advanced programming skills. In addition to descriptor computation, ChemDes offers auxiliary tools for chemical file format conversion, MOPAC-based molecular geometry optimization, and fingerprint similarity analysis. These capabilities enhance the platform's versatility, making it suitable for a wide range of applications in computer-aided drug design, especially in the early stages of QSAR modeling, virtual screening, and lead optimization. ChemDes has proven particularly useful for researchers seeking a centralized, web-accessible solution to generate chemically meaningful data for machine learning-based modeling workflows.

Mordred is a powerful, open-source molecular descriptor calculator written in Python, designed to support QSAR modeling and other cheminformatics applications. Developed by Moriwaki et al. (2018), Mordred is capable of computing more than 1,800 molecular descriptors, including both 2D topological descriptors and 3D geometric descriptors [54]. This makes it one of the most comprehensive descriptor engines available for cheminformatics research. Mordred features a flexible architecture that supports use as a Python library, command-line tool, or through a web interface. It employs RDKit for molecular parsing and basic computation, with additional optimized implementations to enhance performance. Benchmarking results indicate that Mordred operates at least twice as fast as PaDEL-Descriptor and can successfully process larger and more structurally complex molecules that may challenge other descriptor tools. The software supports batch processing, parallel computation, and customizable output formats (e.g., CSV, NumPy arrays, Pandas DataFrames), enabling seamless integration into high-throughput QSAR pipelines. Researchers can tailor descriptor sets to specific modeling tasks and choose whether to incorporate 3D conformational data. In QSAR modeling workflows, Mordred plays a vital role in generating detailed molecular feature matrices used for both regression and classification tasks, such as predicting biological activity, toxicity, solubility, or ADMET properties. Recent studies have demonstrated that Mordred-generated descriptors, either alone or in combination with molecular fingerprints, frequently outperform simpler descriptor sets in modeling challenging end points associated with pharmacologically active compounds [54, 55].

MolFormer is a state-of-the-art, transformer-based molecular language model developed by IBM, designed to learn chemically meaningful representations from SMILES (Simplified Molecular Input Line Entry System) strings using a self-supervised learning framework. Trained on over 1.1 billion molecules, MolFormer employs an efficient linear attention mechanism along with rotary positional embeddings to handle long sequences while maintaining scalability [56]. The core principle of MolFormer lies in masked language modeling: selected tokens in a SMILES string are hidden, and the model is trained to predict them based on context. This allows it to learn both chemical syntax and implicit structural patterns, including atom connectivity, functional motifs, and even 3D spatial relationships despite the inherently one-dimensional nature of SMILES inputs. Through this pretraining, MolFormer generates rich molecular embeddings that can be used in two modes:

1. **Frozen mode**, where embeddings are passed through a downstream neural network.
2. **Fine-tuning mode**, where the entire model is updated based on a specific prediction task.

Empirical benchmarks show that MolFormer consistently outperforms GNN baselines on a variety of molecular property prediction tasks from quantum mechanical properties to physiological end points on datasets such as those in the MoleculeNet benchmark suite. In QSAR and QSPR modeling, MolFormer offers a paradigm shift by replacing handcrafted molecular descriptors with task-specific learned representations. These embeddings act as adaptive descriptors that encode multiscale chemical information, ranging from local substructures to global context, thus eliminating the need for manual feature selection or engineering. As a result, MolFormer facilitates the rapid development of predictive models for ADMET profiling, toxicity prediction, and pharmacokinetic property estimation, especially in scenarios with limited experimental data. The pretrained nature of MolFormer enables few-shot and transfer learning, reducing the demand for large labeled datasets. Moreover, interpretability studies have shown that its embeddings retain spatial and motif-level features, such as ring systems, hydrogen bond donors and acceptors, and electrophilic centers, attributes crucial for mechanistic understanding in QSAR research [56].

DOPtools (descriptors and optimization tools) is a versatile Python-based software library designed for calculating molecular descriptors, performing hyperparameter optimization, and building validated QSPR models. It integrates multiple types of descriptors, including:

- structural descriptors, such as RDKit fingerprints;
- physicochemical descriptors, via the Mordred library;
- reaction-based descriptors, using condensed graphs of reactions (CGRs).

This broad descriptor coverage enables detailed analyses of both molecular structures and chemical reactions. DOPtools features a scikit-learn-compatible API, facilitating seamless integration into ML pipelines for regression and classification tasks. Its modular architecture enables easy incorporation of custom descriptors or ML algorithms, making it highly adaptable to evolving research needs in cheminformatics and QSAR modeling.

PaDEL-Descriptor is a stand-alone, Java-based application designed to compute a wide range of molecular descriptors and fingerprints. It calculates 797 descriptors, including 663 one-dimensional (1D) and 2D descriptors and 134 3D descriptors, as well as 10 types of molecular fingerprints [57].

The software is built upon the **Chemistry Development Kit (CDK)** and extends it by incorporating additional descriptors such as:

- atom-type electro-topological state indices;
- McGowan volume;
- molecular linear free energy relationship (MLFER) descriptors.

PaDEL-Descriptor is commonly integrated into cheminformatics pipelines and platforms, offering batch processing, user-friendly GUI, and scripting support for large-scale descriptor calculation. Its high efficiency and compatibility with diverse file formats (e.g., SMILES, SDF, MOL) make it a preferred choice for researchers working on QSAR/QSPR modeling and chemical database profiling.

Open Babel is one of the earliest and most widely used open-source cheminformatics toolkits, designed to facilitate the interconversion of chemical data across diverse formats. It acts as a "translator" between chemical software, enabling interoperability in computational workflows. Open Babel supports reading and writing over 110 chemical file formats, including SMILES, InChI, MOL, MOL2, SDF, and

CIF, among others [58]. Beyond format conversion, Open Babel provides a rich suite of functionalities that support cheminformatics research and development:

- molecular fingerprint generation for similarity searching and clustering;
- substructure searching using SMARTS patterns;
- molecular mechanics calculations and geometry optimization;
- descriptor calculations, including molecular weight, log$P$, and rotatable bonds.

These features make Open Babel a powerful and flexible tool for tasks such as compound library preparation, virtual screening, and molecular database curation. Its command-line interface and scripting capabilities also enable integration into automated workflows and large-scale batch processing, making it highly suitable for academic and industrial applications alike.

Chemicalize, developed by ChemAxon, is a comprehensive online platform designed for chemical property prediction, structure-based search, and chemical text mining. Users can input molecular structures through various means, including structure drawing, SMILES, InChI, or by uploading structure files, and the system returns a wide array of computed physicochemical properties [59]. In addition to property prediction, Chemicalize supports:

- **Structure-based searching**, enabling users to identify related molecules in large databases.
- **Text-based chemical recognition**, allowing extraction of chemical entities from unstructured documents.
- **Web service integration**, which allow seamless incorporation of its functionalities into third-party applications and custom cheminformatics workflows.

Chemicalize's browser-based interface and programmatic accessibility make it a valuable resource for medicinal chemists, QSAR practitioners, and bioinformaticians seeking fast and reliable chemical analysis without the need for local installations.

Notable AI and Chemometrics Tools

Ai Metrix®

Although primarily used in spectroscopic calibration and process chemistry, Ai Metrix employs an automated pipeline and chemometric rigor that are instructive for QSAR modeling. Its methods, such as robust outlier trimming, factor optimization, and nested cross-validation, mirror QSAR best practices in carefully curating descriptor space to avoid overfitting [4]. By enforcing standardized, reproducible model generation and monitoring, Ai Metrix exemplifies how automation boosts predictive model reliability. However, it lacks native support for molecular descriptor calculation or ML modules tailored to end points such as biological activity or ADMET. To maximize its value in QSAR, Ai Metrix must be combined with cheminformatics platforms (e.g., RDKit, PaDEL-Descriptor) and ML libraries (e.g., scikit-learn, DeepChem), thus creating a hybrid workflow that merges chemometric strength with QSAR-specific predictive analytics.

Chemometric Brain is a cloud-based platform for real-time, nondestructive analysis of complex chemical data, especially from near-infrared (NIR), Raman, and mid-infrared (MIR) spectroscopy. It integrates multivariate statistical methods including PCA, PLS, and discriminant analysis to build predictive calibration models for correlating spectral signatures with chemical or physical properties. The platform features a user-friendly interface, transparent modeling processes, and compatibility with portable spectrometers, enabling real-time model training, validation, and deployment across industrial environments. Applications include pharmaceutical raw-material verification and blend uniformity control [60] and food authenticity testing in olive oil and honey [61]. With support for ISO 17025 and CFR 21 Part 11 compliance, Chemometric Brain is well suited for regulated settings.

The Unscrambler® X, developed by CAMO Software (now AspenTech), is a leading multivariate analysis software used extensively in spectroscopy and process analytics. It supports PCA, PLS regression, and MLR, along with classification models such as SIMCA and PLS-DA, aimed at extracting hidden patterns and constructing predictive models for product quality. The platform features preprocessing tools (e.g., SNV, MSC, derivatives), real-time visualization, and batch modeling, key for Quality by Design (QbD) and Process Analytical Technology (PAT) initiatives. It integrates seamlessly with NIR, Raman, and UV–vis instruments, and is widely used in pharmaceutical development for blend uniformity and material verification [62] and in agricultural quality assessment via hyperspectral imaging.

MLatom 3 is a Python-based platform integrating ML with quantum chemistry, ideal for enhancing QSAR/QSPR modeling with mechanistic depth. It learns quantum observables (e.g., energies, dipoles) via techniques like kernel ridge regression (KRR), Gaussian processes, and neural networks. MLatom 3 supports advanced 3D descriptors such as Coulomb matrices, SOAP, and SLATM and integrates with quantum chemistry tools (e.g., ORCA, Q Chem) for data generation. Its modular pipeline enables automated dataset creation, active learning, cross-validation, uncertainty estimation, and transfer learning, features critical for building reliable predictive models. MLatom 3 has been successfully applied to predict electronic properties like HOMO–LUMO gaps and polarizabilities relevant to bioactivity and reactivity end points [63, 64]. By leveraging high-fidelity quantum descriptors, MLatom 3 enhances chemical space coverage and mechanistic interpretability in AI-driven QSAR workflows.

Current Limitations of AI in QSAR Modeling

Despite the significant advancements brought by AI to QSAR modeling, several limitations still temper its widespread application (Table 13.1). Challenges such as data quality issues, model interpretability, overfitting, and regulatory acceptance remain critical barriers. Additionally, the lack of standardization and the computational demands of modern AI approaches can hinder reproducibility and scalability. These limitations underscore the importance of combining domain knowledge with rigorous model validation and transparency. As the field matures, addressing these constraints will be essential to fully realize the potential of AI in predictive toxicology, drug discovery, and beyond.

Table 13.1 Current limitations of AI in QSAR modeling and potential mitigation strategies.

Limitation	Description	Potential mitigation strategy
Data quality and scarcity	Datasets are often small, noisy, imbalanced, or experimentally inconsistent	Careful curation, data augmentation, and active learning [14]
Lack of interpretability	AI models, especially deep learning, often act as black boxes	Use interpretable models (e.g., SHAP, LIME), hybrid ML–mechanistic approaches
Overfitting and generalizability	Models may perform well on training data but fail on unseen compounds	Employ cross-validation, external test sets, regularization, and dropout
Applicability domain (AD)	AI models often extrapolate outside the chemical space of training data	Use AD estimation techniques (e.g., distance-based, leverage methods)
Descriptor overload	Redundant or irrelevant features can obscure true structure–activity links	Apply feature selection (PCA, RFE) and domain knowledge filtering
High computational demand	Deep models (e.g., GNNs) require significant computational resources	Optimize architectures, use transfer learning, or employ cloud computing
Reproducibility issues	Lack of standardized workflows hinders benchmarking and regulatory adoption	Follow OECD QSAR principles, use open-source tools, and maintain documentation
Ethical and legal concerns	Issues with data ownership, bias, and liability in automated decisions	Ensure data privacy, bias audits, and transparent reporting

Conclusion

The integration of AI into QSAR modeling marks a transformative shift in drug discovery and chemical informatics. Traditional QSAR approaches, though valuable, have often been constrained by limited descriptor selection, oversimplified linear models, and the inability to fully exploit complex chemical–biological relationships. AI, particularly ML and DL, has overcome many of these limitations, enabling the automated learning of nonlinear, high-dimensional, and structure–activity patterns with unprecedented accuracy. Modern AI-based QSAR platforms now leverage advanced techniques such as GNNs, transformer architectures, active learning, transfer learning, and ensemble modeling. These approaches allow for better generalizability, higher predictive performance, and the ability to work effectively with limited or imbalanced datasets. Moreover, tools such as AutoML, XAI, and generative models have democratized model development and interpretation, empowering both chemists and data scientists. Data quality, model interpretability, reproducibility, and regulatory acceptance are persistent concerns. AI systems are only as reliable as the data they are trained on; therefore, curated, standardized, and FAIR (findable, accessible, interoperable, and reusable chemical datasets are foundational to robust model development. Furthermore, transparency in algorithmic decision-making is essential to ensure trust and broader adoption in pharmaceutical settings. Looking forward, the convergence of AI with quantum chemistry, molecular dynamics, and systems biology promises to expand QSAR from molecular prediction to mechanistic insight. Collaborative, open-source ecosystems and interdisciplinary education will be key drivers in shaping the next generation of AI-powered QSAR tools. In conclusion, AI is not just enhancing QSAR modeling but redefining its boundaries. As we move toward a more automated, data-driven, and interpretable future, AI will play an indispensable role in accelerating drug discovery, guiding experimental design, and ultimately improving therapeutic outcomes.

References

1 Rang HP, Dale MM, Ritter JM, et al. *Rang and Dale's Pharmacology*. 8th ed. Elsevier Health Sciences; 2015.
2 Hughes JP, Rees S, Kalindjian SB, Philpott KL. Principles of early drug discovery. *Br. J. Pharmacol.* 2011;162(6):1239–49.
3 Paul SM, Mytelka DS, Dunwiddie CT, et al. How to improve R&D productivity: The pharmaceutical industry's grand challenge. *Nat. Rev. Drug Discov.* 2010;9(3):203–14.
4 Ekins S, Mestres J, Testa B. In silico pharmacology for drug discovery: methods for virtual ligand screening and profiling. *Br. J. Pharmacol.* 2007;152(1):9–20.
5 Liu J, Wang R, Wu W. Molecular interaction field-based virtual screening: understanding structure-activity relationships through molecular interaction patterns. *J. Chem. Inf. Model.* 2007;47(1):166–75.
6 Kitchin DB, Decornez H, Furr JR, Bajorath J. Docking and scoring in virtual screening for drug discovery: methods and applications. *Nat. Rev. Drug. Discov.* 2004;3(11):935–49.
7 Lipinski CA, Lombardo F, Dominy BW, Feeney PJ. Experimental and computational approaches to estimate solubility and permeability in drug discovery and development settings. *Adv. Drug. Deliv. Rev.* 1997;23(1–3):3–25.
8 Waring MJ. Lipophilicity in drug discovery. *Expert Opin. Drug Discov.* 2009;4(4):429–40.
9 Pardridge WM. The blood–brain barrier: bottleneck in brain drug development. *NeuroRx* 2005;2(1):3–14.
10 Hansch C, Fujita T. ρ–σ–π analysis. A method for the correlation of biological activity and chemical structure. *J. Am. Chem. Soc.* 1964;86(8):1616–26.
11 Todeschini R, Consonni V. *Molecular Descriptors for Chemoinformatics*. 2nd ed. Wiley-VCH; 2009.
12 Cherkasov A, Muratov EN, Fourches D, et al. QSAR modelling: where have you been? Where are you going to? *J. Med. Chem.* 2014;57(12):4977–5010.
13 Zhou Z, Kearnes S, Li L, et al. Optimization of molecules via deep reinforcement learning. *Sci. Rep.* 2020;9:10752.

14 Fourches D, Muratov E, Tropsha A. Trust, but verify: On the importance of chemical structure curation in cheminformatics and QSAR modelling research. *J. Chem. Inf. Model.* 2010;50(7):1189–204.

15 Tropsha A. Best practices for QSAR model development, validation, and exploitation. *Mol. Inform.* 2010;29(6–7):476–88.

16 Gramatica P, Cassani S, Chirico N. QSAR modelling of the aquatic toxicity of perfluorinated compounds using theoretical molecular descriptors. *Environ. Int.* 2014;66:150–9.

17 Jaworska J, Nikolova-Jeliazkova N, Aldenberg T. QSAR applicability domain estimation by projection of the training set in descriptor space: a review. *Altern. Lab. Anim.* 2005;33(5):445–59.

18 OECD. *OECD Principles for the Validation, for Regulatory Purposes, of (Quantitative) Structure-Activity Relationship Models.* OECD; 2007.

19 Wang Y, Bryant SH, Cheng T, et al. PubChem BioAssay: 2017 update. *Nucleic Acids Res.* 2020; 45(D1):D955–63. https://doi.org/10.1093/nar/gkw1118

20 Duvenaud D, Maclaurin D, Aguilera-Iparraguirre J, et al. Convolutional networks on graphs for learning molecular fingerprints. In: *Proceedings of the 29th International Conference on Neural Information Process Systems.* vol. 2. Curran Associates; 2015. pp. 2224–32.

21 Irwin R, Wang Y, Renz P, et al. MolFormer: Self-supervised transformer for molecular representation learning. arXiv Preprint, 2022, arXiv:2206.15439. Available from: https://arxiv.org/abs/2206.15439

22 Feurer M, Klein A, Eggensperger K, et al. Efficient and robust automated machine learning. In: *Proceedings of the 29th International Conference on Neural Information Process Systems.* vol. 2. Curran Associates; 2015. pp. 2755–63.

23 Altae-Tran H, Ramsundar B, Pappu AS, Pande V. Low data drug discovery with one-shot learning. *ACS Cent. Sci.* 2017;3(4):283–93.

24 Paul D, Sanap G, Shenoy S, et al. Artificial intelligence in drug discovery and development. *Drug Discov. Today* 2021;26(1):80–93. https://doi.org/10.1016/j.drudis.2020.10.010

25 Scalia G, Grambow CA, Pernici B, et al. Evaluating scalable uncertainty estimation methods for deep learning-based molecular property prediction. *J. Chem. Inf. Model.* 2020;60(6):2697–717.

26 Zhavoronkov A, Ivanenkov YA, Aliper A, et al. Deep learning enables rapid identification of potent DDR1 kinase inhibitors. *Nat. Biotechnol.* 2019;37:1038–40.

27 Rajan K, Zielesny A, Steinbeck C. DECIMER: Deep learning for chemical image recognition. *Sci. Data.* 2021;8(1):1–9.

28 Rajan K, Zielesny A, Steinbeck C. DECIMER: Towards deep learning for chemical image recognition. *J. Cheminform.* 2021;13(1):20.

29 Rajan K, Zielesny A, Steinbeck C. Benchmarking the performance of modern OCSR tools. *J. Cheminform.* 2023;15:36.

30 BioChemInsight: Automated extraction of chemical structures and bioactivities using vision-language models. arXiv Preprint, 2024, arXiv:2403.12345. Available from: https://arxiv.org/abs/2403.12345

31 Guha R, Howard MT, Hutchison GR, et al. The Blue Obelisk Interoperability Project. *J Chem. Inf. Model.* 2011;51(3):614–9.

32 Karulin B, Kozhevnikov M. Ketcher: web-based chemical structure editor. *J. Cheminform.* 2011; 3(Suppl 1):P3. https://10.1186/1758-2946-3-S1-P3

33 Qian Y, Guo J, Tu Z, et al. MolScribe: robust molecular structure recognition with image-to-graph generation. *J. Chem. Inf. Model.* 2023;63(7):1925–34.

34 Rajan K, Brinkhaus HO, Zielesny A, Steinbeck C. Advancements in hand-drawn chemical structure recognition through an enhanced DECIMER architecture. *J. Cheminform.* 2024;16(1):78. https://10.1186/s13321-024-00872-7

35 Dosovitskiy A, Beyer L, Kolesnikov A, et al. An image is worth 16×16 words: Transformers for image recognition at scale. arXiv Preprint, 2021, arXiv:2010.11929. Available from: https://arxiv.org/abs/2010.11929

36 Fan V, Qian Y, Wang A, et al. OpenChemIE: an information extraction toolkit for chemistry literature. *J. Chem. Inf. Model.* 2024;64(14):5521–34.

37 Namdeti R. Artificial intelligence in chemical engineering: past, present, and future perspectives. *J. Chem. Health Risks* 2023;13(6):2051–61.

38 Schneider N, Stiefl N, Landrum GA. ChemScraper: Automating chemical data collection from the web. *J. Chem. Inf. Model.* 2022;62(1):125–35.

39 Chun S, Kim J, Jo A, et al. MolMole: molecule mining from scientific literature. arXiv Preprint, 2025, arXiv:2505.03777v2. Available from: https://doi.org/10.48550/arXiv.2505.03777

40 Caufield H, Kroll C, O'Neil ST, et al. CurateGPT: a flexible language-model assisted biocuration tool. arXiv Preprint, 2024, arXiv:2411.00046. Available from: https://doi.org/10.48550/arXiv.2411.00046

41 Wilkinson MD, Dumontier M, Jan Aalbersberg I, et al. Addendum: the fair guiding principles for scientific data management and stewardship. *Sci. Data.* 2019;6(1):6. http://10.1038/s41597-019-0009-6

42 Tharatipyakul A, Numnark S, Wichadakul D, Ingsriswang S. ChemEx: information extraction system for chemical data curation. *BMC Bioinformatics* 2012;13(Suppl 17):S9. http://10.1186/1471-2105-13-S17-S9

43 ChemAxon. Chemicalize – Predict, search and analyze chemical structures online. 2023. Available from: https://chemicalize.com/

44 Boiteau RM, Hoyt DW, Nicora CD, et al. Structure elucidation of unknown metabolites in metabolomics by combined NMR and MS/MS prediction. *Metabolites* 2018;8(1):8. http:// 10.3390/metabo8010008

45 Tatonetti NP, Ye PP, Daneshjou R, Altman RB. Data-driven prediction of drug effects and interactions. *Sci. Transl. Med.* 2012;4(125):125ra31.

46 Landrum G. RDKit: Open-source cheminformatics. 2013. Available from: http://www.rdkit.org

47 Riniker S, Landrum GA. Open-source platform to benchmark fingerprints for ligand-based virtual screening. *J. Cheminform.* 2015;7:52.

48 Ramsundar B, Eastman P, Walters P, Pande V. *Deep Learning for the Life Sciences.* O'Reilly Media; 2019.

49 Wu Z, Ramsundar B, Feinberg EN, et al. MoleculeNet: a benchmark for molecular machine learning. *Chem. Sci.* 2017;9(2):513–30. http://10.1039/c7sc02664a

50 Swain MC, Cole JM. ChemDataExtractor: a toolkit for automated extraction of chemical information from the scientific literature. *J. Chem. Inf. Model.* 2016;56(10):1894–904.

51 Kleinoeder T, Segler MHS, Heider M. MolScribe: a high-performance deep learning tool for optical chemical structure recognition. *J. Cheminform.* 2023;15(1):6.

52 Gramatica P. Principles of QSAR models validation: internal and external. *QSAR Comb. Sci.* 2007;26(5):694–701.

53 Dong J, Cao D-S, Miao H-Y, et al. ChemDes: An integrated web-based platform for molecular descriptor and fingerprint computation. *Nucleic Acids Res.* 2015; 43(W1):W65–70.

54 Moriwaki H, Tian Y-S, Kawashita N, Takagi T. Mordred: a molecular descriptor calculator. *J. Cheminform.* 2018;10(1):4.

55 Dong J, Wang N-N, Yao Z-J, et al. ADMETLab: a platform for systematic ADMET evaluation based on a comprehensively collected ADMET database. *Chemometr. Intell. Lab. Syst.* 2020;200:104033.

56 Soleimany AP, Amini A, Goldman S, et al. Evidential deep learning for guided molecular property prediction and discovery. *ACS Cent. Sci.* 2021;7(8):1356–67. http://10.1021/acscentsci.1c00546.

57 Yap CW. PaDEL-Descriptor: an open source software to calculate molecular descriptors and fingerprints. *J. Comput. Chem.* 2011;32(7):1466–74.

58 O'Boyle NM, Banck M, James CA, et al. Open Babel: an open chemical toolbox. *J. Cheminform.* 2011;3:33.

59 ChemAxon. ChemAxon Suite Documentation. 2022. Available from: https://chemaxon.com/products

60 Sulub Y, Konigsberger M, Cheney J. Blend uniformity end-point determination using near-infrared spectroscopy and multivariate calibration. *J. Pharm. Biomed. Anal.* 2011;55(3):429–34.

61 Grundy H, Hird H, Bailey-Horne V, et al. Review of current and emerging analytical methods for the testing of edible oil for authenticity. *FSA Res. Evid.* 2025. http://10.46756/001c.145026.

62 Kourti T, MacGregor JF. Process analytical technology in pharmaceutical manufacturing. *J. Process Anal. Control.* 1996;3:87–98.

63 Zaspel P, Bhagavatula V, Delgutte B. Quantum-informed reactivity predictions in drug discovery. *Chem. Sci.* 2019;10:6234–44.

64 Hui W, von Lilienfeld OA. Machine learning electron density from quantum data. *J. Chem. Phys.* 2020;153:044105.

14

CADD in Drug Discovery: From Molecular Modeling to AI-powered Insights

Sweta Kamboj[1], Pankaj Kumar[2], Vandana Chaudhary[3], Rohit Kamboj[1], Shikha Kamboj[1], Rohit Dutt[4], and Kumar Guarve[1]*

[1]*Guru Gobind Singh College of Pharmacy, Yamuna Nagar, Haryana, India*
[2]*Anschutz Medical Campus, University of Colorado, Aurora, CO, USA*
[3]*Antaria Pty. Ltd., Brisbane, QLD, Australia*
[4]*Gandhi Memorial National College, Ambala Cantt, Haryana, India*

*Corresponding author: pankajsharma2707@gmail.com

Introduction

The process of discovering new drugs is an intricate and multifaceted journey, traditionally involving years of research, extensive laboratory work, and significant financial investment. In the past few decades, the integration of computational technologies into pharmaceutical research has transformed this landscape. A potent tool that extends and supplements conventional drug discovery techniques, computer-aided drug design (CADD) enables the quicker, more precise, and more economical creation of new treatments. Considering the recent explosion of technology related to artificial intelligence (AI), the role of computational methods has grown even more profound, providing intelligent systems capable of analyzing massive datasets, predicting molecular behaviors, and even designing new drug candidates with minimal human intervention [1]. Drug discovery is the initial phase in the development of new therapeutic agents. It involves identifying active compounds, or "hits," which can modulate a biological target associated with a disease. These hits are further optimized into "leads" through iterative cycles of synthesis and testing, eventually culminating in a candidate suitable for clinical trials. Traditionally, this process relied heavily on high-throughput screening (HTS) of vast chemical libraries and laborious experimental testing. While effective, this method is resource-intensive, time-consuming, and often limited in its ability to identify truly novel chemical entities. Furthermore, the typical cost of launching a new medication can reach billions of dollars, yet the likelihood of success is still low since biological systems are complicated and can have unanticipated adverse effects [2]. CADD emerged as a solution to these problems by introducing computational methods to expedite several phases of the drug discovery process.

CADD is the process of designing and assessing possible drug candidates using computer tools and algorithms. With the development of structural biology and the capacity to determine the three-dimensional (3D) structures of proteins using methods like X-ray crystallography and NMR (nuclear magnetic resonance) spectroscopy, the strategy gained traction in the 1980s. Key techniques in CADD include:

- molecular modeling and simulation;
- molecular docking;
- QSAR (quantitative structure–activity relationship) modeling;
- pharmacophore modeling;
- molecular dynamics (MD) simulations.

These tools enable researchers to predict how molecules will interact with biological targets, assess their binding affinities, and optimize their chemical structures for better efficacy, safety, and bioavailability. CADD significantly reduces the number of compounds that need to be synthesized and tested, thereby saving time and cost. A revolutionary development in pharmaceutical sciences and biomedical research is the use of AI into medication discovery. Conventional drug development methods are frequently costly, time-consuming, and prone to significant failure rates. It often takes more than 10 years and billions of dollars to bring a new medication to market. AI offers the potential to significantly accelerate this process by automating complex tasks, enhancing data analysis, and predicting outcomes with remarkable accuracy [3].

Fundamentals of CADD

In the field of contemporary pharmaceutical sciences, CADD is an advanced and revolutionary methodology that uses computational tools and techniques to improve and expedite the drug discovery and development process. By offering a logical foundation for drug design, CADD not only expedites the discovery of new therapeutic agents but also lowers expenses and experimental mistakes. In the age of AI and large data, CADD is evolving rapidly, enabling more accurate predictions and integration with other drug discovery tools. This chapter outlines the fundamental principles, workflow, advantages, and limitations of CADD.

Principles and Workflow

The core principle of CADD lies down in its ability to simulate and analyze interaction between small molecules (ligands) and biological targets (typically proteins or nucleic acids). CADD can be broadly categorized into two major strategies.

- **Structure-based Drug Design (SBDD):** utilizes biological target's three-dimensional (3D) structure to create compounds that can attach to it and change its activity.
- **Ligand-based Drug Design (LBDD):** knowledge of molecules that bind to a target, even if the target's structure is unknown, is the foundation of LBDD.

Each approach offers unique strengths and is chosen based on the availability of structural and experimental data [4].

Structure-based Drug Design

SBDD is dependent on the biological target's 3D structure. The first step in the procedure is to resolve the target structure, usually using NMR spectroscopy, cryo-electron microscopy (cryo-EM), or X-ray crystallography. In order to create tiny molecules that can fit into and work well with the target protein's active or binding site, that site is first examined. By forecasting the ideal orientation and conformation of molecules in the binding pocket, SBDD plays a crucial role in discovering high-affinity ligands. This method frequently uses methods like fragment-based design, molecular docking, and MD.

Ligand-based Drug Design

When the 3D structure of the target is unknown, LBDD provides an alternative by relying on information from known active molecules. This method includes examining the molecular characteristics that give rise to the biological activity, such as hydrophobicity, electronic characteristics, and spatial organization. Two popular methods in LBDD include pharmacophore modeling and QSAR research. These models aid in the discovery of novel molecules with comparable activity profiles by establishing a correlation between a compound's chemical structure and biological activities [5].

General Workflow of CADD

The workflow of CADD integrates multiple computational steps to optimize drug discovery systematically. The general workflow can be summarized as follows:

Target identification and validation: finding a pertinent biological target linked to the condition of interest is the first step in the procedure. Bioinformatics tools, genomic data, and disease association studies help validate the target's relevance and druggability.

Structural determination or homology modeling: if the structure of the target is not experimentally available, homology modeling can be used to predict its 3D conformation based on known templates with high sequence similarity. The structural model is essential for further simulation and docking studies.

Molecular modeling and preparation: the structural model undergoes optimization to ensure correct protonation states, tautomeric forms, and charge distributions. Software tools such as Maestro (Schrödinger), MOE, and Chimera are used to prepare both the target and the ligand molecules.

Virtual Screening

- The computer counterpart of high-throughput screening is virtual screening (VS). To find possible "hits," molecules that are likely to bind to the target, a chemical library including hundreds to millions of compounds is computationally screened against the target structure:
  - **Structure-based screening** uses docking algorithms.
  - **Ligand-based screening** uses similarity or pharmacophore searches [6].

Molecular docking: docking entails inserting a ligand into the binding site of the target and utilizing scoring methods to determine the binding affinity. Docking programs like AutoDock, Glide, GOLD, and DOCK offer varying levels of precision and computational cost.

Scoring and ranking: the docked molecules are evaluated and ranked using scoring functions that estimate binding free energy. Top-scoring compounds are shortlisted for further evaluation.

Hit-to-lead optimization: hits identified during screening are subjected to iterative chemical modifications and docking cycles to improve potency, selectivity, and ADMET (absorption, distribution, metabolism, excretion, and toxicity) properties.

MD and free energy calculations: MD simulations are used to evaluate the stability of the ligand–target complex in order to improve predictions. To compute more precise binding free energies, methods such as MM/PBSA and MM/GBSA are employed [7].

In silico ADMET prediction: CADD platforms often include in silico models to evaluate pharmacokinetic and toxicological properties. These models guide chemists in modifying compounds to improve drug-likeness and minimize off-target effects.

Advantages and Limitations of CADD

Despite its broad applicability, CADD is not without challenges. Below is a detailed discussion of its advantages and limitations.

Advantages of CADD

Time and cost efficiency: one of the most significant advantages of CADD is its ability to dramatically reduce the time and cost associated with drug discovery. Traditional drug discovery can take over a decade and billions of dollars, while CADD can screen millions of compounds in silico within days at a fraction of the cost [8].

Rational design approach: CADD promotes a rational design strategy where drug candidates are developed based on an understanding of the target's structure and the underlying disease mechanism. This contrasts with empirical approaches, reducing the trial-and-error element of discovery.

Early elimination of poor candidates: CADD allows for the early prediction of poor ADMET profiles and off-target effects. This helps in eliminating compounds that would likely fail in preclinical or clinical stages, thus saving resources.

Virtual libraries and chemical space exploration: researchers can explore large virtual chemical libraries and even generate novel compounds de novo using AI-integrated platforms. This expands the chemical space far beyond what is feasible in laboratory synthesis.

High hit rate: compared to traditional screening, CADD often yields higher hit rates due to the pre-selection of molecules based on structural complementarity and physicochemical properties.

Integration with AI and machine learning (ML): modern CADD benefits from ML algorithms for predicting binding affinities, toxicity, and pharmacokinetic properties. These tools enhance prediction accuracy and reduce the reliance on exhaustive simulations [9].

Limitations of CADD

Dependency on structural data quality: the success of CADD heavily relies on the accuracy of input data. Poorly resolved crystal structures, incorrect ligand conformations, or inaccurate homology models can lead to misleading results.

Limitations in modeling protein flexibility: many docking algorithms assume a rigid target structure, which doesn't account for induced fit or conformational flexibility. This can reduce the accuracy of binding predictions, especially for allosteric sites.

Challenges with scoring functions: current scoring functions are not perfect. They often fail to accurately predict absolute binding affinities and can produce false positives or negatives. ML-based scoring functions are promising but still under development [10].

Computational cost: while cheaper than experimental screening, accurate simulations (e.g., MD, free energy calculations) are computationally expensive and time-consuming, especially for large biomolecular systems.

Limited predictive power for novel scaffolds: when dealing with novel chemical scaffolds or new biological targets, CADD may not always predict activity accurately due to the lack of representative data in training models.

Incomplete integration with experimental validation: CADD is a powerful predictive tool, but it must be complemented with experimental studies. Overreliance on computational predictions without empirical validation can lead to poor outcomes in clinical development [11].

Molecular Modeling Techniques

A crucial technique in contemporary drug development is molecular modeling. In order to simulate or model the behavior of molecules, a wide range of computer methods are used. Researchers can use these techniques to visualize, predict, and investigate molecular interactions between potential therapeutic molecules and their biological targets. With the exponential growth of computational power and the increasing availability of structural and biochemical data, molecular modeling has become more accurate, efficient, and widely used across both academic and industrial settings. This technique explores three core modeling approaches: SBDD, LBDD, and homology modeling, each of which plays a crucial role in rational drug design [2].

Structure-based Drug Design

SBDD is a logical design approach that finds and optimizes ligands that bind precisely and efficiently to a specified active site by using the 3D structure of biological macromolecules, usually proteins or nucleic acids. The foundation of SBDD lies in the accurate structural characterization of the target protein. High-resolution experimental methods including NMR spectroscopy, X-ray crystallography, and, more recently, cryo-EM are most frequently used to produce these structures [5]. Once the 3D structure is available, computational tools are employed to explore the interaction of small molecules within the target's active site. SBDD's primary benefit is its capacity to direct medicinal chemists in the creation of compounds with enhanced binding affinity and selectivity. It enables detailed visualization of the binding pocket, allowing identification of key residues involved in ligand binding and catalysis. This insight aids in the rational optimization of lead compounds by enhancing their interaction with critical amino acids, thereby improving their potency and therapeutic index. One of the most commonly used tools in SBDD is molecular docking, which predicts the preferred orientation of a ligand when bound to a target and estimates binding affinity using various scoring functions. Docking tools such as AutoDock, GOLD, Glide, and MOE provide a means to virtually screen large libraries of compounds against the target, identifying promising leads for further development. Another important aspect of SBDD is structure-based virtual screening, which applies docking techniques at a large scale. Instead of experimentally testing thousands of compounds, researchers can screen millions of molecules in silico, dramatically accelerating the hit identification phase. Advanced SBDD methods also integrate MD simulations to assess the flexibility of the target and ligand, thus providing a dynamic view of the binding process that complements static docking models. Despite its strengths, SBDD also faces challenges, particularly with targets that exhibit high conformational flexibility or have poorly defined active sites. In such cases, modeling the induced fit upon ligand binding can be difficult. Nevertheless, recent advances in AI and simulation techniques are addressing these limitations by improving the predictive power and accuracy of SBDD models [3].

Ligand-based Drug Design

When the structural details of the biological target are inadequate or unavailable, an alternate approach known as LBDD is employed. Instead of relying on the 3D structure of the target, LBDD utilizes information from known bioactive compounds to predict and design new molecules with similar biological activity. This approach assumes that molecules with similar structural or physicochemical properties will exhibit similar biological behavior. One of the most powerful techniques in LBDD is pharmacophore modeling [12]. A pharmacophore is a theoretical representation of molecular features that are necessary for biological activity, such as hydrogen bond acceptors/donors, hydrophobic centers, aromatic rings, and ionizable groups. By analyzing a set of active compounds, researchers can develop a pharmacophore model that encapsulates the essential features responsible for activity. This model can then be used to screen chemical libraries for new compounds that match the pharmacophore, thereby identifying potential hits. QSAR modeling is another popular technique in LBDD. Creating mathematical models that connect chemical structure to biological activity is an essential part of QSAR. By computing molecular descriptors, numerical values that capture aspects of a molecule's geometry, electronic distribution, hydrophobicity, and more, researchers can build predictive models using statistical or machine learning techniques. These models are then applied to predict the activity of new compounds prior to synthesis and biological testing [13]. When used in conjunction with ADMET predictions, QSAR models have demonstrated a high degree of efficacy in lead chemical optimization. They can help identify modifications to enhance potency, reduce toxicity, and improve pharmacokinetic profiles. However, the reliability of QSAR models depends on the quality of the training data and the diversity of the chemical space covered. Poorly curated or limited datasets can lead to inaccurate predictions and model overfitting. The key advantage of LBDD lies in its applicability even in the absence of target structural information, making it particularly valuable for new or poorly characterized targets. It is also computationally less intensive compared to SBDD, allowing rapid screening and model development. On the downside, LBDD is limited by the need for a robust and diverse dataset of known actives and inactives, and it may not effectively capture novel modes of action or interactions [14].

Homology Modeling

When experimental data is unavailable, homology modeling also referred to as comparative modeling is a structural modeling method used to forecast a protein's 3D structure. The foundation of this approach is the idea that proteins with comparable sequences typically take on comparable 3D structures. As a result, the structure of the target protein (the "query") may be modeled using a homologous protein (a "template") that has a known structure [15]. The process of homology modeling typically involves several key steps:

- **Template identification:** the amino acid sequence of the target protein is searched against structural databases (e.g., PDB) using tools like BLAST to find a suitable template with high sequence similarity.
- **Sequence alignment:** the query sequence is aligned with the template sequence to identify conserved regions and gaps, which are critical for accurate model construction.
- **Model building:** the aligned regions are modeled directly based on the template, while loop regions and insertions/deletions are modeled using either ab initio methods or loop libraries.
- **Model refinement:** the generated model is optimized using energy minimization and, in some cases, MD simulations.
- **Validation:** the final model is evaluated using tools like Ramachandran plots, ProSA, and Verify3D to assess stereochemical quality and structural integrity.

Homology modeling is particularly useful for targets that lack crystallographic or NMR structures but share significant similarity with proteins that have been structurally resolved. It allows researchers to perform docking, virtual screening, and mutation analysis on proteins that otherwise would be inaccessible to structure-based methods. This technique has been widely applied in the discovery of new drugs, especially for G-protein coupled receptors (GPCRs), kinases, and ion channels, many of which have conserved structural features among homologous proteins. Pharmaceutical companies often use homology modeling to build structural models of entire protein families to explore off-target effects and drug selectivity. However, homology modeling comes with limitations [16]. The accuracy of the predicted model depends heavily on the sequence identity between the target and the template. Models built on templates with <30% identity can be unreliable, particularly in loop and active site regions. Moreover, errors in sequence alignment can propagate into the final model, affecting its utility in downstream applications. Despite these limitations, ongoing improvements in modeling algorithms, alignment tools, and integration with AI are enhancing the reliability and usability of homology models. Coupled with structural refinement and validation methods, homology modeling remains a valuable tool in molecular modeling.

Molecular Docking and Virtual Screening

Molecular docking and virtual screening are essential methodologies in CADD, enabling the rapid and economical discovery of bioactive molecules. These methods are pivotal for exploring the interactions between small molecules (potential drugs) and their biological targets (usually proteins), helping researchers understand binding mechanisms and prioritize lead compounds for synthesis and testing. This chapter explores the principles of docking algorithms, scoring functions, real-world applications, and the integration of virtual screening with AI in modern drug discovery workflows [17].

Docking Algorithms and Scoring Functions

Molecular docking is a computer method used to anticipate the optimal orientation of a ligand when it connects with a receptor's active site, with the aim of identifying compounds that exhibit high binding affinity and specificity. The primary goal is to simulate the binding process and evaluate the strength and stability of the resulting complex. Docking involves two main components: the search algorithm, which generates possible binding conformations of ligand, and the scoring function, which evaluates and ranks these poses based on predicted binding affinity.

Docking algorithms: docking algorithms explore the conformational space of the ligand and its possible orientations within the target binding site. They vary in complexity, speed, and accuracy, and can be broadly categorized into rigid docking, where both ligand and receptor are treated as inflexible, and flexible docking, which allows some degree of movement in one or both molecules [18]. Several widely used docking programs implement different algorithmic strategies:

AutoDock: one of the most popular open-source docking tools, AutoDock employs a Lamarckian genetic method to investigate ligand conformations and binding modalities. It allows for partial receptor flexibility and is widely used in both academia and industry.

Glide (from Schrödinger): a commercial software known for its accuracy and efficiency. It employs a hierarchical algorithm combining exhaustive search with energy minimization. Glide offers different precision levels.

GOLD: uses a genetic algorithm and emphasizes flexibility of both ligands and protein side chains. It is particularly effective in exploring multiple binding modes.

DOCK: one of the earliest docking programs developed, it focuses on shape complementarity and uses sphere-based matching techniques. Each algorithm balances computational cost with the quality of results. High-throughput settings may require faster algorithms with less precision, while lead optimization stages may benefit from more detailed and time-consuming approaches [19].

Scoring functions: scoring functions are mathematical models used to estimate the binding affinity of a ligand to its target. They typically incorporate terms that account for:

- van der Waals interactions
- hydrogen bonding
- electrostatic interactions
- desolvation effects
- conformational strain or torsional energy

Scoring functions fall into three main categories:

- Force-field-based scoring functions: these calculate interaction energies using physics-based molecular mechanics. They are accurate but computationally intensive.
- Empirical scoring functions: derived from regression models trained on experimental binding affinities, they are faster but depend on the quality of training data.
- Knowledge-based scoring functions: these are statistical potentials obtained from large structural databases (e.g., PDB). They reflect observed frequencies of interactions.

Recent advances in AI have led to ML-based scoring functions, which outperform traditional methods in specific contexts by learning complex interaction patterns directly from data. Examples include RF-Score, NNScore, and DeepDock. Despite their utility, scoring functions have limitations. They may struggle to accurately rank ligands with similar binding energies or capture entropy and solvation effects comprehensively. As a result, researchers often combine multiple scoring functions or use consensus scoring strategies to improve reliability [20].

Case Studies of Docking in Drug Discovery

Molecular docking has played a crucial role in the discovery and optimization of many FDA-approved drugs. Its application across diverse therapeutic areas illustrates the power of this computational approach.

HIV protease inhibitors: the isolation and development of HIV protease inhibitors was one of the first and most well-reported achievements of structure-based drug design and molecular docking. HIV protease is essential for viral replication, and its structure was among the first to be resolved at high resolution. Docking studies enabled researchers to design inhibitors that fit precisely into the active site, blocking enzymatic activity. Drugs such as saquinavir, indinavir, and ritonavir were developed using iterative cycles of docking, synthesis, and biological testing.

Kinase inhibitors: protein kinases are implicated in various cancers and inflammatory diseases. The ATP-binding pocket of kinases is a well-defined target for small molecule inhibitors. Molecular docking has been instrumental in designing selective inhibitors that exploit subtle differences in the binding sites of different kinases. For instance, imatinib (Gleevec), a BCR-ABL kinase inhibitor used in chronic myeloid leukemia, was developed using structure-guided design principles that relied heavily on docking and modeling [21].

COVID-19 drug repurposing: molecular docking was extensively utilized during the COVID-19 pandemic to quickly virtually screen for SARS-CoV-2 targets, including the major protease (Mpro), spike protein, and RNA-dependent RNA polymerase. These efforts led to the identification of potential repurposing candidates like remdesivir and favipiravir, accelerating the response to the health crisis. While not all docked compounds translated into clinical efficacy, these case studies demonstrated the speed and scalability of docking in emergent situations. Molecular docking has also facilitated the discovery of drugs for Alzheimer's disease (e.g., beta-secretase inhibitors), anti-tuberculosis agents targeting InhA and DprE1 enzymes, and antibiotics against resistant bacterial strains. Its integration into early-stage drug discovery pipelines has now become standard practice across pharmaceutical R&D [22].

VS Applications

The computational analysis of sizable chemical libraries to find compounds that are likely to bind to a particular biological target is known as VS. It significantly lowers the number of chemicals that must be manufactured and examined, making it a high-throughput, economical substitute for experimental screening. VS can be categorized into two main approaches: Structure-based virtual screening (SBVS): uses docking algorithms to evaluate how well each compound in a library fits into the 3D structure of the target binding site. Ligand-based virtual screening (LBVS): involves similarity searching, pharmacophore modeling, or ML based on known active molecules when the structure of the target is not available. The success of virtual screening depends on several factors, including the quality of the target structure, chemical diversity of the library, accuracy of docking algorithms, and reliability of scoring functions. Recent efforts have focused on building curated and diverse libraries such as ZINC, ChemBridge, and Enamine REAL for high-quality VS campaigns [23].

AI in VS

AI is increasingly being integrated into virtual screening workflows to improve hit rates. Deep learning models can be trained to predict binding affinity, drug-likeness, and selectivity based on molecular representations such as SMILES strings, fingerprints, and 3D conformers. Reinforcement learning (RL) and generative models also assist in generating novel compounds that satisfy predefined pharmacological and physicochemical criteria. For example, Deep Docking, a deep learning-based tool developed by Insilico Medicine and collaborators, was able to reduce the virtual screening time of over a billion compounds to a few days by pre-filtering likely binders before docking. Similarly, AtomNet, a CNN model, uses structural data to predict binding and has shown success in prioritizing drug candidates. These AI-powered VS platforms are transforming how pharmaceutical companies approach lead generation, shifting from brute-force screening to more intelligent, targeted exploration of chemical space [10].

Quantitative Structure–Activity Relationship

A key element of computational drug design is QSAR modeling, which uses mathematical models to connect chemistry and biology. QSAR offers a prediction framework for creating novel molecules with desired pharmacological characteristics by quantitatively relating a compound's chemical structure to its biological activities. It saves time, money, and resources in drug development by helping to select compounds for production and testing. As computational power and machine learning techniques have evolved, QSAR has also grown from basic statistical approaches to more complex, data-driven models integrated with AI.

QSAR Modeling Techniques

QSAR is based on the principle that the biological activity of a compound is a function of its molecular structure. To build a QSAR model, chemical compounds are described numerically using molecular descriptors, which represent various physicochemical, topological, geometrical, and electronic features. These descriptors may include hydrophobicity (e.g., log P), molecular weight, polar surface area, counts of specific atoms or functional groups, and 3D shape parameters. The first step in QSAR modeling involves data collection: a dataset of compounds with known biological activity (such as IC_{50}, EC_{50}, K_i, or MIC values) against a particular target. These compounds are then encoded into descriptors using specialized software tools like PaDEL, Dragon, MOE, or RDKit. Next, feature selection algorithms help eliminate redundant or noninformative descriptors, improving model interpretability and performance. The QSAR model is developed using statistical or ML approaches after the descriptors have been determined. The first tools for QSAR were conventional linear regression methods like partial least squares (PLS) and multiple linear regression (MLR). Although these models are straightforward and easy to understand, they could not account for the nonlinear correlations present in biological data [24]. To handle complex interactions, nonlinear modeling techniques are employed. These include:

- Decision trees and random forests: tree-based models that divide data into branches based on descriptor thresholds. Random forests are ensembles of decision trees and are popular for their robustness and accuracy.
- Support vector machines (SVMs): these find hyperplanes that separate active and inactive compounds in a high-dimensional descriptor space, effective in classification-based QSAR.
- *k*-Nearest neighbors (*k*-NN): a straightforward technique that groups compounds according to the descriptor space activities of their closest neighbors.

- Artificial neural networks (ANNs): these models learn intricate, nonlinear correlations between activity and descriptors by simulating the architecture of the brain.
- Deep learning: newer developments in deep neural networks (DNNs), such as recurrent neural networks (RNNs) and convolutional neural networks (CNNs), have enabled modeling from raw molecular representations such as SMILES strings or graphs, reducing dependence on predefined descriptors.

Validation is a critical step in QSAR modeling. Internal validation methods such as cross-validation and bootstrapping, and external validation using an independent test set, are used to assess the reliability of the model. Statistical parameters such as R^2, RMSE, Q^2 (cross-validation coefficient), and external predictive R^2 (R^2_pred) help evaluate model performance. Moreover, applicability domain (AD) analysis defines the chemical space for which a model is reliable. Only compounds that fall within the AD should be considered for prediction to avoid misleading results.

Applications in Lead Optimization

QSAR has found extensive applications in lead optimization, a critical stage in the drug discovery pipeline where promising compounds are refined to improve efficacy, selectivity, and pharmacokinetic profiles while minimizing toxicity. By using QSAR models, researchers can predict how structural changes to a lead molecule will affect its biological activity before synthesis, reducing trial-and-error experimentation. One of the key contributions of QSAR is in structure–activity optimization. After initial hits are identified through screening or virtual screening, QSAR models guide chemists in making structural modifications to enhance binding affinity, potency, or specificity. For example, by identifying molecular features associated with improved target interaction, chemists can add or substitute functional groups to achieve better activity. QSAR is also widely used in predicting ADMET properties. These predictions are invaluable in eliminating compounds with unfavorable pharmacokinetics or safety profiles early in the pipeline. Moreover, multitarget QSAR models can be built to predict activity across different biological targets. This is particularly useful in identifying dual inhibitors or repurposing existing compounds for new therapeutic uses [25]. In polypharmacology, where the goal is to modulate multiple targets simultaneously (e.g., in cancer or CNS disorders), QSAR provides insights into off-target effects and potential drug interactions. QSAR also plays a role in green chemistry and regulatory toxicology. Regulatory agencies such as the OECD and FDA accept validated QSAR models for screening chemicals for environmental toxicity and human risk assessment, especially when in vivo testing is limited or prohibited.

MD Simulations

Strong computer methods for examining the physical motions of atoms and molecules over time are MD simulations. Unlike static models derived from crystallography or docking, MD simulations provide a dynamic view of biomolecular systems, revealing conformational flexibility, transient interactions, and long-timescale behaviors that are crucial in understanding biological function and drug efficacy. In the realm of drug discovery, MD plays an essential role in refining structural hypotheses, validating docking poses, and predicting the stability and dynamics of drug–target complexes [26].

Role in Drug Discovery

In drug discovery, MD simulations are commonly used to observe how small molecules bind to their targets and how the protein–ligand complex behaves under physiological conditions. These simulations can provide insights into induced fit mechanisms, water-mediated interactions, and protein allostery phenomena often missed by rigid-body docking approaches. Moreover, MD can reveal intermediate conformational states that are important for function but may not be observable in crystallographic snapshots. For example, in kinase inhibitors or GPCR ligands, MD simulations help in identifying the "active" vs. "inactive" conformations of the target protein. These structural insights guide medicinal chemists in designing molecules that selectively stabilize desirable conformational states. Furthermore, MD aids in evaluating the binding free energy between the ligand and receptor using advanced post-processing techniques like molecular mechanics Poisson–Boltzmann surface area (MM-PBSA) or molecular mechanics generalized Born surface area (MM-GBSA), providing a more accurate estimation of binding affinity [27].

Force Fields and Simulation Tools

The success of MD simulations depends heavily on the quality of the force fields, which define the potential energy functions governing atomic interactions. Popular force fields include:

- AMBER (Assisted Model Building with Energy Refinement): commonly used in simulating proteins and nucleic acids.
- CHARMM (Chemistry at HARvard Macromolecular Mechanics): suitable for large biomolecular systems with extensive parameter sets.
- GROMOS (GROningen MOlecular Simulation): widely applied in studies of solvent systems and biomolecular dynamics.

These force fields use classical mechanics to calculate the forces and energies based on parameters such as bond lengths, angles, dihedrals, and nonbonded interactions like van der Waals forces and electrostatics. The actual simulations are carried out using specialized software platforms:

- GROMACS: a fast and flexible MD package commonly used in academic research.
- NAMD: suitable for large-scale simulations on parallel computing architectures.
- AMBER Suite: a comprehensive toolkit designed for biomolecular modeling, particularly useful for free energy calculations.

- Desmond: developed by D.E. Shaw Research, known for its high performance and use in drug discovery pipelines. These tools allow researchers to model systems ranging from simple peptide–ligand interactions to complex membrane-bound receptors embedded in lipid bilayers [28].

Introduction to AI And Machine Learning

A kind of AI called ML allows computers to learn from data without explicit programming. To forecast the characteristics of novel compounds, ML models are trained on historical data, including chemical structures and biological activity, in the drug development process. Decision trees, random forests, SVMs, and ensemble approaches are examples of ML techniques [29]. A more complex subset of ML called deep learning models nonlinear interactions using multilayered neural networks. When examining unstructured data, such as pictures, chemical graphs, or sequences, it works very well. Another component of AI is natural language processing, which helps in hypothesis development and medication repurposing by extracting knowledge from patents and clinical trial data; AI's capacity to handle large, complex datasets makes it a valuable tool for integrating multi-omics data, predicting target–drug interactions, and designing new molecules with desired properties [30].

AI-driven Virtual Screening and Predictive Modeling

AI significantly enhances virtual screening, where large chemical libraries are searched for compounds likely to bind a target of interest. Traditional virtual screening relies on docking, but AI-based methods like random forests, SVMs, and neural networks can predict binding affinity directly from molecular descriptors or structural fingerprints. AI models are particularly useful when trained on diverse and high-quality datasets such as ChEMBL or PubChem. These models can prioritize candidates with higher probabilities of activity, improving hit rates and reducing experimental validation costs. Multitask learning, a form of deep learning where a model learns several related tasks simultaneously, is used to predict multiple pharmacological endpoints (e.g., potency, selectivity, toxicity) in parallel. AI also facilitates predictive modeling for physicochemical and ADMET properties. For instance, CNNs can analyze molecular images to predict lipophilicity or solubility, while RNNs can forecast time-dependent phenomena such as drug metabolism or clearance.

Deep Learning (CNNs, RNNs) for Molecular Prediction

Deep learning models have shown superior performance in molecular property prediction tasks. CNNs excel at recognizing spatial patterns and are commonly used for analyzing molecular grids or 3D voxelized data representing molecular structures. They are effective in tasks like protein–ligand binding prediction or molecular docking rescoring. Popular RNN variations include gated recurrent units (GRUs) and long short-term memory (LSTM) networks used in drug discovery for sequence-based property predictions, including bioactivity, toxicity, and metabolic stability. Graph neural networks (GNNs) are another deep learning innovation, particularly powerful for analyzing molecular graphs. GNNs treat atoms as nodes and bonds as edges, learning embeddings that capture the topology and chemistry of molecules. Tools like DeepChem and Chemprop have demonstrated high accuracy in property prediction and scaffold hopping [31].

Generative Models for De Novo Drug Design

De novo drug design, which involves creating completely new molecules with desired features from start, is made possible by generative models, which represent a revolutionary change in drug development. These models learn from existing chemical space and generate novel, synthetically accessible structures that meet predefined pharmacological criteria.

RL Approaches

Reward and penalty-based feedback and agent-environment interactions constitute RL, a subfield of ML. In drug design, RL is used to iteratively optimize molecules by modifying their structure in ways that improve their predicted activity, selectivity, or ADMET properties. For example, the REINVENT framework uses an RNN as a generative model and applies RL to fine-tune it toward molecules with favorable predicted outcomes. The agent is trained to "score" each generated molecule, guiding the system to explore promising regions of chemical space. This method allows for the optimization of multiple objectives simultaneously, such as potency and solubility.

Variational Autoencoders

Generative models called variational autoencoders (VAEs) learn to encode molecules into a continuous latent space from which sampled points may be decoded to produce new molecules. This smooth latent representation allows for interpolation between molecules and systematic exploration of chemical diversity. In practice, VAEs are trained on large datasets like ZINC or ChEMBL, where each molecule is encoded into a compressed vector. During training, the encoder learns to capture meaningful chemical features, while the decoder reconstructs valid SMILES strings from the latent space. VAEs are particularly useful for scaffold hopping and generating analogs of active compounds.

Designing Molecules with Desired Properties

The integration of VAEs and RL provides a powerful framework for property-optimized molecule generation. By combining the latent space representation of VAEs with the reward-based optimization of RL, researchers can design molecules that fulfill specific criteria, including:

- high binding affinity to a particular target;
- good oral bioavailability;

- low toxicity and off-target effects;
- synthetic accessibility.

AI for ADMET and Toxicity Prediction

One of the most critical stages in drug progress is evaluating the ADMET profiles of candidate molecules. Failures due to poor pharmacokinetics or unexpected toxicity account for a large percentage of drug attrition. AI offers a proactive strategy for early ADMET prediction, allowing researchers to prioritize compounds with better chances of clinical success [32].

Pharmacokinetic and Pharmacodynamic Modeling

AI models trained on in vitro and in vivo ADME data can predict pharmacokinetic behaviors such as oral bioavailability, BBB penetration, volume of distribution, and half-life and clearance. These models often use ensemble techniques like random forests, which are robust against noisy data. More recently, deep learning models have been developed to process high-dimensional descriptor sets or even raw chemical representations. In pharmacodynamic modeling, AI is used to predict dose–response curves, receptor binding kinetics, and multitarget activity profiles. These insights inform dosing regimens and therapeutic index estimation, improving drug safety and efficacy.

In Silico Toxicology Tools

Numerous AI-based in silico tools have emerged for toxicity prediction, enabling early screening of compounds for mutagenicity, hepatotoxicity, cardiotoxicity, and other adverse effects. Prominent tools include:

- **DeepTox:** a deep neural network trained on Tox21 datasets to predict multiple toxicity endpoints.
- **ProTox-II:** a web-based platform that predicts toxicity class, LD50, and organ-specific toxicity using ML models.
- **ADMETlab:** an integrated platform offering predictions for over 50 ADMET-related properties. These tools reduce reliance on animal models, support regulatory submissions, and enhance preclinical safety profiling. They are also useful in chemical risk assessment and environmental toxicology [29].

Successful Application of CADD in Drug Discovery and Development

The importance of these strategies in the fields of drug discovery and development has been confirmed by successful programs that used CADD. More than 70 clinically authorized medications were discovered using CADD approaches, according to recent studies. For instance, pharmacophore modeling and SBVS techniques were used to find the medicines norfloxacin for urinary tract infections and isoniazid for TB. Using a combination of molecular docking and MD, vaborbactam was identified as a strong inhibitor of serine β-lactamases in carbapenem-resistant Enterobacteriaceae, and the US FDA authorized it in 2017. Other commercially accessible medications, such as oseltamivir and zanamivar, which are inhibitors of the influenza virus's neuraminidase, were also discovered thanks to CADD technology. Similar to this, other studies have thoroughly examined the contributions of CADD techniques to the development of several medications, including fedratinib hydrochloride (tyrosine kinase inhibitors), axitinib, lorlatinib, adagrasib, entrectinib, zanubrutinib, ritonavir, indinavir, saquinavir, and nelfinavir (blocking HIV-1 protease), and captopril (targeting angiotensin-converting enzymes) [33].

Conclusion and Future Perspectives

Computing power and technology advancements, along with advances in genomics, transcriptomics, proteomics, and metabolomics, are quickly enhancing the potential of CADD to transform the creation of novel medications, particularly antimicrobial treatments. From an "omics" standpoint, CADD can significantly enhance antimicrobial drug discovery by leveraging advanced techniques that identify and prioritize drug targets based on insights into bacterial genomes and metabolic pathways. Methods such as subtractive genomics, which identifies proteins unique to pathogens but absent in the host, structural bioinformatics, which focuses on predicting and analyzing the 3D structures of macromolecules, and metabolic pathway analysis, which links metabolite clusters to cellular signaling pathways, are gaining prominence in tackling antimicrobial resistance. These strategies support the development of sophisticated tools and databases that help researchers select promising protein targets for drug development. Notable resources include TDR Targets, TuberQ, targetTB, Biopcy, BLASTp, BioCyc, and MetaCyc, which assist in identifying potential drug targets. In terms of computational advancements, the integration of quantum computing into drug design and generative chemistry is becoming increasingly relevant. Hybrid quantum-classical ML models can be employed in generative AI systems to design novel chemical compounds with drug-like properties. These compounds can be evaluated for attributes such as uniqueness, potency, metabolic stability, drug-likeness, and safety, enabling faster and more accurate virtual screening, particularly for emerging pathogens and antimicrobial resistance threats. In summary, CADD is poised to become more intelligent and impactful across all stages of drug discovery and development by combining cutting-edge "omics" knowledge, innovative AI algorithms, and increasingly powerful computational capabilities. This evolution marks a transformative step in the pursuit of new and effective therapies.

References

1 Schneider G. Automating drug discovery. *Nat. Rev. Drug Discov.* 2018; 17(2):97–113.

2 Hughes JP, Rees S, Kalindjian SB, Philpott KL. Principles of early drug discovery. *Br. J. Pharmacol.* 2011;162(6):1239–49.

3 Lionta E, Spyrou G, Vassilatis DK, Cournia Z. Structure-based virtual screening for drug discovery: principles, applications and recent advances. *Curr. Top. Med. Chem.* 2014;14(16):1923–38.

4 Lounkine E, Keiser MJ, Whitebread S, et al. Large-scale prediction and testing of drug activity on side-effect targets. *Nature* 2012;486(7403):361–7.

5 Kitchen DB, Decornez H, Furr JR, Bajorath J. Docking and scoring in virtual screening for drug discovery: methods and applications. *Nat. Rev. Drug Discov.* 2004;3(11):935–49.

6 Sliwoski G, Kothiwale S, Meiler J, Lowe Jr EW. Computational methods in drug discovery. *Pharmacol. Rev.* 2014;66(1):334–95.

7 Wang Y, Xiao J, Suzek TO, et al. PubChem: a public information system for analyzing bioactivities of small molecules. *Nucleic Acids Res.* 2009; 37(suppl_2):W623–33.

8 Jumper J, Evans R, Pritzel A, et al. Highly accurate protein structure prediction with AlphaFold. *Nature* 2021;596(7873):583–9.

9 Ekins S, Puhl AC, Zorn KM, et al. Exploiting machine learning for end-to-end drug discovery and development. *Nat. Mater.* 2019;18(5):435–41.

10 Vamathevan J, Clark D, Czodrowski P, et al. Applications of machine learning in drug discovery and development. *Nat. Rev. Drug Discov.* 2019;18(6):463–77.

11 Yu W, MacKerell Jr AD. Computer-aided drug design methods. In: *Antibiotics: Methods and Protocols.* Humana Press; 2016. pp. 85–106.

12 Cherkasov A, Muratov EN, Fourches D, et al. QSAR modeling: where have you been? Where are you going to? *J. Med. Chem.* 2014;57(12):4977–5010.

13 Genheden S, Ryde U. The MM/PBSA and MM/GBSA methods to estimate ligand-binding affinities. *Expert Opin. Drug Discov.* 2015;10(5):449–61.

14 Leelananda SP, Lindert S. Computational methods in drug discovery. *Beilstein J. Org. Chem.* 2016;12(1):2694–718.

15 Cavasotto CN, Orry AJW. Ligand docking and structure-based virtual screening in drug discovery. *Curr. Top. Med. Chem.* 2007;7(10):1006–14.

16 Yuriev E, Holien J, Ramsland PA. Improvements, trends, and new ideas in molecular docking: 2012–2013 in review. *J. Mol. Recognit.* 2015;28(10):581–604.

17 Pagadala NS, Syed K, Tuszynski J. Software for molecular docking: a review. *Biophys. Rev.* 2017;9(2):91–102.

18 Ouma RB, Ngari SM, Kibet JK. A review of the current trends in computational approaches in drug design and metabolism. *Discov. Public Health* 2024;21(1):108.

19 Wang K, Huang Y, Wang Y, et al. Recent advances from computer-aided drug design to artificial intelligence drug design. *RSC Med. Chem.* 2024;15(12): 3978–4000.

20 Wu Z, Chen S, Wang Y, et al. Current perspectives and trend of computer-aided drug design: a review and bibliometric analysis. *Int. J. Surg.* 2024;110(6):3848–78.

21 Chua HM, Moshawih S, Kifli N, et al. Insights into the computer-aided drug design and discovery based on anthraquinone scaffold for cancer treatment: a systematic review. *PLoS One* 2024;19(5):e0301396.

22 Chandrika BR, Subramanian J, Sharma SD. Managing protein flexibility in docking and its applications. *Drug Discov. Today* 2009;14(7–8):394–400.

23 Zhavoronkov A, Vanhaelen Q, Oprea TI. Will artificial intelligence for drug discovery impact clinical pharmacology? *Clin. Pharmacol. Ther.* 2020;107(4):780–5.

24 Chen H, Engkvist O, Wang Y, et al. The rise of deep learning in drug discovery. *Drug Discov. Today* 2018;23(6):1241–50.

25 Gao KY, Fokoue A, Luo H, et al. Interpretable drug target prediction using deep neural representation. *InIJCAI* 2018;2018:3371–7.

26 Olivecrona M, Blaschke T, Engkvist O, Chen H. Molecular de-novo design through deep reinforcement learning. *J. Cheminform.* 2017;9:1–4.

27 Fang Y, Pan X, Shen HB. De novo drug design by iterative multiobjective deep reinforcement learning with graph-based molecular quality assessment. *Bioinformatics* 2023;39(4):btad157.

28 Fralish Z, Chen A, Skaluba P, Reker D. DeepDelta: predicting ADMET improvements of molecular derivatives with deep learning. *J. Cheminform.* 2023;15(1):101.

29 Özçelik R, van Tilborg D, Jiménez-Luna J, Grisoni F. Structure-based drug discovery with deep learning. *ChemBioChem* 2023;24(13):e202200776.

30 Aksamit N, Tchagang A, Li Y, Ombuki-Berman B. Hybrid fragment-SMILES tokenization for ADMET prediction in drug discovery. *BMC Bioinformatics* 2024;25(1):255.

31 Hu X, Liu G, Zhao Y, Zhang H. Activity cliff-aware reinforcement learning for de novo drug design. *J. Cheminform.* 2025;17(1):54.

32 Kant S, Roy S. Artificial intelligence in drug discovery and development: transforming challenges into opportunities. *Discov. Pharm. Sci.* 2025;1(1):1–4.

33 Martinelli DD. Generative machine learning for de novo drug discovery: A systematic review. *Comput. Biol. Med.* 2022;145:105403.

15

From Target to Candidate: A Stepwise Guide to Modern Computational Drug Discovery

*Titirsha Kayal[1], Soumyadip Ghosh[2], Reetika Debroy[3], Sudha Ramaiah[2,4], Anand Anbarasu[2], and Soumya Basu[5]**

[1]*Department of Microbiology, Indira Gandhi Government Medical College and Hospital (IGGMCH), Nagpur, Maharashtra, India*
[2]*Medical and Biological Computing Laboratory, School of Biosciences and Technology (SBST), Vellore Institute of Technology (VIT), Vellore, Tamil Nadu, India*
[3]*Department of Biotechnology, Jaypee Institute of Information Technology, Noida, Uttar Pradesh, India*
[4]*Department of Bio Sciences, School of Bio Sciences and Technology, Vellore Institute of Technology (VIT), Vellore, Tamil Nadu, India*
[5]*Department of International Health, Bloomberg School of Public Health, Johns Hopkins University, Baltimore, MD, USA*

*Corresponding author: sbasu12@jh.edu

Introduction

"One of the discoveries being recognised this year concerns the construction of spectacular proteins. The other is about fulfilling a 50-year-old dream: predicting protein structures from their amino acid sequences. Both of these discoveries open up vast possibilities," Heiner Linke

> *(Chair of the Nobel Committee for Chemistry; during the 2024 announcement of the 2024 Nobel Prize in Chemistry being awarded to David Baker, Demis Hassabis, and John Jumper for their pioneering work in Computational Chemistry).*

Computational drug discovery (CDD) has revolutionized how drugs are discovered today. CDD combines computer models, large amounts of data, and rapid data analysis to quickly identify and optimize potential new medicines. This chapter outlines current and emerging computational drug discovery methods, tools, and workflows that can be used to apply theoretical concepts to drug development and serves as a resource for drug developers to understand the process from target identification to lead optimization. The growth of CDD is characterized by several recent, groundbreaking studies that have demonstrated the interrelationship between computational prediction and experimental validation, which are changing long-standing drug development paradigms, and also helping to address long-standing challenges in drug development. The emergence of the large-scale virtual screening systems illustrates these improvements. An example of such a scale is VirtualFlow, an open-source platform that has allowed for the efficient docking of greater than 1.4 billion compounds, ultimately resulting in the identification of nanomolar-affinity KEAP1 inhibitors, which is a significant accomplishment within the realm of protein–protein interactions [1]. As such, scalable docking will be important as it relates to the exploration of vast chemical libraries, especially when paired with machine learning (ML) models that can predict compound binding affinity and off-target activity. A recent study by Sztain et al. (2025) illustrated the combination of ML models and DNA-encoded library screens to accelerate the identification of active compounds while minimizing dependence on high-resource experimental assays [2]. These advancements also highlight the shift toward what could be described as "gigascale" screening systems utilizing cloud computing and distributed architectures to provide broad accessibility to high-performance drug discovery pipelines [3]. Target identification, a foundational step in CDD, has undergone significant advancements through structural biology and deep learning (DL). Although primarily focused on protein structures, AlphaFold3 has increased the number of potentially druggable targets beyond proteins and other RNA targets [1]. Additional tools such as MIT's DiffDock use diffusion models to predict ligand binding poses at a 38% success rate greater than that obtained using traditional docking models or methods [4]. For poorly characterized targets, integrative approaches that combine molecular dynamics simulation and binding site detection algorithms, such as the University of Central Florida's BindingSite-AugmentedDTA, increase the accuracy of predicted affinities by identifying additional potential cryptic binding sites [4]. These tools are particularly valuable for novel targets, including non-coding RNAs and allosteric sites, where traditional structure-based drug design encounters limitations. The lead optimization process utilizes iterative computational/experimental workflows that optimize for potency, selectivity, and pharmacokinetics. The study published in 2024 by Wu et al. demonstrated one of the major limitations of virtual screening, eliminating "cheating molecules" via improved solvation energy calculations, and provided an example of how physics-based models may be used to improve hit selection [1]. Additionally, covalent drug design has advanced using tools such as those developed by ChemRxiv (2023), which mapped reactive residues and warhead compatibility to reduce the risk associated with off-target effects of irreversible inhibitors [5]. This approach is supported by ML-driven predictive models such as Insilico Medicine's InClinico, which demonstrated a 79% accuracy in predicting clinical trial outcomes based on analysis of over 55,000 phase 2 clinical trials based on multimodal data [4]. Such tools enable researchers to triage compounds earlier in development, mitigating attrition risks. Examples of these types of models allow researchers to prioritize compounds earlier in the development cycle, thus mitigating the risks of compound failure during development. Public databases and cheminformatics tools represent the foundation on which these workflows are constructed. Libraries such as ZINC20 and ChEMBL contain hundreds of millions of synthesizable compounds that are publicly available. Tools such as those described in the 2024 *Frontiers in Drug Discovery* review support ADMET (absorption, distribution, metabolism, excretion, and toxicity) profiling, which provides a means to screen compounds for ADMET prior to synthesis. The review also highlights integrative platform technologies that utilize molecular docking, QSAR modeling, and synthetic accessibility scoring to enable complete optimization cycles [6]. For example, Pharmacia utilized an integrative platform technology to perform a virtual screening campaign for tyrosine phosphatase-1B inhibitors, which resulted in a 35% hit rate, demonstrating the potential for synergistic efficiencies gained when computational and experimental validation approaches are closely coupled [7]. While these advances have significantly enhanced the effectiveness of CDD, there remain several challenges to overcome. The ability to accurately predict interactions between RNA and small molecule ligands remains hindered due to a lack of structural information regarding the complex and dynamic conformations of RNA. Additionally, the lack of interpretability of many deep learning models hinders their adoption in regulated environments [1]. However, the development of hybrid models that combine coarse-grained simulations with machine learning is rapidly closing these knowledge gaps. As computational power increases and model sophistication improves, CDD will move away from its current role as a supporting tool and into a key driver of first-in-class therapeutic agents, especially for under-drugged targets in oncology, neurodegenerative disease, and infectious disease. Through the utilization of open-source tools, collaborative frameworks, and reproducible workflows, future generations of researchers can accelerate this transformation to deliver safer and more effective treatment options for patients.

Historical Perspective and Current Trends: An Overview

The history of computer use in drug discovery has undergone significant transformations over the years. The most notable changes occurred with the introduction of AI/ML and, more recently, quantum computing. Historically, drug discovery was a process based on trial and error that relied heavily on chance or serendipity. Drug discovery can be expensive and take many years to complete without the assurance that a successful drug will result. In the latter part of the 20th century, the rise of computer-assisted drug design (CADD) allowed researchers to utilize the scientific discoveries made in structural biology and to apply this information in order to predict potential drug–target interactions. A major paradigmatic shift occurred when researchers moved from traditional hypothesis-based research methodologies to those that utilize large amounts of data as illustrated in Figure 15.1. With the application of CADD, researchers have been able to rationally design therapeutics [8]. Recent developments involving AI/ML have significantly impacted the field of drug discovery by allowing researchers to rapidly evaluate large amounts of data, design new molecules, and predict the pharmacological characteristics of these molecules; thus, saving time and cost relative to evaluating new drug candidates [9, 10]. Quantum computing represents an emerging technology that will allow researchers to exponentially increase their computational abilities; therefore, it may enable them to model and simulate aspects of molecular chemistry that are currently beyond the limits of classical computational power [8]. As a direct result of these technological

1950 - PRESENT

CADD: A Historical Perspective

Year	Discovery/Advancement	Key Scientists	Notable Awards
1950	Foundation of molecular modeling concepts	Robert Mulliken, Linus Pauling	Nobel Prize in Chemistry (Mulliken, 1966; Pauling, 1954)
1953	Double-helix structure of DNA discovered	James Watson, Francis Crick	Nobel Prize in Physiology or Medicine (1962)
1964	First use of molecular mechanics	Martin Karplus, Michael Levitt	Nobel Prize in Chemistry (2013)
1970	Introduction of quantum chemistry in drug design	Roald Hoffmann, Kenichi Fukui	Nobel Prize in Chemistry (1981)
1976	Concept of molecular docking introduced	Irwin "Tack" Kuntz	
1977	First high-resolution protein structure (myoglobin)	John Kendrew	Nobel Prize in Chemistry (1962)
1981	Application of structure-based drug design (SBDD) for HIV protease inhibitors	Multiple research groups	
1984	Development of molecular dynamics (MD) simulations	Martin Karplus, Levitt, Warshel	Nobel Prize in Chemistry (2013)
1985	Homology modeling introduced	Michael Levitt	Nobel Prize in Chemistry (2013)
1987	Discovery of Captopril through SBDD	David Cushman, Miguel Ondetti	Lasker-DeBakey Clinical Medical Research Award (1999)
1990	Virtual screening for drug discovery	Jürgen Drews, Paul Ehrlich	
1993	Launch of DOCK, first automated docking software	Irwin Kuntz	
1994	Introduction of pharmacophore modeling	Jürgen Drews	
1995	HIV protease crystal structure solved	Stephen Kent, Arthur Olson	
2001	Human Genome Project completed	Francis Collins, Craig Venter	
2002	Integration of high-throughput screening (HTS)	Multiple research groups	
2004	Fragment-based drug discovery (FBDD) introduced	Astex Pharmaceuticals	
2007	In silico ADME prediction tools	Various computational chemists	
2010	Machine learning in drug-target interactions	Andrew Ng, Daphne Koller	
2012	AlphaFold for protein structure prediction	Demis Hassabis, John Jumper	Breakthrough Prize in Life Sciences (2023)
2015	Deep learning for de novo drug design	David Shaw, Geoffrey Hinton	
2018	CRISPR-Cas9 in computational target validation	Jennifer Doudna, Emmanuelle Charpentier	Nobel Prize in Chemistry (2020)
2019	AlphaFold 2 revolutionizes protein folding	Demis Hassabis, John Jumper	Breakthrough Prize in Life Sciences (2023)
2020	AI platforms (Atomwise, Exscientia) predict new drugs	Atomwise, Exscientia	
2021	Quantum computing in molecular simulation	Google AI Quantum Team	
2022	AlphaFold releases nearly all known protein structures	DeepMind	
2023	Generative AI models design novel drug-like molecules	Insilico Medicine, Exscientia	
2024	Computational protein design & structure prediction	David Baker, Demis Hassabis, John Jumper	Nobel Prize in Chemistry (2024)

(Salient achievement years for the advancement of CADD are highlighted in bold)

Figure 15.1 The development of CADD has been a meticulous conglomeration of decades of classical chemistry and computational discoveries, many of which silently paved the way for subsequent breakthroughs that received global accolades.

improvements, the pharmaceutical industry has experienced measurable advantages including lower costs, reduced timelines, and higher success rates for new therapeutic agents. As an example, researchers using CADD and AI/ML have been able to screen hundreds of millions of compounds; thereby, rapidly identifying promising drug candidates and substantially reducing the failure rate during clinical trials [3, 11]. Additionally, there are numerous examples of successful applications of computational methods in drug discovery. For example, researchers have used computational methods to design a class of kinases to treat cancer and also to rapidly repurpose existing drugs to address recent emergent public health threats such as COVID-19 [11]. Central to these achievements is the integration and validation of diverse biological and chemical data, which ensure the reliability and scalability of computational predictions. For instance, in a 2023 review article, researchers attempted to formulate potential therapeutic interventions against biofilm-associated microbial infections and projected a bioinformatics-based approach for deriving treatment solutions. They highlighted the existing web databases and servers, as well as the contemporary computational methods dedicated towards anti-biofilm research, and proposed "knowledge-guided computational pipeline" for favorable discovery of anti-biofilm drug candidates. The proposed pipeline encompasses the prominent methodologies in genomics, transcriptomics, interactomics, and proteomics accompanied by ML–AI algorithmic validations for the successful detection of potent biofilm-inhibitors. They also recommended the *in silico* pipelines to be strengthened by *in vitro* validations for an efficient anti-biofilm drug discovery [12]. The "lab in a loop" strategy, for instance, exemplifies how experimental data from laboratories and clinical studies are used to train and refine AI models, creating a feedback loop that enhances both predictive performance and experimental outcomes [13]. The necessity for rigorous model validation using cross-validation as well as other robust performance metrics is necessary to ensure computational predictions not only accurately predict experimental results but also generalize to real-world applications of computational modeling in drug development [14]. The collaboration between computational innovation and experimental validation will continue to be an important component of future developments in CDD allowing for drug discovery to be faster, more precise and more patient-centric [8, 15].

Core Workflows

Robust target identification and validation workflows, which are foundation to modern computational drug discovery, are supported by large-scale databases and sophisticated computational tools. Recently, scientists have used large-scale screening databases, such as the Library of Integrated Network-based Cellular Signatures (LINCS) to screen tens of thousands of potential drugs at once. In a study published by the University of Cincinnati, the authors demonstrated that an integration of LINCS-based screening with docking simulation could be used to accelerate the drug candidate screening process from months down to just a few hours; therefore, reducing the time it takes to identify new lead candidates [15]. This dual-method approach shows how computational tools and curated databases can synergistically streamline early-stage drug discovery. Furthermore, the increasing adoption of deep learning (DL) for protein structure prediction, is enhancing the accuracy of virtual screens for novel biological targets, especially where traditional ligand discovery methods are limited by sparse chemical information [1, 16]. These kinds of advances are also facilitated by the growing number of publicly available compound libraries, and the growing number of algorithms capable of screening up to a trillion compounds per week, thereby greatly expanding the chemical space available for exploration and hit identification [1, 17, 18]. These advances are changing the way drug discovery is completed by allowing researchers to perform drug discovery more efficiently, through more scalable and data-driven workflows.

Hit Discovery

The hit discovery stage has seen a significant increase with the use of both structural and ligand-based virtual screening along with AI-based methodologies. With the use of the docking method for structural-based virtual screening, hundreds of millions to over a billion compounds are being docked onto protein targets on an ongoing basis. For example, Lyu et al. were able to dock 138 million compounds to the dopamine D4 receptor and 99 million to AmpC beta lactamase while identifying multiple diverse new hits [1]. Additionally, Gorgulla et al. reported a 12% hit rate after screening 1.3 billion compounds against the KEAP1 redox sensor using surface plasmon resonance (SPR) to confirm binding [1, 18]. Ligand-based methods, such as QSAR models and pharmacophores, are being improved through the use of deep learning in low data scenarios. Research recently published in *Nature Computational Science*, in 2024, found that active deep learning methods can improve the hit discovery rate by up to six times relative to traditional methods, even when there is limited experimental data. In addition, the use of AI for drug discovery is revolutionizing the process of clinical translation, with a study showing that the drugs identified by AI and that are entering Phase I clinical trials have an 80–90% success rate vs. a 40–65% success rate for traditionally identified drugs [19]. These research findings clearly demonstrate how the convergence of virtual screening, advanced modeling, and AI is reshaping hit discovery, making it faster, more accurate, and more productive than ever before.

Hit-to-lead Optimization

The hit-to-lead (H2L) stage represents the most crucial step in the transition from an initial hit compound to one that has enhanced efficacy, specificity, and "drug-likeness." High-throughput experimentation (HTE), combined with the application of deep learning and molecular property optimization, can be utilized to accelerate this H2L transition. A recent 2025 study created a large dataset of 13,490 novel C–H alkylations and trained Deep Graph Neural Networks (DGNNs) on this data to predict the accuracy of these chemical reactions. Following scaffold-based enumeration and virtual library screening, 212 potential inhibitors were identified and 14 synthetic ligands were prepared that showed 4,500-fold increases in potency over the initial hits as well as good pharmacology [20]. Another method, Site Identification and Next Choice (SINCHO) uses molecular dynamics (MD) ensembles to take into consideration the dynamic movement of proteins, allowing medicinal chemists to visualize and design modifications to active site functional groups for the purpose of optimizing their H2L transition. The protocol demonstrated high prediction accuracy in both experimentally determined structures and MD ensembles, facilitating rational optimization of lead candidates. These approaches, coupled with *in silico* ADMET prediction tools, enable early identification of compounds with optimal pharmacokinetic and safety profiles, reducing attrition rates and resource expenditure in downstream development [21–23].

Lead Validation: Network Pharmacology

Lead validation increasingly employs network pharmacology to assess the broader biological impact of candidate compounds by mapping their interactions across complex cellular pathways. This systems-level approach enables the identification of key genes, proteins, and pathways involved in disease modulation, as well as potential off-target effects and synergistic interactions. Network pharmacology has been widely used to screen novel targets and drug candidates for infectious and systemic diseases and cancer. Comprehensive analysis of host–pathogen interactions for theragnostic development can also be devised through network-based studies [24–34]. Figure 15.2 gives an inclusive overview of network analyses pipeline with mention of specific tools that are widely used in disease prognosis. A network pharmacology study constructed a compound-target network for 14 active phytocompounds and 283 gene targets, using topological parameters such as network density and centralization to elucidate the complexity and functional properties of the network [35]. The study identified five key genes (EGFR, MAPK1, MAPK3, PTGS2, and ESR1) as central nodes in breast cancer, with in silico docking confirming strong ligand–target interactions. Such analyses provide quantitative and visual insights into the organization and influence of specific nodes within biological networks, supporting the rational prioritization of leads for experimental validation [35]. By integrating multi-omics data and computational modeling, network pharmacology enhances the robustness of lead validation and aligns with the growing emphasis on precision medicine and the development of safer and more effective therapeutics.

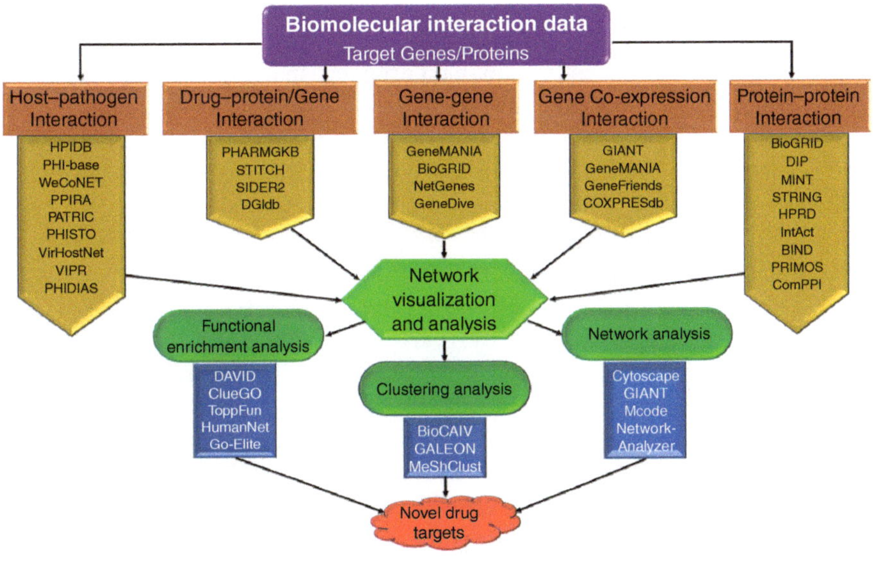

Figure 15.2 The basic network pharmacology workflow.

Databases and Repositories

The foundation of each stage of the process of computational drug discovery is an ecosystem of numerous databases and libraries of chemicals. These libraries and databases provide investigators with the means to access and integrate large quantities of molecular, biological, and pharmacological data. ChEMBL is one of these resources, a database of curated bioactive compounds that have the characteristics of a drug. ChEMBL contains three types of data: chemical, bioactivity, and genomic, all of which are needed to carry out virtual screening, QSAR modeling, and target validation [36]. Another key resource is DrugBank, which contains detailed information about approved drugs and investigational drugs, including the target proteins for those drugs, the mechanism by which they work, and the pharmacokinetic properties of those drugs. This information is important for drug repositioning and understanding the interaction of a drug with its target [37]. PubChem and ChemSpider are two major databases that contain structural, property, and linking information to literature, patents, and biological assays for millions of compounds, making them useful for both early-stage hit identification and cheminformatics analysis [38, 39]. Finally, the Protein Data Bank (PDB) contains a very large collection of 3D structures of proteins that can be used to support structure-based drug design through molecular docking and homology modeling [40]. In addition to the general databases mentioned above, there are specialized databases that add more depth to the landscape of drug discovery. An example of this is BindingDB, which is a curated database of binding affinities of small molecule ligands for macromolecules. This type of data can be useful for identifying inhibitors and carrying out structure–activity relationship (SAR) studies [41]. Additionally, the Chemical Probes Portal is a database of expert-reviewed chemical probes that assists researchers to identify high-quality chemical probes that may otherwise cause problems during biological investigations [42]. For pharmacogenomics and personalized medicine, PharmGKB is another example of a specialized database that helps investigators understand the effect of genetic variation on drug response. Open Targets is a database that integrates genomics and functional data to help researchers prioritize disease relevant targets [43, 44]. Clinical trials databases, such as ClinicalTrials.gov and ClinicalTrialsRegister.eu, are important for tracking the current status and results of clinical trials of investigational drugs. Other specialized databases such as the FDA's Orange Book, SureChEMBL, and PATENTSCOPE provide information regarding drug patents, regulatory status, and intellectual property that are useful for investigators to plan strategically for drug development. For ADMET prediction and safety profiling, tools like SwissADME, ADMETlab 3.0, and vNN-ADMET offer computational assessments of pharmacokinetic and toxicity properties. Collectively, these databases and libraries – ChEMBL, DrugBank, PubChem, ChemSpider, PDB, BindingDB, Chemical Probes Portal, PharmGKB, Open Targets, ClinicalTrials.gov, ClinicalTrialsRegister.eu, FDA Orange Book, SureChEMBL, SwissADME, PATENTSCOPE, ADMETlab 3.0, and vNN-ADMET – form the backbone of modern computational drug discovery, enabling efficient data-driven research and informed decision-making throughout the drug development pipeline [45, 46]. Table 15.1 summarizes all important databases and repositories that are conventionally used.

Table 15.1 Various tools, web servers, software, and databases involved in CDD.

CDD steps	Analysis type or description	Tools/software/platform/database	Accessibility	Features and description	Reference URL
Target identification and validation	Gene–gene interaction, protein–protein interaction, host–pathogen interaction, drug–gene/protein interaction	BioGrid	Free	PPI data with genetic interactions	https://thebiogrid.org/
		InnateDB	Free	Experimentally verified interactions and signaling pathways	https://www.innatedb.com/
		TRRUST v2	Free	A manually curated database of human and mouse transcriptional regulatory networks	https://www.grnpedia.org/trrust/
		DIP	Free	PPI interactions database	https://dip.doe-mbi.ucla.edu/dip/Main.cgi
		MINT	Free	Molecular interaction database	https://mint.bio.uniroma2.it/
		STRING	Free	PPI networks, functional enrichment analysis	https://string-db.org/
		PATRIC	Free	Vital pathogen information with extensive data and analysis tools	https://www.patricbrc.org/
		HPIDB 3.0	Free	Host–pathogen interaction database	https://cales.arizona.edu/hpidb/

(Continued)

Table 15.1 (Continued)

CDD steps	Analysis type or description	Tools/software/platform/database	Accessibility	Features and description	Reference URL
		GeneMania	Free	Predicts the function of different genes and gene sets	https://genemania.org/
		DGIdb	Free for academia and research	Drug–gene interactions and the druggable genome	https://dgidb.org/
		IntAct	Free	Molecular interactions database	https://www.ebi.ac.uk/intact/
		PHARMGKB	Free	Information on how human genetic variation affects response to medications	https://www.pharmgkb.org/
		BindingDB	Free	Chemical and biological information sources and tools for data exploration	https://www.bindingdb.org/rwd/bind/index.jsp
		NetworkAnalyst 3.0	Free and open-source platform	Comprehensive gene expression profiling via network visual analytics	https://www.networkanalyst.ca/NetworkAnalyst/home.xhtml
	Target predictions and drug pathway	Open Targets	Free and open-source tool	Pre-competitive insights for systematic drug target selection	https://www.opentargets.org/
		BindingSite-AugmentedDTA	Open-source	Next-gen pipeline for interpretable prediction models in drug repurposing	https://github.com/yazdanimehdi/BindingSite-AugmentedDTA
		PHAROS	Open-source	Illuminating the druggable genome	https://pharos.nih.gov/

		MetaDrug	Open-source	Information on biological effects of small molecule compounds	https://library.medicine.yale.edu/collections/title/metadrug
		STITCH	Open-source	Known and predicted interactions of chemicals and proteins	http://stitch.embl.de/
		TargetNet	Open web server	Netting or predicting the binding of multiple targets for any given molecule	http://targetnet.scbdd.com/home/index/
		Chem2Bio2RDF	Free	A semantic framework for linking and data mining chemogenomic and systems chemical biology data	http://cheminfov.informatics.indiana.edu/
		DrugCIPHER	Open-source	A linear regression framework to predict drug–target relations	https://github.com/judyboon/drugCIPHER
		DeepDTA	Open-source	A deep-learning-based approach to predict drug–target binding affinity using only sequences of proteins and drugs	https://github.com/hkmztrk/DeepDTA
Protein structure retrieval, prediction, and analysis	Retrieval of 3D protein structure	PDB	Free and open-source	Experimentally determined 3D protein structures	https://www.rcsb.org/
	3D structure prediction and modeling	AlphaFold3	Open-source	Accurate protein structure predictions	https://github.com/google-deepmind/alphafold3
		RoseTTAFold2	Open-source	Protein structure prediction	https://github.com/uw-ipd/RoseTTAFold2

(Continued)

Table 15.1 (Continued)

CDD steps	Analysis type or description	Tools/software/platform/database	Accessibility	Features and description	Reference URL
		ESMFold	Free software	ML-based models to predict 3D protein structures from amino acid sequences	https://github.com/mit-ll/ESMFold
		OmegaFold	Commercial	High-resolution de novo structure prediction from primary sequence	https://cosmic-cryoem.org/tools/omegafold/
		Robetta	Free	Protein structure prediction server	https://robetta.bakerlab.org/
		I-TASSER	Free	Protein structure prediction and structure-based function annotation	https://zhanggroup.org/I-TASSER/
		Swiss-Model	Free	Automated protein structure homology-modeling server	https://swissmodel.expasy.org/
		RaptorX	Free for noncommercial use	Prediction of 3D structures for protein sequences without close homologs in the PDB	http://raptorx.uchicago.edu/
		MODELLER	Free for academic use	Homology or comparative modeling of protein 3D structures	https://salilab.org/modeller/
	3D Protein structure optimization	3Drefine	Free	Protein structure refinement server	https://3drefine.mu.hekademeia.org/
		GalaxyRefine	Free	A web server for protein structure prediction and refinement	https://galaxy.seoklab.org/cgi-bin/submit.cgi?type=REFINE

	YASARA	Requires a license fee	Computer program for molecular visualization, modeling, and dynamics	https://www.yasara.org/
	Swiss-PdbViewer	Free for academic use	Swiss-PdbViewer is an application that provides a user-friendly interface, allowing users to analyze several proteins at the same time	https://spdbv.unil.ch/
3D protein structure validation	PSIPred	Free	Investigates protein structure	https://bioinf.cs.ucl.ac.uk/psipred/
	ProtSAV	Free	A meta-server that assesses the quality of protein structures	https://www.scfbio-iitd.res.in/software/proteomics/protsavzip.jsp
	ProSA-Web	Free	Protein structure analysis	https://prosa.services.came.sbg.ac.at/prosa.php
	SAVES v6.1	Free	Protein structure analysis	https://saves.mbi.ucla.edu/
	DynaMine	Free	Fast predictor of protein backbone dynamics	https://bio2byte.be/dynamine/#
	PDBSum	Free	Atlas of proteins for 3D structural model	https://www.ebi.ac.uk/thornton-srv/databases/pdbsum/
	MolProbity	Free	Structure validation web service	https://phenix-online.org/documentation/reference/molprobity_tool.html
	DISOPRED3	Free	ML-based approach to detect disordered regions	http://bioinf.cs.ucl.ac.uk/disopred

(Continued)

Table 15.1 (Continued)

CDD steps	Analysis type or description	Tools/software/platform/database	Accessibility	Features and description	Reference URL
	Druggability prediction or drug-binding pockets prediction	CASTp 3.0	Free	Computed atlas of surface topology of proteins	http://sts.bioe.uic.edu/castp/index.html?2cpk
		DoGSiteScorer	Free	Binding site prediction	https://proteins.plus/help/dogsite
		PockDrug	Free	Predicts druggability probability for a protein pocket	https://pockdrug.rpbs.univ-paris-diderot.fr/cgi-bin/index.py?page=home
		CavityPlus	Free	Precise and robust protein cavity detection and functional analysis	http://www.pkumdl.cn:8000/cavityplus/index.php#/
		PrankWeb	Free	ML-based method for prediction of ligand-binding sites from protein structure	https://prankweb.cz/
Ligand screening	Selection of ligand library	PubChem	Free	Chemical information in the world's largest free chemistry database	https://pubchem.ncbi.nlm.nih.gov/
		DrugBank	Free	Comprehensive and reliable drug data, structured for immediate use or easy software integration	https://go.drugbank.com/
		IMPPAT 2.0	Free	Manually curated database of Indian medicinal plants, phytochemistry, and therapeutics information	https://cb.imsc.res.in/imppat/

		ZINC15/ZINC20	Free	A database of commercially available compounds for virtual screening and other information	https://zinc15.docking.org/
		ChEMBL	Free	Manually curated database of bioactive molecules with drug-like properties	https://www.ebi.ac.uk/chembl/
		PhytAMP	Free	Plant AMP sequence information and physicochemical or biological data	https://webs.iiitd.edu.in/raghava/satpdb/catalogs/phytamp/
		APD3	Free	Antimicrobial Peptide Database	https://aps.unmc.edu/
	QSAR and other ligand-based models	QSAR toolbox	Free software	It supports reproducible and transparent chemical hazard assessment	https://qsartoolbox.org/
		PaDEL-Descriptor	Open-source	Calculates molecular descriptors and fingerprints	http://yapcwsoft.com/dd/padeldescriptor/
		Chemprop	Open-source	Message passing neural networks for molecule property prediction	https://github.com/chemprop/chemprop
		DeepChem	Open-source	Democratizing deep learning for drug discovery, quantum chemistry	https://github.com/deepchem/deepchem
Hit discovery	Virtual screening of ligands	AutoDock Vina	Open-source	Virtual screening and molecular docking	https://vina.scripps.edu/
		GOLD	Commercial	To predict how small molecules bind to protein targets	https://www.ccdc.cam.ac.uk/solutions/software/gold/

(Continued)

Table 15.1 (Continued)

CDD steps	Analysis type or description	Tools/software/platform/database	Accessibility	Features and description	Reference URL
		MTiOpenScreen	Free	Facilitates the identification of new bioactive compounds for drug discovery and chemical biology	https://bioserv.rpbs.univ-paris-diderot.fr/services/MTiOpenScreen/
		DockBlaster	Free	Structure-based ligand discovery	https://blaster.docking.org/
Chemoinformatics		RDKit	Open-source	Collection of cheminformatics and machine learning software	https://github.com/rdkit/rdkit
		DataWarrior	Free	Data visualization and analysis with chemical intelligence	https://openmolecules.org/datawarrior/
		MOE	Free	Integrated computer-aided molecular design platform	https://www.chemcomp.com/en/Products.htm
Molecular docking	Protein–ligand docking	AutoDock tools	Open-source	Suite of automated docking tools	https://autodock.scripps.edu/
		AutoDock Vina	Open-source	Virtual screening and molecular docking	https://vina.scripps.edu/
		Dock6	Free for academic use	Genetic algorithms and de novo design for fragment-based ligand searching	https://dock.compbio.ucsf.edu/DOCK_6/index.htm
		SwissDock	Free	Predicts the molecular interactions likely to occur between a target protein and a small molecule	https://www.swissdock.ch/
		GNINA	Open-source	A deep learning framework for molecular docking	https://github.com/gnina/gnina

	Glide (Schrödinger)	Commercial	Reliable ligand–receptor docking	https://www.schrodinger.com/platform/products/glide/
	DiffDock	Open-source	Diffusion steps, twists, and turns for molecular docking	https://github.com/gcorso/DiffDock
	FlowDock	Open-source	A geometric flow matching model for generative protein–ligand docking and affinity prediction	https://github.com/BioinfoMachineLearning/FlowDock
	DeepDock	Open-source	A geometric deep learning approach to predict binding conformations of bioactive molecules	https://github.com/OptiMaL-PSE-Lab/DeepDock
Protein–protein docking	ClusPro	Free for noncommercial use	Protein–protein docking	https://cluspro.org/help.php
	HADDOCK 2.4	Free for noncommercial use	High Ambiguity Driven protein–protein DOCKing	https://rascar.science.uu.nl/haddock2.4/
	HDock	Free	Predicts the interaction interface of two interacting proteins or protein–DNA/RNA	http://hdock.phys.hust.edu.cn/
	ZDock	Free for noncommercial use	An automated protein docking server	https://zdock.wenglab.org/
	DockQ	Open-source	Single continuous quality measure for protein, nucleic acids, and small molecule docking models	https://github.com/bjornwallner/DockQ
	FRODOCK	Open-source	Fast Rotational DOCKing	https://github.com/chaconlab/FRODOCK

(Continued)

Table 15.1 (Continued)

CDD steps	Analysis type or description	Tools/software/ platform/database	Accessibility	Features and description	Reference URL
		GRAMM	Free	Template-based protein docking	https://gramm.compbio.ku.edu/
		InterEvDock2	Free	Predicts the structure of protein–protein interactions using evolutionary information	https://bioserv.rpbs.univ-paris-diderot.fr/services/InterEvDock2/
		HawDock	Free	Global macromolecular docking	http://cadd.zju.edu.cn/hawkdock/
HTL optimization	MD and other simulations	GROMACS	Free and open-source	Suite for high-performance MD and output analysis	https://www.gromacs.org/
		AMBER	Commercial	Suite of biomolecular simulation programs	https://ambermd.org/
		NAMD	Free	Parallel MD software for simulating large biomolecular systems	https://www.ks.uiuc.edu/Research/namd/
		DESMOND (Schrödinger)	Free for noncommercial use	GPU-powered high-performance MD engine for simulating biological systems	https://www.schrodinger.com/platform/products/desmond/
		MDweb	Free for academic use	It is based on well-known simulation programs like Amber, NAMD, and GROMACS, and a series of preparation and analysis tools, joined together in a common interface	https://mmb.irbbarcelona.org/MDWeb2/

		Vienna-PTM 2.0	Free	Resource for exploring protein post-translational modifications (PTMs) using MD simulations	http://vienna-ptm.univie.ac.at/
		CABS-flex 3.0	Free	For efficient simulations of protein and peptide structural flexibility	https://lcbio.pl/cabsflex3/
PK/PD analysis	PK/PD modeling	Phoenix WinNonlin	Commercial	Analyze PK and PD data, particularly noncompartmental analysis (NCA)	https://www.certara.com/software/phoenix-winnonlin/
		NONMEM	Commercial	Statistical software package for conducting population PK and PD analyses	https://www.iconplc.com/solutions/technologies/nonmem
		MATLAB-Simbiology	Commercial	Tools for modeling, simulating, and analyzing dynamic systems, focusing on quantitative systems pharmacology, physiologically based pharmacokinetic PK/PD applications	https://www.mathworks.com/products/simbiology.html
		GastroPlus	Commercial	Mechanistically based simulation software package that simulates biopharmaceutics, PK, and PD in humans and animals	https://www.simulations-plus.com/software/gastroplus/
		PKPDsim	Open-source	Library for numerical integration of ODE systems, in particular PK–PD mixed-effects models	https://insightrx.github.io/PKPDsim/
		ADAPT	Free for academic use	A computational modeling platform developed for PK and PD applications	https://bmsr.usc.edu/software/adapt/

(Continued)

Table 15.1 (Continued)

CDD steps	Analysis type or description	Tools/software/platform/database	Accessibility	Features and description	Reference URL
		Nlmxir	Open-source	An R package for population PK–PD modeling	https://nlmixrdevelopment.github.io/nlmixr/
	ADME and toxicity prediction	ADMETlab 3.0	Free	A systematical evaluation of ADMET properties and screening	https://admetlab3.scbdd.com/
		admetSAR 3.0	Free	A comprehensive platform for compound ADMET property search, prediction, and optimization	https://lmmd.ecust.edu.cn/admetsar3/predict.php
		SwissADME	Free	To compute physicochemical descriptors as well as to predict ADME parameters, pharmacokinetic properties, drug-like nature, and medicinal chemistry friendliness	http://www.swissadme.ch/about.php
		pkCSM	Free	Predicting small-molecule pharmacokinetic properties using graph-based signatures	https://biosig.lab.uq.edu.au/pkcsm/
		PreADMET	Free	A web-based application for predicting ADME data and building drug-like library using in silico method	https://preadmet.webservice.bmdrc.org/
		ProTox 3.0	Free	A virtual lab for the prediction of toxicities of small molecules	https://tox.charite.de/protox3/
		ToxinPred	Free	Designing and prediction of toxic peptides	http://crdd.osdd.net/raghava/toxinpred/

		ADMET-AI	Free and open-source	Web interface for predicting the ADMET properties of molecules using machine learning models	https://admet.ai.greenstonebio.com/
AI–ML-based analysis tools	Optimization of small molecules	OpenChem	Open-source	Deep learning toolkit for computational chemistry and drug design research	https://github.com/Mariewelt/OpenChem
		Datamol	Open-source	It simplifies molecular processing and featurization workflows for ML scientists in drug discovery	https://datamol.io/
		ProLIF	Open-source	A Python library used to generate interaction fingerprints for molecular complexes	https://github.com/cbouy/ProLIF
		DiffSBDD	Open-source	A Euclidean diffusion model for structure-based drug design	https://github.com/arneschneuing/DiffSBDD
		TorchDrug	Free	It is a PyTorch-based ML toolbox designed for several purposes in drug discovery	https://torchdrug.ai/
		Chemprop	Open-source	Message passing neural networks for molecule property prediction	https://github.com/chemprop/chemprop
		Targetdiff	Open-source	3D equivariant diffusion for target-aware molecule generation and affinity prediction	https://github.com/guanjq/targetdiff
		DiffLinker	Open-source	Equivariant 3D-conditional diffusion model for molecular linker design	https://github.com/igashov/DiffLinker
	Dynamics of interactions	OpenMM	Open-source	A toolkit for molecular simulation using high-performance GPU code	https://github.com/openmm/openmm

(Continued)

Table 15.1 (Continued)

CDD steps	Analysis type or description	Tools/software/platform/database	Accessibility	Features and description	Reference URL
		TorchMD	Open-source	End-to-end MD engine using PyTorch	https://github.com/torchmd/torchmd
		SAnDReS	Open-source	Computational environment for the development of ML models for the prediction of ligand-binding affinity	https://github.com/azevedolab/sandres
		Paccmann_predictor	Open-source	PyTorch implementation of bimodal neural networks for drug–cell (pharmarcogenomics) and drug–protein (proteochemometrics) interaction prediction	https://github.com/PaccMann/paccmann_predictor
		PLUMED	Open-source	Community-developed library that provides a wide range of different methods, such as enhanced-sampling algorithms, free energy methods, and tools to analyze the vast amounts of data produced by MD simulations	https://www.plumed.org/
	AI/ML generative platforms	REINVENT4	Open-source	AI molecular design tool for de novo design, scaffold hopping, R-group replacement, linker design, and molecule optimization	https://github.com/MolecularAI/REINVENT4

	MolBERT	Open-source	A novel DL framework that combines an effective molecular representation with a pretrained BERT model tailored for molecular property prediction	https://github.com/cxfjiang/MolBERT
	MolGAN	Open-source	An implicit generative model for small molecular graphs	https://github.com/nicola-decao/MolGAN
	InClinico (In silico)	Commercial	Data-driven multimodal forecast of single clinical trial's probability of success	https://pharma.ai/inclinico
	DeepDrive MD	Free	A deep learning-driven adaptive molecular simulation	https://deepdrivemd.github.io/
Quantum computing and hybrid tools	OpenVQE	Open-source	An extension of the Variational Quantum Eigensolver (VQE) for quantum chemistry, building on the Quantum Learning Machine (QLM) and developed using tools	https://github.com/OpenVQE/OpenVQE
	QAOA	Open-source	A flexible Python implementation of the Quantum Approximate Optimization Algorithm/Quantum Alternating Operator Ansatz (QAOA) aimed at researchers to readily test the performance of a new ansatz, a new classical optimizer, etc.	https://github.com/OpenQuantumComputing/QAOA
	Rigetti Forest	Commercial	Future pipeline for molecular dynamics	https://github.com/rigetti/forest-software

Classical and New Tools

CDD incorporates multifaceted approach for selecting and developing novel drug candidates. Unlike the traditional method of discovering drugs via their active ingredients, contemporary methods involve rather interdisciplinary route employing knowledge from biological sciences, chemistry, and computational sciences [47].

Classical Tools in CADD: Molecular Docking and Simulations

Among the many conventional methods in CDD, molecular docking and molecular dynamics simulations (MDS) are two of the most exploited techniques. Molecular docking determines the ideal conformations with which a ligand binds to target to form a stable complex, showing only a static representation of the binding event. Meanwhile, MDS, often performed as a post-docking approach, investigate the dynamic behavior of that complex over a time duration, thereby evaluating the dynamic stability of the docked conformations [48]. Docking tools such as AutoDock, GOLD, and Glide employ physics-based or empirical scoring functions to predict ligand binding poses and estimate affinities. These tools are particularly valuable during the virtual screening and lead optimization phases. While certain assumptions were made to simplify this method (i.e., that the receptor is a rigid body), these methods can be very fast at screening for ligands, especially after experimental validation of results or after refinement using consensus scoring. Atomically detailed MDS provide insights into the interactions between proteins and small molecules, including receptor motion, ligand stability, sampling of different conformations, motions in the binding site, and effects of the solvent on ligand–protein interaction [49–51]. These simulations have been instrumental in identifying binding site plasticity and cryptic pockets, thereby supporting more accurate binding free energy calculations.

Cheminformatics: Structure–Property Integration

Cheminformatics platforms such as RDKit, Open Babel, and MOE facilitate the manipulation and standardization of chemical structures, the generation of molecular descriptors, and the prediction of properties. Their integration with QSAR models enables the automated prediction of pharmacological and ADMET properties, thereby enhancing compound prioritization prior to synthesis. These platforms support rapid property calculations and serve as foundational tools for virtual libraries and ML input [17, 46, 52, 53]. Intensive efforts have been undertaken to optimize CADD pipelines, and a comprehensive framework utilizing phytochemicals for therapeutic breakthroughs has recently been reported; this framework can be universally applied to other compounds and disease targets (Figure 15.3) [17]. This can also be considered as subsequent steps for the systems pharmacology pipeline.

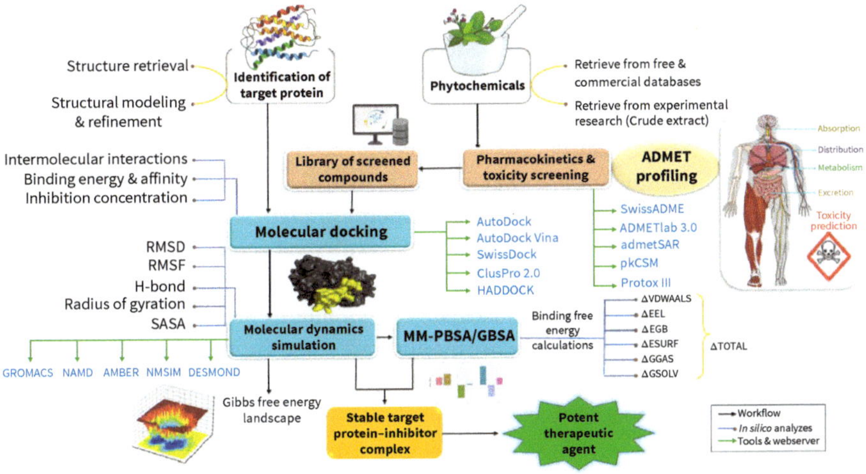

Figure 15.3 *In silico* structure-based drug-design pipeline (here, phytochemicals were used as an exemplary class of potent therapeutic agents). *Source:* [17]. Springer Nature/ CC BY 4.0.

AI and ML Platforms

Contemporary AI and ML models have substantially advanced the processes of hit discovery and lead optimization. Structure-based drug design or rational drug design has not only acted as the starting point toward lead identification for several experiments, but has also boosted up research in synthetic chemistry [31]. Researchers have also referred to this approach as a "human-engineered process with multiple independently optimized steps" [32]. On the other hand, in the modern scenario, integration of state-of-the-art computational techniques and AI-based algorithms for efficient and rapid screening of leads has become a powerful option to save time and resources. This notion was applied in a review article (2003) where authors discuss the conventional and contemporary AI–ML-aided methods and tools to be integrated at various stages for structure-based drug discovery pipelines. For example, AlphaFold 2.0, a DL-based *ab initio* protein structure prediction method has been claimed to be the "solution to a 50-year-old grand challenge in biology." DL-based RoseTTAFold platform has also been promising in the same endeavor for rapid and accurate protein structure and interaction predictions. ML-based methods such as DeepDrug3D, DeepSite, BionoiNet, and DeeplyTough efficiently predict the binding sites in targets; whereas "Deep Docking," a recently developed DL-based platform, has the power to rapidly dock billions of compounds with optimized accuracy and predicts docking scores using deep QSAR models generated from docking scores of a training set compound library [54]. On the contrary, in certain scenarios where 3D structures of macromolecules are unavailable, ligand-based virtual screening has opened new doorways for continuing research in formulating therapeutic interventions instead of conventional methods. This approach works on the "similarity principle," hypothesizing that similar molecules may exhibit similar biological properties [55]. For instance, in an *in silico* study conducted in 2023, authors employed multidisciplinary approach to validate drug targets and identify leads against MDR *Enterobacter cloacae* str. ATCC 13047. Based on network topology, ribosomal subunit S5 (RpsE) protein was selected as a therapeutic biomarker, and its structure was predicted, meanwhile nucleoside analogs were identified as bioavailable drug-like compounds through ligand-based virtual screening against a standard inhibitor molecule. While molecular docking and MDS confirmed the binding efficacy and protein–ligand complex stability, binding free energy (G_{bind}) calculations, principal component (PC) and free energy landscape (FEL) analyses affirmed the hypothesis [56]. AI–ML platforms have been instrumental in escalating the diagnosis of pediatric tuberculosis, which will significantly reduce a global healthcare burden [57]. Similar computational methods were employed in bladder cancer study where a library of disease-specific inhibitors was selected from various databases and was screened for drug-like criteria by a QSAR and pharmacophore model generated from a library of approved drugs. The final drug candidates were subjected to molecular docking and MDS for accessing their interaction profiles with FGFR3 (a favorable drug target in cancer) and complex stability, respectively [58]. ML-based frameworks, including DeepChem, Chemprop, and MolBERT, enhance predictive modeling by utilizing deep neural networks, graph-based learning, and structure–activity relationships derived from extensive datasets to forecast bioactivity, ADMET properties, and retrosynthetic pathways. Transformer-based architectures like MolBERT and graph neural networks (GNNs) represent a scalable alternative to traditional methods used in de novo design, target prediction, and retrosynthesis planning. Improved performance in low-data settings has been achieved with the introduction of transfer learning and active learning methodologies, which are especially advantageous for rare disease-related applications. Companies utilize autoencoders, GNNs, and reinforcement learning agents to survey the chemical space of potential molecules, improve molecular design, and project clinical outcomes for new drugs [16, 59]. Uses of AI increase in predicting toxicological risk via transcriptomics and simulating compound pharmacokinetics, these integrations enable compounds to be ranked and prioritized at an early stage in the drug development process. By doing so, they reduce the need for experimentation and minimize the associated risks.

Cloud-based and Open-source Platforms

The use of present-day platforms such as Google Cloud, AWS, and Azure for High-Performance Computing (HPC), enables scalable drug design using docking and simulation. Platforms such as VirtualFlow, DeepDriveMD, and ColabFold are examples of how drug discovery has been democratized via cloud access and have enabled researchers all over the world to utilize advanced work flows without requiring the purchase or maintenance of sophisticated computer systems. VirtualFlow is a unique, open-source platform that offers a high level of automation and flexibility and has demonstrated significant scale-up capabilities for the automated preparation and screening of very large compound libraries. DeepDriveMD employs deep learning to enable adaptive molecular simulations that provide a quantifiable measure to demonstrate how the combination of deep learning methods with molecular dynamics simulations can improve speed and decrease solution times when using supercomputers. ColabFold accelerates the drug discovery process by making it easier to predict the 3D protein structure rapidly and accurately [60].

CADD and Clinical Breakthroughs

Computational drug discovery has completely changed the way new medicines are developed. It has had positive results in fighting diseases that were thought to be impossible to fight with antibiotics, cancer, and pandemics like COVID-19. These results show that combining high-throughput screening, AI, and omics-based techniques will allow us to discover, develop, and bring new medicines into clinical use much faster than before [3, 6, 8]. In antimicrobial drug discovery, the urgent threat presented by multidrug-resistant (MDR) and extensively drug-resistant (XDR) pathogens has prompted a shift from traditional high-throughput screening methods to more sophisticated, computationally driven strategies [61, 62]. The use of AI-powered virtual screening has been successful in discovering novel antibiotics. For example, using a deep learning model trained on an extensive dataset of antibacterial activity, researchers discovered halicin, a structurally unique antibiotic effective against a broad spectrum of MDR bacteria [62]. This approach also facilitated the repurposing of RS102895 as abaucin, a selective agent against *Acinetobacter baumannii*, and the identification of PHAR(A), a new bacterial growth inhibitor, via *in silico* screening of the ZINC natural product library [61, 62]. Chemical genomic profiling and structure-based optimization also provided researchers with new targets, such as the efflux pump Efpa in *Mycobacterium tuberculosis*, which led to potent inhibitors like BRD-8000.3 [3]. Similarly, the discovery of clovibactin, an antibiotic derived from previously uncultivated microbes, demonstrates how advanced cultivation techniques, solid-state NMR spectroscopy, confocal microscopy, and computational workflows can be synergistically used to elucidate unusual mechanisms of action and resistance evasion [3]. The COVID-19 pandemic was a global testbed for computational drug discovery, leading to rapid innovation and collaboration at a scale not seen before [19, 50, 63]. Researchers utilized molecular docking, molecular dynamics simulations, and AI-driven knowledge mining to screen existing libraries of compounds against SARS-CoV-2 proteins and identify new compounds that may have activity against the same [63]. Signature-based approaches, including connectivity mapping using transcriptomic data, have been used to identify existing drugs that had expression profiles that were antagonistic to viral pathogenesis [63]. The rapid nomination and clinical testing of several agents, including remdesivir and dexamethasone, were direct results of the computational efforts made during the COVID-19 pandemic [64]. The integration of omics data and machine learning classifiers has also facilitated the prioritization of potential drug candidates and identification of key gene modules involved in COVID-19. Furthermore, the integration of computational predictions with experimental validation and open science and data sharing practices established a new standard for research and development in drug discovery [63]. Similarly, cancer drug discovery has also benefited significantly from the use of computational methods, with several examples moving from *in silico* design to clinical trials and regulatory approval [8, 65]. Structure-based drug design and pharmacophore modeling have provided targeted therapies such as gefitinib, erlotinib, sorafenib, lapatinib, abiraterone, and crizotinib. Crizotinib is a selective c-Met/ALK dual inhibitor that was developed using computational methods and approved for the treatment of non-small cell lung cancer (NSCLC) and other cancers [3, 65]. The optimization of SR13668, a compound derived from indole-3-carbinol, using computational models provided strong anticancer effects in phase I clinical trials. Perhaps most notably, the development of BBO-8520 – a first-in-class medication targeting KRAS mutations, which were long considered "undruggable," was achieved through collaboration using supercomputing and AI-driven design. Thousands of compounds have been modeled *in silico*, and only those compounds showing the greatest promise have been selected for human clinical trials, dramatically reducing timelines and costs [3, 8, 65]. These case studies illustrate the revolution in drug discovery being driven by computational methods. The use of AI, high-throughput virtual screening, and integrative omics enables researchers to rapidly identify, optimize, and validate drug candidates, providing patients with safer and more effective treatments at unprecedented speeds. In addition to addressing urgent public health challenges, these advances establish a new era of precision medicine and therapeutic innovation.

Best Practices and Pitfalls

CDD expedites the drug discovery and development process, but it is imperative to adequately address the inherent challenges associated with this approach. One of the most faced obstacles is the lack of experts being proficient in the AI–ML element of the contemporary CDD pipeline. This challenge becomes more evident especially if the AI models generated are far too complex, referred to as "black boxes," and the predictions become exceedingly difficult to interpret [66]. Thus, a skilled workforce is crucial for harnessing advanced *in silico* methods to incorporate multi-omics data into an ideal data-driven systematic drug design pipeline. Hence, targeted recruitment and specialized training programs need to be held for a better understanding of the complex scientific data and making accurate predictions [8]. Modern AI-aided CDD practices are highly dependent on the accuracy of computational models built, for example, several scoring algorithms used for predicting the binding affinity between leads and their targets play pivotal role in drug discovery [67]. Since the key predictions entirely rely on the theoretical models, it is essential to estimate the evaluation matrices and performing cross validations to reduce the number of false positive and false negative results. Prior to model evaluation and validation, a cleaned, well-curated, and preprocessed dataset must be fed to build the training set to ensure accurate predictions. Integration of diverse molecular descriptors, appropriate calibration of scoring parameters, and repeated validations against experimental data can significantly improve the model performance and ensure the results to be reliable [8]. The increasing autonomy of AI models necessitates the implementation of transparency, reproducibility, and accountability measures. It is imperative to integrate frameworks, such as FAIR (Findable, Accessible, Interoperable, and Reusable), and emerging regulatory guidance, including EMA/AI White Papers, into CDD pipelines. Ethical issues include concerns about data privacy, ensuring that ML processes can be replicated, and providing fair access to AI-based tools [68]. Since AI–ML must aid the drug discovery process, overreliance on the model predictions without considering experimental validations can possibly result in misguided efforts. Hence, maintaining an ideal synergy between *in silico* predictions and experimental evidence is crucial for a successful CDD pipeline [69].

Future Directions

CDD involves the use of information technologies for identifying, on a rational ground, potential leads which exhibit a desired set of biological and physicochemical features. While traditional methods focus on the molecular determinants of the ligand–target interactions, current pipelines seem more complex with multiparameter optimization tasks due to the diverse, yet specific, property requirements [70].

Generative AI for Molecule Design

Generative models – VAEs, GANs, and reinforcement learning-based agents – enable de novo molecular generation with tailored pharmacophores [59]. Recent platforms like REINVENT and MolGAN can propose drug-like compounds conditioned on binding site constraints, synthetic accessibility, ADMET profiles and predicted bioactivity. Combining generative outputs with retrosynthetic planning and synthetic feasibility filters is now a standard best practice [71]. Beyond the several computational methods discussed in this chapter, other supporting cutting-edge technologies may help in the modern-day CDD. Language models, such as GPT models like GPT 3.5, may be applied in automating literature reviews, formulating research hypotheses, and deducing complex biological data. For example, T5Chem can be used to study the target–ligand interaction patterns [72].

Quantum Computing: Emerging Frontier

In addition, quantum computing has emerged as a promising field for solving intricate molecular simulations and optimizing troubleshooting problems, as well as recognizing highly complex associations or patterns hidden in the data. Prior to molecular docking, quantum-mechanical (QM) simulations can be utilized to optimize the atomic structures of both macromolecules and ligands, where the electronic interactions are treated quantum mechanically. Thus, QM simulations have proven to be more accurate than force field simulations, which use empirical potentials. There exist a few toolkits, such as a Python-based builder "MCPB.py" and a DL package "DeePMD-kit," that can build force fields from the output data of QM calculations [72]. Quantum computing also promises breakthroughs in the field of CDD, especially in solving Schrödinger's equation for medium-sized drug-like molecules. Quantum ML models have shown early advantages in data-constrained QSAR modeling. Chiang et al. demonstrated that quantum-enhanced classifiers outperformed classical classifiers when the training data and feature dimensionality were reduced, suggesting near-term value in rare disease or orphan indication modeling. Companies like IBM and Rigetti are advancing quantum algorithms such as VQE - and -QAOA for binding free energy prediction. While still early, hybrid quantum-classical algorithms are already outperforming traditional methods in niche simulations [73, 74]. Ideally, both force-field methods and QM simulations can be combined for simulating larger biochemical systems with satisfactory accuracy for diverse atomic interactions. Although QM simulations happen to be more accurate, but it can be less efficient and practical for larger compounds at finite temperature. Hence, the hybrid method called as QM/MM method can be incorporated to provide both aspects, i.e., accuracy and efficiency, by exclusively treating the active sites quantum mechanically while the other regions are treated classically [75]. The integration of AI for QM calculations has been currently initialized and, hence, the routine use of quantum computing in CDD can be foreseen [48].

Toward Fully Integrated AI-powered Pipelines

Visionary drug discovery ecosystems will combine target discovery, virtual screening, de novo design, ADMET profiling, and clinical trial prediction into seamless, end-to-end pipelines integrating AI-powered platforms. These "closed-loop" systems – currently in prototype form at organizations like *in silico* medicine – promise to shrink drug development timelines from years to months. Interoperability between cheminformatics, bioinformatics, and health informatics will define next-generation precision drug design [76].

Conclusion

CDD has evolved from being an additional tool used within the drug discovery process to one of the key tools used today as a leading driver of innovation. This chapter provides an overview of all of the major components involved in the process of drug design from identifying targets to validating leads, and it illustrates how molecular docking and dynamics, when combined with ML, AI, generative models, and quantum computing, will allow researchers to design drugs more effectively than ever before. The integration of virtual screening platforms, AI-enhanced pharmacokinetic modeling, and large databases of known compounds has greatly decreased the time required to find potential hits using virtual screening methods and has increased the number of potential hits found per search. The use of cloud-based infrastructure has made access to the same CDD workflow available to researchers globally and has enabled global collaborations between researchers. The integration of AI, quantum computing, and cloud-based infrastructure will enable a complete transformation of therapeutic R&D. AI-enabled pipelines designed to perform de novo design, ADMET profiling, and trial simulations will propel the drug discovery process toward a predictive, personalized model. It further illustrates that the future of drug discovery is based on augmenting researcher's scientific intuition by using computational precision. Through innovation and integration, researchers are well positioned to address some of the most complex biological challenges and develop precision medicine. This chapter provides a review of the various strategies and approaches available to researchers in order to improve and expedite the modern-day drug discovery process timeline.

Declaration of Interest

The authors have no conflict of interest.

Data Availability

No new data generated. Figure 15.3 has been included from our group's recent publication, which is available as an open access article (CC by 4.0) at http://dx.doi.org/10.1007/s42452-025-06772-1.

Author Contributions

TK, SG and RD: data curation, literature review, visualization, writing-original draft; SR: data validation, writing-review and editing; AA: conceptualization, supervision, writing- review and editing; SB: study design, visualization, writing- review and editing.

Funding

None to declare

Use of Software and Generative AI Tools

Grammarly was used for language, grammar, and sentence corrections/reconstructions wherever necessary.

List of Abbreviations

ADMET	Absorption, distribution, metabolism, excretion, and toxicity
AI	Artificial intelligence
AWS	Amazon Web Services
CADD	Computer-aided drug design

CDD	Computational drug discovery
DL	Deep learning
EGFR	Epidermal growth factor receptor
ESR1	Estrogen receptor 1
FAIR	Findable, Accessible, Interoperable, and Reusable
FDA	Food and Drug Administration
FEL	Free energy landscape
FGFR3	Fibroblast growth factor receptor 3
G_{bind}	Binding free energy
GNN	Graph neural network
H2L	Hit-to-lead
HPC	High-performance computing
HTE	High-throughput experimentation
LINCS	Library of Integrated Network-based Cellular Signatures
MAPK1	Mitogen-activated protein kinase 1
MAPK3	Mitogen-activated protein kinase 3
MD	Molecular dynamics
MDR	Multi-drug resistant
MDS	Molecular dynamics simulation
ML	Machine learning
MM	Molecular mechanics
NSCLC	Non-small cell lung cancer
PC	Principal component
PDB	Protein Data Bank
PTGS2	Prostaglandin-endoperoxide synthase 2
QAOA	Quantum Approximate Optimization Algorithm
QM	Quantum-mechanical
QSAR	Quantitative structure–activity relationship
SINCHO	Site Identification and Next Choice
SPR	Surface plasmon resonance
VQE	Variational Quantum Eigensolver
XDR	Extensively-drug resistant

References

1 Xu W. Current status of computational approaches for small molecule drug discovery. *J. Med. Chem.* 2024;67(21):18633–6.

2 Shears T. Sztain deploys computational methods to advance drug discovery, protein engineering. *The Apothecary*. September 25, 2024. Available from: https://pharmacy.umich.edu/blog/sztain-deploys-computational-methods-to-advance-drug-discovery-protein-engineering/

3 Sadybekov AV, Katritch V. Computational approaches streamlining drug discovery. *Nature* 2023;616(7958):673–85.

4 Buntz B. A year in review: AI's evolving role in drug discovery and development in 2023. 2023. Available from: https://www.drugdiscoverytrends.com/ai-drug-discovery-2023-trends/

5 Yu W, Weber DJ, MacKerell AD. An integrated covalent drug design workflow using site-identification by ligand competitive saturation. *J. Chem. Theory Comput.* 2023;19:3007–21.

6 Naithani U, Guleria V. Integrative computational approaches for discovery and evaluation of lead compound for drug design. *Front. Drug Discov.* 2024. https://doi.org/10.3389/fddsv.2024.1362456

7 Bakare OS. Advances in computational methods for drug design: a revolution in pharmaceutical development. *J. Angiother.* 2024;8(7):1–10.

8 Niazi SK, Mariam Z. Computer-aided drug design and drug discovery: a prospective analysis. *Pharmaceuticals* 2023;17(1):22.

9 Rashid MBMA. Artificial intelligence effecting a paradigm shift in drug development. *SLAS Technol.* 2021;26(1):3–15.

10 Bassaganya-Riera J. The AI paradigm shift: how cutting-edge tech is fast-tracking the future of medicine. *Pharmaceutical Executive.* 2024. Available from: https://www.pharmexec.com/view/ai-paradigm-cutting-edge-tech-fast-tracking-future-medicine

11 Vlasiou MC. Success stories in computer-aided drug discovery. In: *Computer-Aided Drug Discovery Methods: A Brief Introduction.* Bentham Science Publishers; 2024. pp. 112–25.

12 Debroy R, Ramaiah S. Consolidated knowledge-guided computational pipeline for therapeutic intervention against bacterial biofilms: a review. *Biofouling* 2023; 39 (9–10):928–47.

13 Regev A. AI and machine learning: Revolutionising drug discovery and transforming patient care. 2025. Available from: https://www.roche.com/stories/ai-revolutionising-drug-discovery-and-transforming-patient-care

14 BioDawn Innovations. Foundations of AI models in drug discovery series: Step 4 of 6 – Model evaluation and validation in drug discovery. 2024. Available from: https://biodawninnovations.com/foundations-of-ai-drug-discovery-series-step-4-of-6-model-evaluation-and-validation

15 Elhadi A. Bridging the gap between computational biology and experimental validation in drug discovery. 2024. Available from: https://www.linkedin.com/pulse/bridging-gap-between-computational-biology-validation-ahmed-elhadi-bafnc

16 Carracedo-Reboredo P, Liñares-Blanco J, Rodríguez-Fernández N, et al. A review on machine learning approaches and trends in drug discovery. *Comput. Struct. Biotechnol. J.* 2021;19:4538–58.

17 Ghosh S, Basu S, Kayal T, et al. Computational advancements to facilitate therapeutic application of phytochemicals: where do we stand? *Discov. Appl. Sci.* 2025;7(5):491.

18 Malathi K, Ramaiah S. Bioinformatics approaches for new drug discovery: a review. *Biotechnol. Genet. Eng. Rev.* 2018;34(2):243–60.

19 Colwell NA. Harnessing artificial intelligence in drug discovery and development. 2024. Available from: https://www.accc-cancer.org/view/harnessing-artificial-intelligence-in-drug-discovery-and-development

20 Nippa DF, Atz K, Stenzhorn Y, et al. Expediting hit-to-lead progression in drug discovery through reaction prediction and multi-objective molecular optimization. ChemRxiv. 2025. Available from: https://chemrxiv.org/doi/full/10.26434/chemrxiv-2025-0lxhw-v2

21 Joshi T, Vijayakumar S, Ghosh S, et al. Identifying novel therapeutics for the resistant mutant F533L in PBP3 of *Pseudomonas aeruginosa* using ML techniques. *ACS Omega* 2024;9(26):28046–60.

22 Basu S, Varghese R, Debroy R, et al. Non-steroidal anti-inflammatory drugs ketorolac and etodolac can augment treatment against pneumococcal meningitis by targeting penicillin-binding proteins. *Microb. Pathog.* 2022;170:105694.

23 Ghosh S, Basu S, Anbarasu A, Ramaiah S. A comprehensive review of antimicrobial agents against clinically important bacterial pathogens: prospects for phytochemicals. *Phytother. Res.* 2025;39(1):138–61.

24 Basu S, Naha A, Veeraraghavan B, et al. In silico structure evaluation of BAG3 and elucidating its association with bacterial infections through protein–protein and host–pathogen interaction analysis. *J. Cell Biochem.* 2022;123(1):115–27.

25 Naha A, Ramaiah S. Structural chemistry and molecular-level interactome reveals histidine kinase EvgS to subvert both antimicrobial resistance and virulence in *Shigella flexneri* 2a str. 301. *3 Biotech.* 2022;12(10):258.

26 Naha A, Miryala SK, Debroy R, et al. Elucidating the multi-drug resistance mechanism of *Enterococcus faecalis* V583: a gene interaction network analysis. *Gene* 2020;748:144704.

27 Mishra A, Soundharya H, Naha A, et al. Designing novel serine protease inhibitors against potential biomarkers specific to lung adenocarcinoma. In: *Handbook of Proteases in Cancer.* CRC Press; 2024. pp. 401–15.

28 Miryala SK, Ramaiah S. Cellular and molecular level host–pathogen interactions in *Francisella tularensis*: a microbial gene network study. *Comput. Biol. Chem.* 2022;96:107601.

29 Miryala SK, Anbarasu A, Ramaiah S. Discerning molecular interactions: a comprehensive review on biomolecular interaction databases and network analysis tools. *Gene* 2018;642:84–94.

30 Ashok G, Ramaiah S. FN1 and cancer-associated fibroblasts markers influence immune microenvironment in clear cell renal cell carcinoma. *J. Gene. Med.* 2023. https://doi.org/10.1002/jgm.3556

31 Ashok G, Miryala SK, Anbarasu A, Ramaiah S. Integrated systems biology approach using gene network analysis to identify important pathways and new potential drug targets for neuroblastoma. *Gene. Rep.* 2021;23:101101.

32 Priyamvada P, Debroy R, Anbarasu A, Ramaiah S. A comprehensive review on genomics, systems biology and structural biology approaches for combating antimicrobial resistance in ESKAPE pathogens. *World J. Microbiol. Biotechnol.* 2022;38(9):153.

33 Ashok G, Priyamvada P, Miryala SK, et al. Systems biology tools for identification of potential drug targets and biomarkers for cancer therapeutics. In: *Computational Methods in Drug Discovery and Repurposing for Cancer Therapy*. Elsevier; 2023. pp. 259–92.

34 Naha A, Banerjee S, Debroy R, et al. Network metrics, structural dynamics and DFT calculations identified a novel ursodeoxycholic acid derivative against Parkin. *Comput. Struct. Biotechnol. J.* 2022;20:4271–87.

35 Wan S, Potterton A, Husseini FS, et al. Hit-to-lead and lead optimization binding free energy calculations for G protein-coupled receptors. *Interface Focus* 2020;10(6):20190128.

36 Zdrazil B, Felix E, Hunter F, et al. The ChEMBL Database in 2023: a drug discovery platform spanning multiple bioactivity data types and time periods. *Nucleic Acids Res.* 2024; 52(D1):D1180–92.

37 Knox C, Wilson M, Klinger CM, et al. DrugBank 6.0: the DrugBank Knowledgebase for 2024. *Nucleic Acids Res.* 2024; 52(D1):D1265–75.

38 Kim S, Chen J, Cheng T, et al. PubChem 2019 update: improved access to chemical data. *Nucleic Acids Res.* 2019; 47(D1):D1102–9.

39 Pence HE, Williams A. ChemSpider: an online chemical information resource. *J. Chem. Educ.* 2010;87(11):1123–24.

40 Kouranov A. The RCSB PDB information portal for structural genomics. *Nucleic Acids Res.* 2006;34:D302–5.

41 Gilson MK, Liu T, Baitaluk M, et al. BindingDB in 2015. *Nucleic Acids Res.* 2016; 44(D1):D1045–53.

42 Antolin AA, Sanfelice D, Crisp A, et al. The Chemical Probes Portal. *Nucleic Acids. Res.* 2023; 51(D1):D1492–502.

43 Barbarino JM, Whirl-Carrillo M, Altman RB, Klein TE. PharmGKB. *WIREs Syst. Biol. Med.* 2018;10(4):e1417.

44 Carvalho-Silva D, Pierleoni A, Pignatelli M, et al. Open Targets Platform. *Nucleic Acids Res.* 2019; 47(D1):D1056–65.

45 One Health Drugs. Databases list and access. https://onehealthdrugs.com/working-area/database-collections-open-sources-and-vendors/databases-list-and-access/

46 Ashok G, Ramaiah S. A critical review of datasets and computational suites for improving cancer theranostics and biomarker discovery. *Med. Oncol.* 2022;39(12):206.

47 Lavanya C, Niranjan V, Kanagarajan D. The technological leap in drug discovery with cutting-edge solutions. In: *Molecular Modeling and Docking Techniques for Drug Discovery and Design*. IGI Global; 2024. pp. 473–504.

48 Bassani D, Moro S. Past, present, and future perspectives on computer-aided drug design methodologies. *Molecules* 2023;28(9):3906.

49 Shukla R, Tripathi T. Molecular dynamics simulation of protein and protein–ligand complexes. In: *Computer-Aided Drug Design*. Springer; 2020. pp. 133–61.

50 Basu S, Ramaiah S, Anbarasu A. In-silico strategies to combat COVID-19: a comprehensive review. *Biotechnol. Genet. Eng. Rev.* 2021;37(1):64–81.

51 Basu S, Ashok G, Ghosh S, et al. Cefiderocol susceptibility endows hope in treating carbapenem-resistant *Pseudomonas aeruginosa. RSC Adv.* 2024;14(30):21328–41.

52 Basu S, Veeraraghavan B, Anbarasu A. Impact of PmrB mutations on *Klebsiella pneumoniae* with variable colistin susceptibilities. *Chem. Biol. Drug Des.* 2024;103(1):e14381.

53 Basu S, Debroy R, Kumar H, et al. Bioactive phytocompounds against *Borrelia recurrentis. Med. Vet. Entomol.* 2023;37(2):213–18.

54 Choudhury C, Arul Murugan N, Priyakumar UD. Structure-based drug repurposing: traditional and AI/ML-aided methods. *Drug Discov. Today* 2022;27(7):1847–61.

55 Ajjarapu SM, Tiwari A, Ramteke PW, et al. Ligand-based drug designing. In: *Bioinformatics*. Elsevier; 2022. pp. 233–252.

56 Debroy R, Ramaiah S. Translational protein RpsE as an alternative target for novel nucleoside analogues. *World J. Microbiol. Biotechnol.* 2023;39(7):187.

57 Basu S, Chakraborty S. Diagnostics for pediatric tuberculosis. *Microorganisms* 2025;13(1):178.

58 Ganji M, Bakhshi S, Shoari A, Ahangari Cohan R. Discovery of potential FGFR3 inhibitors. *J. Transl. Med.* 2023;21(1):111.

59 Gangwal A, Lavecchia A. Unleashing the power of generative AI in drug discovery. *Drug Discov. Today* 2024;29(6):103992.

60 Gorgulla C, Boeszoermenyi A, Wang ZF, et al. An open-source drug discovery platform enables ultra-large virtual screens. *Nature* 2020;580(7805):663–8.

61 Kolar B. Virtual screening for new antibiotics: A machine learning approach to Lpxc inhibition. 2024. Available from: https://blog.latch.bio/p/virtual-screening-for-gram-negative

62 Durrani J. AI tool screens 107 million molecules, discovers potent new antibiotics. 2020. Available from: https://www.chemistryworld.com/news/ai-tool-screens-107-million-molecules-discovers-potent-new-antibiotics/4011233.article

63 Muratov EN, Amaro R, Andrade CH, et al. Computational approaches employed for COVID-19 drug discovery. *Chem. Soc. Rev.* 2021;50(16):9121–51.

64 Chang CC, Hsu HJ, Wu TY, Liou JW. Computer-aided discovery of COVID-19 therapeutics. *Tzu. Chi. Med. J.* 2022;34(3):276–86.

65 Lin X, Li X, Lin X. A review on applications of computational methods in drug screening and design. *Molecules* 2020;25(6):1375.

66 Ching T, Himmelstein DS, Beaulieu-Jones BK, et al. Opportunities and obstacles for deep learning in biology and medicine. *J. R. Soc. Interface* 2018;15(141):20170387.

67 Warren GL, Andrews CW, Capelli AM, et al. A critical assessment of docking programs and scoring functions. *J. Med. Chem.* 2006;49(20):5912–31.

68 Mirakhori F, Niazi SK. Harnessing AI/ML in drug and biological products discovery: regulatory perspective. *Pharmaceuticals (Basel)* 2025;18(1):47.

69 Kitchen DB, Decornez H, Furr JR, Bajorath J. Docking and scoring in virtual screening. *Nat. Rev. Drug Discov.* 2004;3(11):935–49.

70 Talevi A. Computer-aided drug discovery and design: Recent advances and future prospects. In: *Computational Drug Discovery and Design*. Humana Press; 2024. pp. 1–20.

71 Loeffler HH, He J, Tibo A, et al. Reinvent 4: modern AI-driven generative molecule design. *J. Cheminform*. 2024;16(1):20.

72 Pei Z. Computer-aided drug discovery: from traditional simulation methods to language models and quantum computing. *Cell Rep. Phys. Sci*. 2024;5(12):102334.

73 Chiang WY, Kao PY, Yeh TL, et al. Quantum machine learning for QSAR prediction. arXiv. 2025. Available from: http://arxiv.org/abs/2501.13395.

74 Abarbanel OD, Hutchison GR. QupKake: integrating machine learning and quantum chemistry for micro-pKa predictions. *J. Chem. Theory Comput*. 2024;20(15):6946–56.

75 Senn HM, Thiel W. QM/MM methods for biomolecular systems. *Angew. Chem. Int. Ed*. 2009;48(7):1198–229.

76 Zhavoronkov A. Artificial intelligence for drug discovery, biomarker development, and generation of novel chemistry. *Mol. Pharm*. 2018;15(10):4311–3.

16

Case Studies: Successful Drug Discovery Using CADD

Shaik Asha Begum[1], Sk. Abdul Rahaman[2], Anilkumar Adimulapu[3], and Ajay Bommareddy[4]*

[1]*KVSR Siddhartha College of Pharmaceutical Sciences, Vijayawada, Andhra Pradesh, India*
[2]*School of Medical and Allied Sciences, Galgotias University, Greater Noida, Uttar Pradesh, India*
[3]*Department of Pharmaceutics, School of Pharmacy, The Assam Kaziranga University, Jorhat, Assam, India*
[4]*Schmidt School of Medicine, Florida Atlantic University, Boca Raton, FL, USA*

*Corresponding author: abommareddy@health.fau.edu

Introduction

For a long time, finding new drugs has been a difficult, risky, and resource-intensive process. Conventional approaches required a significant amount of time, money, and labor because they mostly relied on trial-and-error testing. However, the field of drug development has seen a significant transformation with the introduction of sophisticated computer technology. The difficulties in drug discovery, the increasing significance of computational methods in contemporary pharmaceutical research, and the development of computer-aided drug design (CADD) as a central instrument in the creation of new treatments are all thoroughly examined in this chapter [1].

Overview of Drug Discovery Challenges

Drug discovery is a complex process with many logistical and scientific obstacles. Finding a viable biological target, typically a protein or enzyme involved in the pathophysiology of a disease, is the first step in the process. The design, production, and optimization of compounds capable of very selectively and effectively modulating the target's function are the next steps. The high attrition rate in drug discovery is one of the biggest obstacles. Only a tiny percentage of medication candidates, roughly 1 in 5,000, are thought to make it from initial discovery to clinical approval. Poor bioavailability, toxicity, ineffectiveness, and unanticipated adverse effects are among of the many causes of failure [2]. Furthermore, it still takes a very long time to develop new drugs; it usually takes 10–15 years from target identification to market introduction. The cost of launching a new medication can reach $2 billion, making the financial strain severe. These difficulties are exacerbated by regulatory obstacles, growing biological complexity, and the growing need for precision medicines, highlighting the necessity of creative approaches to expedite and improve the discovery process.

Role of Computational Methods in Modern Drug Development

Computational techniques have become revolutionary tools in drug research as a result of the shortcomings of conventional methodologies. Researchers can now simulate and anticipate biological interactions with previously unheard-of speed and accuracy by utilizing algorithms, molecular modeling, and artificial intelligence (AI) [3]. Target validation, hit identification, lead optimization, and toxicity prediction are just a few of the phases of drug discovery that computational techniques make easier. Large chemical libraries can be quickly evaluated using methods like molecular docking, virtual screening, quantitative structure–activity relationship (QSAR) modeling, and molecular dynamics (MD) simulations, which help identify promising candidates prior to experimental testing [4]. Furthermore, computational methods greatly lessen the need for expensive and time-consuming wet-lab investigations. Early in the development process, they enable researchers to forecast off-target effects, pick the most promising drugs, and optimize pharmacokinetic and pharmacodynamic characteristics. Computational methods are therefore essential in modern pharmaceutical research since they not only shorten the time it takes to find new drugs but also increase the likelihood that they will be successful in clinical trials [5].

Evolution of CADD

The growth of CADD is a reflection of more general technological developments in structural biology, software development, and processing power. In the 1970s, crude molecular modeling software that offered simple three-dimensional representations of tiny molecules gave rise to CADD. But the discovery of the first high-resolution crystal structures of pharmacological targets, like enzymes and receptors, really gave the research a boost [6]. The majority of early CADD efforts were ligand-based, using comparative analysis to infer the design of novel candidates based on known active compounds. Structure-based drug design (SBDD) techniques became more popular as structural data became more widely available. The ability to create inhibitors with precise molecular interactions has been transformed by methods like structure-based virtual screening and molecular docking, which predicts a ligand's preferred orientation within a target's binding site [7]. CADD has grown over the past few decades to include cutting-edge innovations including de novo drug design, fragment-based drug design (FBDD), and the fusion of AI and machine learning (ML). The accuracy and predictive power of CADD models have been greatly improved by these developments. Additionally, the development of cloud computing and high-performance computational infrastructure has made complex CADD tools more accessible, enabling researchers in academia and industry to fully utilize its potential [8]. These days, CADD is more than just an auxiliary tool; it is a key component of drug development, influencing precision medicine's future and making it possible to rationally design treatments for some of the most difficult illnesses [9].

Fundamentals of CADD

A variety of computational techniques are included in CADD, which aims to improve and expedite the drug development process. CADD techniques enable the rational design of molecules with optimal efficacy, selectivity, and safety profiles by employing structural information about biological targets and/or bioactive chemicals [9]. CADD is based on two main paradigms, SBDD and LBDD (ligand-based drug design), which are bolstered by essential methods such as molecular docking, virtual screening, pharmacophore modeling, and QSAR research.

Structure-based Drug Design

SBDD is the process of creating effective and targeted drug candidates by utilizing the three-dimensional (3D) structure of a biological target, usually a protein. The fundamental idea of SBDD is that the shape, charge distribution, and chemical complementarity of a medication and its target determine how they interact [10]. The first step in the SBDD process is usually to determine the target's three-dimensional structure using experimental techniques like cryo-electron microscopy, nuclear magnetic resonance (NMR) spectroscopy, or X-ray crystallography. Following the resolution of the structure, binding pockets and crucial interaction sites inside the target are found using computational methods. In order to maximize binding affinity and biological activity, molecules can then be engineered or refined to fit precisely into specific binding sites [11]. Key strategies within SBDD include:

- **Molecular docking:** predicting the optimal binding orientation and affinity of small molecules to the target.
- **FBDD:** building larger drug molecules by combining small chemical fragments identified to bind at different regions of the target site.
- **Structure-based virtual screening:** evaluating vast libraries of compounds for their ability to bind to the target structure [12].

SBDD has played a pivotal role in the development of many successful therapeutics, offering a rational, targeted approach to drug discovery that minimizes reliance on empirical trial-and-error [12].

Ligand-based Drug Design

When the biological target's 3D structure is unclear or unavailable, LBDD is used. Rather, LBDD uses the chemical and biological characteristics of molecules that are already known to interact with the target to create novel compounds [13]. Molecules with comparable chemical structures or physicochemical characteristics are expected to have similar biological activities, according to the fundamental tenet of LBDD. Thus, by examining a number of active compounds, scientists are able to pinpoint structural motifs and characteristics that are crucial for activity, often referred to as the "pharmacophore." LBDD's fundamental methods comprise [13]:

- **Pharmacophore modeling:** creating a model that represents the essential features required for molecular recognition at the target site.
- **QSAR:** developing mathematical models that correlate chemical structure with biological activity, enabling prediction of activity for novel compounds.
- **Similarity searching:** identifying new compounds by comparing molecular fingerprints or descriptors of known active molecules [6].

LBDD is particularly valuable in the early stages of drug discovery when structural information is scarce, but biological assay data are available. It enables efficient optimization of hit compounds into more potent and selective leads.

Molecular Docking and Virtual Screening

Molecular docking is a computational technique that predicts the preferred orientation of a small molecule (ligand) when bound to a target (typically a protein) and estimates the strength of this association through scoring functions. Docking involves two primary steps:

- **Pose generation:** sampling possible orientations and conformations of the ligand within the binding site [14].
- **Scoring:** evaluating each pose using a scoring function that approximates binding affinity based on factors like hydrogen bonding, hydrophobic interactions, electrostatics, and desolvation.

Virtual screening (VS) extends molecular docking to high-throughput scales, allowing the rapid evaluation of thousands to millions of compounds against a target structure. Virtual screening can be categorized as:

- **Structure-based virtual screening (SBVS):** based on docking ligands into the known 3D structure of the target.
- **Ligand-based virtual screening (LBVS):** based on similarity to known active compounds when target structures are unavailable [15].

By efficiently narrowing down chemical space, molecular docking and virtual screening greatly enhance the likelihood of identifying promising drug candidates while saving significant experimental resources.

Pharmacophore Modeling and QSAR Analysis

Pharmacophore modeling and QSAR analysis are essential computational approaches that bridge chemical structure with biological function [16]. Pharmacophore modeling involves identifying and abstracting the key molecular features necessary for optimal interactions with the target, regardless of the specific structure of the ligand. Typical pharmacophoric features include:

- hydrogen bond donors and acceptors
- hydrophobic regions

- aromatic rings
- positive or negative ionizable groups

Once a pharmacophore model is constructed, it can be used to screen compound libraries to identify new molecules that fulfill the essential interaction criteria, even if they are structurally diverse from known ligands [17]. QSAR analysis goes a step further by building statistical models that quantitatively relate a set of chemical descriptors (e.g., hydrophobicity, electronic properties, steric effects) to biological activity. Commonly used QSAR techniques include:

- linear regression (e.g., multiple linear regression [MLR]);
- ML approaches (e.g., random forest, support vector machines);
- 3D-QSAR methods like comparative molecular field analysis (CoMFA).

QSAR models not only predict the activity of novel compounds but also help elucidate the underlying molecular determinants of activity, guiding rational design and optimization efforts [18].

Case Studies of Successful Drug Discovery Using CADD

Numerous significant advances in pharmaceutical science have proved the power of CADD. The real-world case studies in this part demonstrate how CADD has been crucial to the development and improvement of effective treatments in a variety of therapeutic domains (Table 16.1) [19, 20].

HIV Protease Inhibitors (Saquinavir, Ritonavir, Darunavir)

Importance of HIV protease as a drug target: HIV protease is a crucial enzyme responsible for the maturation of infectious HIV particles. Inhibiting this protease prevents the virus from producing mature, infectious progeny, making it a prime target for antiretroviral therapy. Targeting HIV protease fundamentally shifted the management of HIV/AIDS, offering a strategic intervention point for therapy [31, 32].

Structure-based design and optimization: the HIV-1 protease crystal structure gave scientists a comprehensive guide for creating new drugs. The active site was modeled, and interactions with potential compounds were optimized using SBDD. The first FDA-approved HIV protease inhibitor created with SBDD principles was saquinavir. Second-generation inhibitors like Ritonavir and Darunavir, which showed improved pharmacokinetic characteristics and increased binding affinity, were developed as a result of additional optimization through computer modeling [33, 34].

HIV treatment and resistance management: HIV protease inhibitors dramatically improved patient survival rates and quality of life. The use of multiple inhibitors in combination therapies like highly active antiretroviral therapy (HAART) reduced viral loads significantly. Resistance mutations were a challenge, but successive iterations of inhibitors, informed by structural modeling, helped manage emerging resistant strains effectively.

Table 16.1 Comparison of case studies [21–30].

Case study	Target	CADD methodologies used	Key drugs developed	Impact
HIV protease inhibitors	HIV-1 protease	SBDD, molecular docking, structure-based optimization	Saquinavir, Ritonavir, Darunavir	Revolutionized HIV treatment; enabled combination therapies; addressed drug resistance through successive designs
Kinase inhibitors in cancer	BCR-ABL, EGFR, ALK kinases	VS, MD simulations, structure-based design	Imatinib, Gefitinib, Crizotinib	Transformed oncology with targeted therapies; highlighted challenges of acquired resistance; continuous innovation needed
Antiviral drug discovery	Viral RNA polymerases, neuraminidase	VS, structure-based optimization, binding affinity prediction	Remdesivir, Oseltamivir	Enabled rapid pandemic response; set precedent for repurposing and quick antiviral development pipelines
Neurological drug development	Glutamate receptors, acetylcholinesterase	LBDD, pharmacophore modeling, QSAR analysis	Riluzole, Donepezil	Offered first therapeutic options for ALS and Alzheimer's; shaped future CNS-targeted drug discovery
Antibiotic and antimicrobial discovery	Bacterial ribosomal RNA, bacterial membrane	SBDD, rational drug design, natural product-inspired CADD	Linezolid, Daptomycin	Addressed antibiotic resistance; introduced new antibiotic classes; invigorated antibacterial research with CADD

Kinase Inhibitors in Cancer Therapy (Imatinib, Gefitinib, Crizotinib)

Role of kinase inhibitors in cancer treatment: protein kinases regulate critical cellular processes such as proliferation and survival. Mutations or dysregulation of kinases often drive oncogenesis, making them attractive therapeutic targets. Small molecule kinase inhibitors selectively block aberrant signaling pathways, offering targeted cancer therapies with improved safety profiles [35].

Virtual screening and MD simulations: CADD techniques such as virtual screening and MD simulations were integral in identifying potent kinase inhibitors. For example, Imatinib (Gleevec) was designed to specifically inhibit the BCR-ABL fusion protein in chronic myeloid leukemia (CML). Gefitinib and Crizotinib emerged from similar computational pipelines targeting EGFR mutations in lung cancer and ALK-positive tumors, respectively [36].

Clinical success and drug resistance challenges: these kinase inhibitors revolutionized oncology by delivering dramatic responses in patients with specific molecular profiles. However, the development of resistance through secondary mutations necessitated the ongoing use of computational strategies to design next-generation inhibitors, illustrating the dynamic interplay between CADD and clinical oncology.

Antiviral Drug Discovery (Remdesivir for COVID-19, Oseltamivir for Influenza)

Computational strategies in targeting viral proteins: the urgent need for antiviral agents, particularly during pandemics, underscores the importance of rapid drug discovery. Viral proteins such as RNA-dependent RNA polymerase (RdRp) and neuraminidase were prioritized targets. CADD methods were crucial in rapidly identifying and optimizing inhibitors against these enzymes.

Structure-based optimization and binding affinity predictions: Remdesivir, originally developed for Ebola, was repurposed for COVID-19 after computational screening demonstrated strong binding to SARS-CoV-2 RdRp. Structure-based design principles also guided the development of Oseltamivir (Tamiflu) for influenza by targeting the neuraminidase enzyme, using crystallographic data and docking simulations to optimize binding affinity and selectivity.

Role of CADD in pandemic preparedness: the rapid identification of antiviral candidates during the COVID-19 pandemic showcased the critical role of CADD in accelerating drug development. Computational modeling reduced the time from target identification to clinical trials, marking a paradigm shift in pandemic response strategies [37].

Neurological Drug Development (Riluzole for ALS, Donepezil for Alzheimer's Disease)

Mechanism of action and target selection: neurodegenerative diseases like ALS and Alzheimer's involve complex pathophysiological mechanisms. For ALS, glutamate-induced excitotoxicity is a major factor, leading to the development of Riluzole. In Alzheimer's disease, the loss of cholinergic neurons guided the selection of acetylcholinesterase as a target, resulting in Donepezil [38].

Ligand-based drug design in CNS drug development: given the complexity of CNS targets and the blood–brain barrier challenges, LBDD approaches were heavily utilized. Pharmacophore modeling and QSAR analysis helped optimize molecules with appropriate CNS penetration and activity profiles. Donepezil's development was a prime example of successfully applying LBDD strategies in designing CNS-active drugs.

Future prospects in neurodegenerative disease treatment [39]**:** advancements in CADD, including AI-driven molecular design, offer new hope for the development of disease-modifying therapies in neurodegenerative disorders, moving beyond symptomatic treatments toward true neuroprotection and neuroregeneration.

Antibiotic and Antimicrobial Drug Discovery (Linezolid, Daptomycin)

Combating antibiotic resistance using CADD: the growing threat of antibiotic resistance necessitates the design of novel antimicrobial agents. CADD strategies have been vital in identifying new bacterial targets and designing molecules with mechanisms distinct from traditional antibiotics.

Rational drug design against bacterial targets [40]**:** Linezolid, an oxazolidinone antibiotic, was the first new class of antibiotics in decades, designed using structure-based methods targeting bacterial ribosomal RNA. Similarly, Daptomycin, a lipopeptide, was optimized for potent bactericidal activity against Gram-positive pathogens through rational design efforts informed by computational analysis.

Advancements in peptide and natural product drug design: CADD is increasingly integrated with natural product research and peptide design, offering new avenues to discover antibiotics with novel mechanisms. Hybrid approaches combining traditional knowledge with computational innovations are expected to revitalize the antimicrobial discovery pipeline.

Integration of AI and Machine Learning in CADD

The integration of AI and ML with CADD has revolutionized the drug discovery landscape. These technologies enhance the efficiency, accuracy, and innovation of traditional CADD methods, enabling faster identification of drug candidates, prediction of molecular properties, and optimization of pharmacokinetic profiles. By learning from massive datasets, AI and ML systems can uncover complex patterns that elude conventional computational techniques, opening new horizons in rational drug design [41].

Deep Learning for Drug Discovery

Deep learning, a subfield of machine learning characterized by multilayered neural networks, has emerged as a transformative tool in drug discovery. It enables the automatic extraction of complex features from raw data, significantly reducing the need for manual intervention in model development. Key applications include [42]:

- **Molecule generation:** generative models such as variational autoencoders (VAEs) and generative adversarial networks (GANs) can create novel chemical structures with desired biological properties.
- **Activity prediction:** deep neural networks (DNNs) can be trained to predict the biological activity of compounds against specific targets by learning from large bioactivity databases.
- **Protein structure prediction:** breakthroughs like AlphaFold have demonstrated the power of deep learning in accurately predicting protein 3D structures, a critical input for SBDD.
- **De novo drug design:** AI algorithms can suggest entirely new chemical scaffolds optimized for target binding and drug-like properties, bypassing traditional synthetic constraints.

Deep learning models have vastly improved the speed and quality of lead identification and optimization, representing a major leap beyond classical CADD methods [43].

Predictive Models for ADMET Profiling

The success of a drug candidate is not solely determined by its binding affinity but also by its absorption, distribution, metabolism, excretion, and toxicity (ADMET) profiles. Poor ADMET properties are among the most common reasons for drug failure in clinical trials. AI and ML models are now extensively used to predict ADMET properties early in the drug development process:

- **Absorption prediction:** ML algorithms can estimate intestinal absorption, blood–brain barrier penetration, and skin permeability from molecular structure [44].
- **Metabolism simulation:** predicting metabolic pathways and sites of metabolism (SOMs) using neural networks trained on enzyme–substrate interaction data.
- **Toxicity forecasting:** deep learning classifiers are deployed to predict cardiotoxicity, hepatotoxicity, and mutagenicity based on chemical structure and physicochemical properties.

- **Pharmacokinetics modeling:** ML approaches help simulate the time-course of drug concentrations in plasma and tissues, informing dosing regimens. By incorporating ADMET predictions early, researchers can prioritize compounds with the highest probability of clinical success, saving time, resources, and costs.

Automation of Drug Screening and Optimization

AI-driven automation is reshaping the traditional workflows of drug discovery, bringing unprecedented speed and scalability to CADD processes. Major advancements [45] include:

- **Virtual screening automation:** AI models rapidly screen vast chemical libraries, intelligently ranking compounds by predicted binding affinity, selectivity, and ADMET profiles, dramatically reducing the number of compounds needing experimental validation.
- **Lead optimization:** ML models suggest chemical modifications to enhance potency, selectivity, and pharmacokinetics, iteratively refining drug candidates with minimal human input [46].
- **Multi-objective optimization:** algorithms can optimize multiple properties simultaneously (e.g., efficacy, safety, manufacturability), balancing trade-offs that are traditionally hard to manage manually.
- **Closed-loop discovery systems:** integration of AI with automated synthesis and biological testing platforms creates a self-driving laboratory system where hypotheses are generated, tested, and refined with minimal human oversight. This degree of automation accelerates the entire drug development pipeline, from hit discovery to clinical candidate nomination, often reducing timelines from years to months.

Challenges and Future Prospects

While CADD and AI/ML have revolutionized the landscape of drug discovery, several significant challenges remain. Understanding these limitations is crucial for advancing the field further. Equally important is envisioning future directions where the synergy between human expertise, computational power, and innovative technologies could lead to unprecedented breakthroughs in therapeutics [47].

Limitations of Current Computational Models

Despite their tremendous value, current computational models are not without limitations:

- **Accuracy and generalizability:** many predictive models, including those based on machine learning, suffer from overfitting and limited generalization

across different chemical spaces or biological contexts. A model trained on one dataset may perform poorly on novel molecules.
- **Data quality and availability:** the success of AI and ML models hinges on the availability of high-quality, diverse, and unbiased datasets. In drug discovery, datasets often suffer from inconsistencies, incomplete annotations, and experimental variability, which can introduce significant errors.
- **Model interpretability:** many deep learning models, especially deep neural networks, operate as "black boxes," making it difficult to understand how input features influence predictions. Lack of interpretability poses challenges for regulatory approval and scientific validation.
- **Complexity of biological systems:** biological processes involve multiscale interactions across molecules, cells, tissues, and organs. Current models often oversimplify these interactions, limiting their ability to accurately predict in vivo outcomes.
- **Computational cost:** despite advances, large-scale simulations such as molecular dynamics and free energy calculations remain computationally intensive and time-consuming, sometimes limiting their practical use in early drug discovery phases.

Overcoming Drug Resistance and Off-target Effects

Drug resistance and unintended off-target effects remain persistent challenges in therapeutic development, and computational methods must evolve to address them more effectively:

- **Predicting resistance mutations:** AI models are increasingly used to predict how mutations in target proteins (such as HIV protease or cancer kinases) may lead to drug resistance. This predictive capability allows for proactive drug design to pre-emptively counteract resistance.
- **Poly pharmacology modeling:** many drugs interact with multiple targets, leading to side effects or unexpected therapeutic benefits. Computational approaches are now focusing on modeling polypharmacology to design drugs that either minimize off-target interactions or deliberately modulate multiple targets for complex diseases.
- **Dynamic target profiling:** integration of MD simulations with AI can capture the flexibility and transient states of biological targets, providing insights into hidden binding sites ("cryptic sites") that can be exploited to avoid resistance.
- **Allosteric modulation and novel targeting strategies:** shifting focus from active site inhibitors to allosteric modulators offers a promising strategy to combat resistance and minimize side effects. Computational tools are evolving to predict and design such novel modulators [48].

Future of AI-driven Drug Discovery

The future of drug discovery will be increasingly shaped by AI-driven innovations, offering transformative possibilities:

- **End-to-end drug design platforms:** fully integrated AI platforms will design, synthesize, test, and optimize drug candidates autonomously, creating a continuous feedback loop that accelerates discovery cycles dramatically.
- **Explainable AI (XAI):** new efforts are focused on making AI models more interpretable, thereby increasing trust, regulatory acceptance, and scientific insights derived from predictions.
- **Quantum computing:** as quantum computing matures, it is expected to revolutionize molecular simulations, enabling the precise modeling of complex chemical systems and reaction pathways that are currently beyond classical computing capabilities [49].
- **Integration of multi-omics data:** combining genomics, proteomics, metabolomics, and transcriptomics data with AI models will allow for a more holistic understanding of disease mechanisms and facilitate truly personalized medicine.
- **Global collaborative AI networks:** future drug discovery efforts may be supported by decentralized AI systems where multiple research groups and pharmaceutical companies collaborate in real time, sharing models and data securely using blockchain technologies.

Impact of CADD on Drug Development

CADD has significantly accelerated and de-risked drug discovery efforts in various ways including:

- **Efficiency:** VS, molecular modeling, and AI-driven predictions have drastically reduced the time and cost associated with identifying promising lead compounds compared to traditional high-throughput screening methods.
- **Precision:** structure-based and ligand-based design approaches have enabled the creation of highly specific drugs minimizing off-target effects and enhancing therapeutic efficacy.
- **Innovation:** CADD has allowed researchers to explore chemical and biological spaces previously deemed inaccessible, leading to the discovery of novel scaffolds, allosteric modulators, and biologics.
- **Adaptability:** throughout emerging crises, such as the COVID-19 pandemic, computational strategies have been pivotal in rapidly repurposing existing drugs and designing new antiviral agents.
- **Integration:** the synergy between CADD, ML, omics technologies, and experimental biology has created a multidisciplinary paradigm for rational drug design.

Collectively, these contributions underscore CADD's role as an indispensable tool in the arsenal of drug developers, enabling the translation of scientific insights into clinical realities more efficiently than ever before.

Conclusion

As the landscape of pharmaceutical research continues to evolve, CADD has established itself as a cornerstone technology. From the discovery of life-saving antivirals to the development of targeted cancer therapies, CADD has transformed how we approach the intricate process of drug discovery and optimization. This chapter provided a comprehensive view of CADD's applications, successes, challenges, and future prospects, highlighting its profound influence on modern medicine. In conclusion, the evolution of CADD, powered by continuous advancements in computational sciences, holds immense promise for overcoming some of the most daunting challenges in drug discovery. The next generation of therapies, faster, safer, and more precisely targeted will undoubtedly owe much of their existence to the strategic integration of CADD into the scientific process.

List of Abbreviations

ADME	Absorption, distribution, metabolism, and excretion
AI	Artificial intelligence
ALS	Amyotrophic lateral sclerosis
API	Active pharmaceutical ingredient
BACE1	Beta-site amyloid precursor protein cleaving enzyme 1
BCR-ABL	Breakpoint cluster region–Abelson tyrosine kinase
CMC	Chemistry, manufacturing, and controls
CML	Chronic myeloid leukemia
COVID-19	Coronavirus disease 2019
CTD	Common technical document
DDL	Deep learning
DDR1	Discoidin domain receptor 1
EMA	European Medicines Agency
FDA	Food and Drug Administration
GCN	Graph convolutional network
HIV	Human immunodeficiency virus
HTS	High-throughput screening
ICH	International Council for Harmonisation
IPR	Intellectual property rights
KPI	Key performance indicator
MD	Molecular dynamics
ML	Machine learning
MND	Motor neuron disease
NEJM	New England Journal of Medicine

PD	Pharmacodynamics
PDB	Protein Data Bank
PK	Pharmacokinetics
QbD	Quality by design
QSAR	Quantitative structure–activity relationship

References

1 Jorgensen WL. The many roles of computation in drug discovery. *Science* 2004;303:1813–8.

2 Kitchen DB, Decornez H, Furr JR, Bajorath J. Docking and scoring in virtual screening for drug discovery: methods and applications. *Nat. Rev. Drug Discov.* 2004;3:935–49.

3 Shoichet BK. Virtual screening of chemical libraries. *Nature* 2004;432:862–5.

4 Hughes JP, Rees S, Kalindjian SB, Philpott KL. Principles of early drug discovery. *Br. J. Pharmacol.* 2011;162:1239–49.

5 Schneider G. Automating drug discovery. *Nat. Rev. Drug Discov.* 2018;17:97–113.

6 Lionta E, Spyrou G, Vassilatis DK, Cournia Z. Structure-based virtual screening for drug discovery: principles, applications, and recent advances. *Curr. Top. Med. Chem.* 2014;14:1923–38.

7 Paul SM, Mytelka DS, Dunwiddie CT, et al. How to improve R&D productivity: the pharmaceutical industry's grand challenge. *Nat. Rev. Drug Discov.* 2010;9:203–14.

8 Ghosh AK, Osswald HL. BACE1 (β-secretase) inhibitors for the treatment of Alzheimer's disease. *Chem. Soc. Rev.* 2014;43:6765–813.

9 Daina A, Michielin O, Zoete V. SwissADME: a free web tool to evaluate pharmacokinetics, drug-likeness and medicinal chemistry friendliness of small molecules. *Sci. Rep.* 2017;7:42717.

10 Leelananda SP, Lindert S. Computational methods in drug discovery. *Beilstein. J. Org. Chem.* 2016;12:2694–718.

11 Zhang L, Tan J, Han D, Zhu H. From machine learning to deep learning: progress in machine intelligence for rational drug discovery. *Drug. Discov. Today* 2017;22:1680–5.

12 De Vivo M, Masetti M, Bottegoni G, Cavalli A. Role of molecular dynamics and related methods in drug discovery. *J. Med. Chem.* 2016;59:4035–61.

13 Lyu J, Wang S, Balius TE, et al. Ultra-large library docking for discovering new chemotypes. *Nature* 2019;566:224–9.

14 Ghosh AK, Anderson DD, Weber IT, Mitsuya H. HIV protease inhibitors: success and challenges. *Adv. Pharmacol.* 2012;67:277–318.

15 Mitsuya H, Broder S. Strategies for antiviral therapy in AIDS. *Nature* 1987;325:773–8.

16 Druker BJ, Talpaz M, Resta DJ, et al. Efficacy and safety of a specific inhibitor of the BCR-ABL tyrosine kinase in chronic myeloid leukemia. *N. Engl. J. Med.* 2001;344:1031–7.

17 Lynch TJ, Bell DW, Sordella R, et al. Activating mutations in the epidermal growth factor receptor underlying responsiveness of non-small-cell lung cancer to gefitinib. *N. Engl. J. Med.* 2004;350:2129–39.
18 Kwak EL, Bang YJ, Camidge DR, et al. Anaplastic lymphoma kinase inhibition in non-small-cell lung cancer. *N. Engl. J. Med.* 2010;363:1693–703.
19 Beigel JH, Tomashek KM, Dodd LE, et al. Remdesivir for the treatment of COVID-19 – final report. *N. Engl. J. Med.* 2020;383:1813–26.
20 Moscona A. Neuraminidase inhibitors for influenza. *N. Engl. J. Med.* 2005;353:1363–73.
21 Patel TH, Hackam DJ. Donepezil and rivastigmine: mechanisms of action. *Curr. Alzheimer Res.* 2002;19:377–83.
22 Miller RG, Mitchell JD, Lyon M, Moore DH. Riluzole for amyotrophic lateral sclerosis (ALS)/motor neuron disease (MND). *Cochrane Database Syst. Rev.* 2003;(2):CD001447.
23 Bozdogan B. Linezolid resistance: a review. *Ann. Clin. Microbiol. Antimicrob.* 2009;8:1.
24 Steenbergen JN, Alder J, Thorne GM, Tally FP. Daptomycin: a lipopeptide antibiotic for the treatment of serious Gram-positive infections. *J. Antimicrob. Chemother.* 2005;55:283–8.
25 Hartenfeller M, Schneider G. Enabling future drug discovery by de novo design. *Wiley Interdiscip. Rev. Comput. Mol. Sci.* 2011;1:742–59.
26 Zhavoronkov A, Ivanenkov YA, Aliper A, et al. Deep learning enables rapid identification of potent DDR1 kinase inhibitors. *Nat. Biotechnol.* 2019;37:1038–40.
27 Jumper J, Evans R, Pritzel A, et al. Highly accurate protein structure prediction with AlphaFold. *Nature* 2021;596:583–9.
28 Vamathevan J, Clark D, Czodrowski P, et al. Applications of machine learning in drug discovery and development. *Nat. Rev. Drug Discov.* 2019;18:463–77.
29 Schneider G, Fechner U. Computer-based de novo design of drug-like molecules. *Nat. Rev. Drug Discov.* 2005;4:649–63.
30 Bishop CM. *Pattern Recognition and Machine Learning.* Springer; 2006.
31 Walters WP, Murcko MA. Prediction of 'drug-likeness'. *Adv. Drug Deliv. Rev.* 2002;54:255–71.
32 Fulle S, Gohlke H. Statics and dynamics of protein–protein binding. *Curr. Opin. Struct. Biol.* 2010;20:194–200.
33 Congreve M, Carr R, Murray C, Jhoti H. A 'rule of three' for fragment-based lead discovery? *Drug. Discov. Today* 2003;8:876–77.
34 Ferreira LG, dos Santos RN, Oliva G, Andricopulo AD. Molecular docking and structure-based drug design strategies. *Molecules* 2015;20:13384–421.
35 Cherkasov A, Muratov EN, Fourches D, et al. QSAR modeling: where have you been? Where are you going to? *J. Med. Chem.* 2014;57:4977–5010.
36 Irwin JJ, Shoichet BK. ZINC – a free database of commercially available compounds for virtual screening. *J. Chem. Inf. Model.* 2005;45:177–82.
37 Gaulton A, Bellis LJ, Bento AP, et al. ChEMBL: a large-scale bioactivity database for drug discovery. *Nucleic Acids Res.* 2012;40:D1100–7.

38 Bajorath J. Integration of virtual and high-throughput screening. *Nat. Rev. Drug Discov*. 2002;1:882–94.

39 Lavecchia A, Di Giovanni C. Virtual screening strategies in drug discovery: a critical review. *Curr. Med. Chem*. 2013;20:2839–60.

40 Lipinski CA. Lead- and drug-like compounds: the rule-of-five revolution. *Drug Discov. Today Technol*. 2004;1:337–41.

41 Mak KK, Pichika MR. Artificial intelligence in drug development: present status and future prospects. *Drug Discov. Today* 2019;24:773–80.

42 Ekins S, Puhl AC. Exploiting machine learning for end-to-end drug discovery and development. *Nat. Mater*. 2019;18:435–41.

43 Greener JG, Sternberg MJE. Structure-based prediction of protein–protein interactions. *Curr. Opin. Struct. Biol*. 2019;58:45–52.

44 Jacob RB, Andersen T, McDougal OM. Accessible virtual screening molecular docking software for students and educators. *PLoS Comput. Biol*. 2012;8:e1002499.

45 Schneider P, Walters WP, Plowright AT, et al. Rethinking drug design in the artificial intelligence era. *Nat. Rev. Drug Discov*. 2020;19:353–64.

46 Grebner C, Matter H, Plowright AT, et al. Virtual screening in the cloud: how big is big enough? *J. Chem. Inf. Model*. 2020;60:4303–15.

47 Fout A, Byrd J, Shariat B, Ben-Hur A. Protein interface prediction using graph convolutional networks. In: *Proceedings of the 31st International Conference on Neural Information Processing Systems*. vol. 30. Curran Associates; 2017. pp. 6533–42.

48 Gupta A, Müller AT, Huisman BJH, et al. Generative recurrent networks for de novo drug design. *Mol. Inform*. 2018;37:1700111.

49 Chodera JD, Noé F. Markov state models of biomolecular conformational dynamics. *Curr. Opin. Struct. Biol*. 2014;25:135–144.

17

Practical Guide to Computational Drug Discovery: Tools, Databases, and Workflows

Arijit Mondal[1], Debashis Metya[2], Sabina Yasmin[3], Suddhasattya Dey[4], and Arindam Maity[1]*

[1]*Department of Pharmaceutical Technology, JIS University, Kolkata, West Bengal, India*
[2]*Department of Pharmacy, Bengal College of Pharmaceutical Technology, Dubrajpur, West Bengal, India*
[3]*Department of Pharmaceutical Chemistry, College of Pharmacy, King Khalid University, Abha, Saudi Arabia*
[4]*Department of Pharmacy, IQ City Institute of Pharmaceutical Sciences, Durgapur, West Bengal, India*

*Corresponding author: sahussain@kku.edu.sa

Introduction

Computer-aided drug design (CADD) is a multifaceted area that uses both computational and experimental approaches to find, design, and optimize novel medicinal molecules [1, 2]. This technology dramatically speeds up the drug development process by lowering the time and cost involved in conventional methods [3]. Such approaches analyze biological data, forecast molecular interactions, and direct the creation of new medications by utilizing a variety of computational tools and algorithms. Computational methods expedite the drug development strategy by understanding and guiding research [4]. The method of developing new drugs requires a lot of time and resources. More efforts are being made to accelerate medication discovery and development by combining computing power in biological and chemical domains. To optimize absorption, hit-to-lead selection, toxicity profiles, metabolism, hit identification, excretion, distribution, metabolism, and safety issues to prevent existing ones, biomedicine uses computational or in silico design [5]. Major computational techniques used in drug design (drug target docking) include quantitative structure, structure-based drug design (SBDD), and ligand-based drug design (LBDD) (Figure 17.1).

This includes the quantitative relationship between activity and structure as well as the association between the two [6]. In order to identify critical interactions and locations that are important to each biological function, the SBDD approach analyzes the 3D structural data of macromolecular sites, which are typically proteins or RNA [7]. This information can be utilized to create antibiotic treatments that

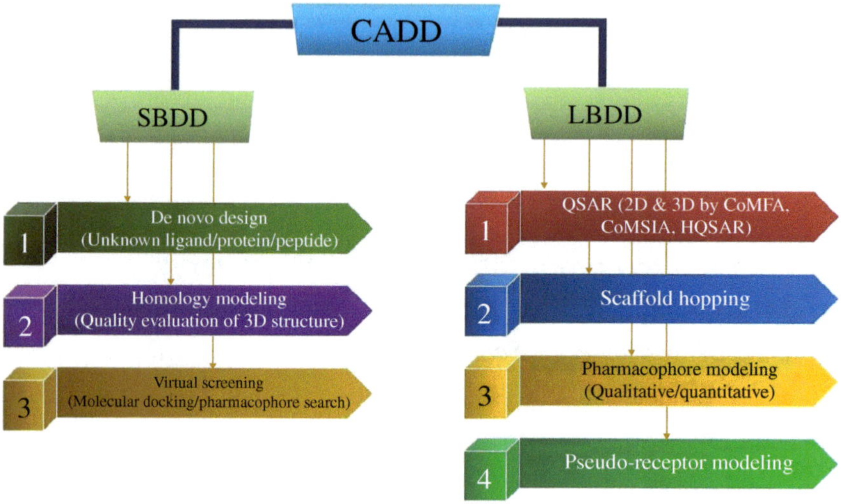

Figure 17.1 Types of CADD.

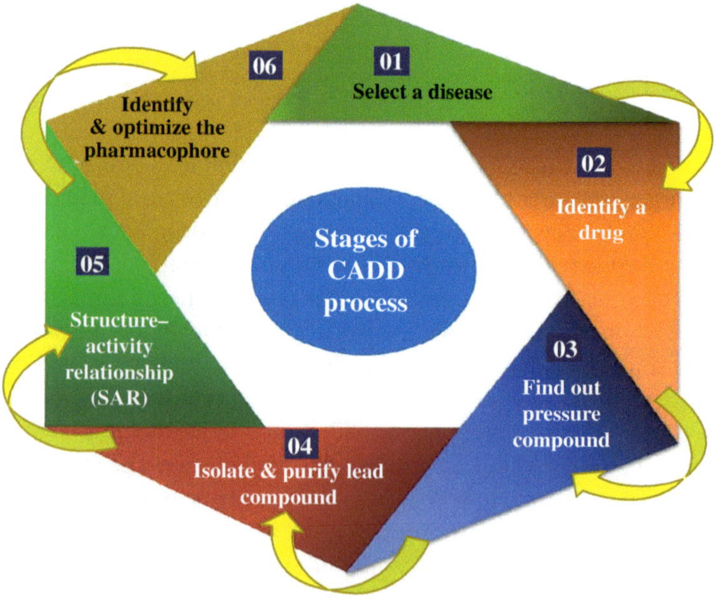

Figure 17.2 Stages of CADD process.

disrupt the pharmacological pathways that allow the bacteria or organisms to thrive by competing with the site's essential interactions. The LBDD method's objective was to identify an already identified antibiotic ligand for a target and establish a structure–activity relationship (SAR) between the antibiotic's impact and its physiochemical properties [8]. Numerous pharmaceutical medications that have been approved by the Food and Drug Administration (FDA) and are presently available for acquisition, were found to be attributed in large part to CADD (Figure 17.2) [9].

There is a continuous development of new methods and technology, and the field of CADD is expanding quickly. In the course of finding new drugs, it holds great promise and possibilities [10]. Combining wet-lab methods with CADD can help identify new antibiotic targets, improve our understanding of drug resistance processes, and create new antibiotics for both known and undiscovered targets. CADD is a helpful tool for finding new molecular configurations [11]. These include developing predictive tools to identify possible ADME, bettering the design and administration of data sources, developing computer programs to generate large libraries of pharmacologically intriguing compounds, and developing new algorithms to evaluate the effectiveness and selectivity of lead candidates [12]. There has been substantial advancement in the utilization of computers in drug development. These techniques are advantageous to do extensive data analysis from clinical and empirical studies. In order to create pharmacologically active drugs, pharmaceutical corporations and academic institutions have recently employed computational methods for finding new drugs, like molecular docking, QSAR analysis, pharmacophore modeling, and virtual screening [13]. CADD, also referred to as "*in silico* screening," has become a potent tactic that can be applied at various stages of drug research and discovery because of its numerous sophisticated characteristics. *In silico* screening also makes it possible to produce and test compounds that have been chosen for improved treatment. Despite its major focus on drug discovery and computational chemistry, this chapter aims to encompass an abundance of computational techniques, including novel approaches and useful components of the field. CADD is essential for *in silico* drug design procedures such as conformation generation, docking, homology modeling, pharmacophore development, multitarget search and design, and quantitative structure–activity connection. It has created clinically approved medications [14]. Furthermore, computational drug development requires utilization of scrupulously managed chemical, biological, and structural resources such as the Protein Data Bank (PDB), Chemical Database of Bioactive Molecules (ChEMBL), and PubChem. These resources assist people in making informed decisions by providing readily accessible and well-organized facts. This chapter offers a thorough analysis of the databases, techniques, and tools necessary for computational drug development. Its objective is to assist academic and industry researchers in improving the efficacy and success rates of drug development. This chapter explores some of the basic computational techniques that aid in drug discovery, including molecular docking, virtual screening, molecular dynamics (MD) simulations, and the combination of artificial intelligence (AI) and machine learning (ML). These methods allow researchers to predict, model, and enhance interactions between drugs and biological targets.

Computational Approach

Molecular Docking

Molecular docking, an *in silico* method that determines the correct binding pose of a protein–ligand complex and evaluates its ability to select the best posture for each molecule generated to a rank order, uses a range of scoring functions [15]. The goal of docking approaches is to smoothly place a ligand inside a target protein's binding region. In order to determine their binding free energy, this entails balancing and optimizing factors including hydrophobic, steric, and electrostatic complementarity [16]. Pose prediction, virtual screening, and binding affinity estimation are the three main goals of molecular docking [17].

Process Involved in Molecular Docking

The essential steps of molecular docking (Figure 17.3), a crucial procedure in structure-based drug design, are described below. These procedures offer an outline for clarifying molecular binding mechanisms and methodically direct the planning, carrying out, and analyzing of ligand–protein interactions.

Preparation of protein and ligand: the 3D structures of the ligand molecule to be docked and the protein receptor can be generated or retrieved. Remove co-crystallized ligands, heteroatoms, and water molecules to create the protein structure [18]. Additionally, look for and fix any structural flaws. Add hydrogen atoms, allocate the appropriate charges, and optimize the ligand's shape to establish its structure [19].

Docking method selection: it depends on what the researcher needs. For example, flexible docking programs may be preferred if the goal is to dock many molecules at a protein's binding site under particular circumstances such as pH, water present, and solubility. Flexible docking techniques, however, would not be feasible when scanning hundreds of chemicals from databases unless there is a fast computer and a lot of computing capacity. Therefore, depending on the capability of their computer and the characteristics of the target, users can choose from a variety of docking techniques [10].

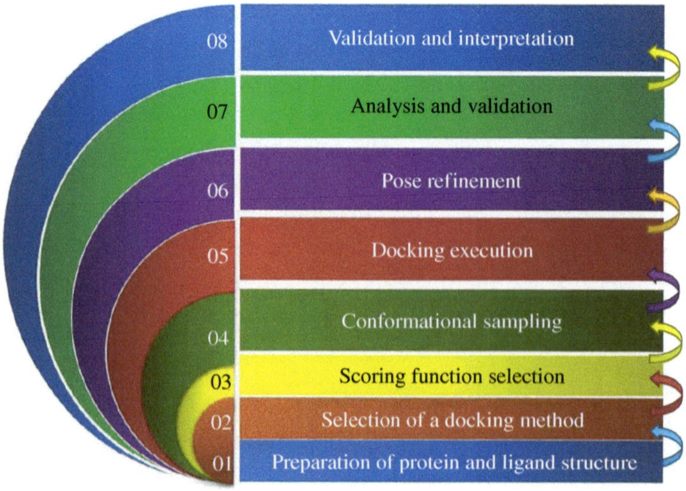

Figure 17.3 Steps of molecular docking.

Picking a scoring function: select a suitable scoring function to assess the ligand poses' suitability in the binding region. The scoring function evaluates the degree of binding using a number of factors, including solvation effects, electrostatics, hydrogen bonding, and van der Waals interactions [20, 21].

Conformational execution: within the specified search space, conduct a methodical search of the ligand conformational space to investigate various potential binding positions. Algorithms such as MD simulations, evolutionary algorithms, or Monte Carlo techniques can be used to generate and refine ligand conformations [22].

Docking performance: employing a docking technique, insert the ligand into the binding site based on the energy scores that have been established. Several popular docking algorithms are AutoDock, DOCK, AutoDock Vina, Glide, and GOLD [23].

Refinement of pose: improve accuracy and dependability by fine-tuning the initial ligand postures that docking algorithms produce. Utilize strategies such as energy minimization, rescoring, or MD simulations to improve the ligand–protein interactions [24].

Visualization and analysis: examine the docked positions to determine possible ligand–protein receptor interactions and binding mechanisms. To understand the results and decide on ligand design or additional experimental validation, see the docked complexes using molecular visualization tools like Discovery Studio [25].

Validation of result: verify the docking results by contrasting them with any accessible experimental data. To comprehend the molecular mechanisms behind ligand binding and to direct future experimental research or drug discovery initiatives, analyze the interactions and binding positions [26].

MD Simulation

A computer simulation technique for examining the actual motions of atoms and molecules is called MD simulation. It allows researchers to see how a system of particles according to Newton's laws of motion changes over time [27]. Protein–ligand complex stability and conformational changes are evaluated with the aid of these simulations. For MD simulations, programs like GROMACS, Assisted Model Building with Energy Refinement (AMBER), and Chemistry at HARvard Macromolecular Mechanics (CHARMM) are frequently used [28].

Workflow of MD Simulation

The essential steps of MD simulation (Figure 17.4) are described below.

Specify the system: select the molecular system that will be examined: first, protein–ligand complex, membrane protein, DNA/RNA fragment, etc., and second, compose the structure using modeling tools or get it using the PDB. Common tools mainly used for this step are RCSB PDB, Chimera, and PyMOL [28].

Select the force field: a force field uses atomic locations to characterize a system's potential energy. Bonded (bond, angle, dihedral) and nonbonded (electrostatics,

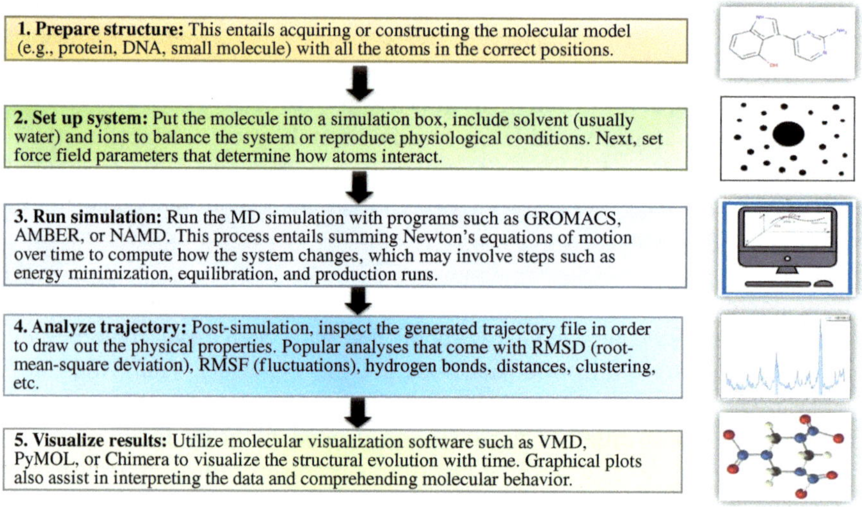

Figure 17.4 Workflow of MD simulation.

van der Waals) interactions make up force fields. Conventionally used force fields are AMBER: biomolecules (proteins, nucleic acids); CHARMM: versatile for biomolecules; GROMOS: simple molecules; and OPLS: appropriate for organic and small molecules [29].

Assemble the basic structure: make the molecular structure presentable and ready. At first, eliminate ligands and water. Then add any unobtainable hydrogens or residues. Specify the simulation box, which is often cubic- or dodecahedron-shaped, and then use water molecules to solve the box. Lastly, to neutralize the system add counterions [30]. Common tools are GROMACS pdb2gmx, CHARMM-GUI, AMBER tleap, and VMD.

Minimization of energy: the objective is to use potential energy optimization to eliminate steric conflicts or improper geometry. Then employ algorithms such as conjugate gradient or steepest descent and continue until a local minimum energy threshold is reached by the system [31].

Finding equilibrium: maintain the system's desired pressure and temperature. There are two common stages: NVT quintet (constant temperature, volume, and number of particles) and NPT quintet (constant temperature, pressure, and particle number). This procedure maintains thermodynamic stability prior to the start of production [32].

Manufacturing run: execute the framework to make real-time observations. The parameters for the execution are time: 1–2 fs and the simulation duration varies from 10 ns to 1 μs based on the study requirements and processing capabilities. Over time, this stage generates trajectories that display atomic motions [32].

Analysis of trajectories: extrapolate practical insights like root-mean-square deviation, radius of gyration, hydrogen bonding, secondary structure, radial distribution function, and free energy landscape [32].

Interpretation and visualization: trajectories can be used to better understand molecular behavior. Monitor diffusion, ligand binding, or conformational changes. VMD, PyMOL, and ChimeraX are helpful tools for this [33].

Tutorial of Pharmacophore Modeling

In drug discovery, pharmacophore modeling (Figure 17.5) is a computational method that determines the spatial arrangement of characteristics necessary for a molecule to cooperate with a particular biological target. It helps with lead optimization, virtual screening, and ligand–target association interpretation [34].

Active Ligand Selection and Preparation

Assemble a list of known active chemicals for a target first. These substances should ideally: have experimental validation, possess a variety of chemical scaffolds, and discuss a variety of topics (e.g., IC_{50}, K_i). Common databases used for active ligand selection are ChEMBL, PubChem, and BindingDB. Utilize cheminformatics tools to create 3D structures from SMILES, reduce the molecules' energy, and maintain appropriate protonation states at the pH of the body. Commonly used tools are Open Bable, LigPrep, etc. [35].

Ligand-based and Structure-based Pharmacophore Model Generation

Ligand-based pharmacophore model is used when protein structure is not available. The first step of this model generation is to set the active components in three dimensions. Then determine common chemical characteristics, such as the hydrophobic center, aromatic ring, and H-bond donor/acceptor, and finally create models or hypotheses [36]. Structure-based pharmacophore model is used when the target's 3D structure is available. The first step of this model generation is to bring in the structure of the protein–ligand combination. Then describe the binding pocket. Next is to find out how the ligand and receptor interact. Final step is to obtain pharmacophoric characteristics. LigandScout, Schrödinger Phase, and Discovery Studio are used for ligand-based and structure-based pharmacophore model generation [37].

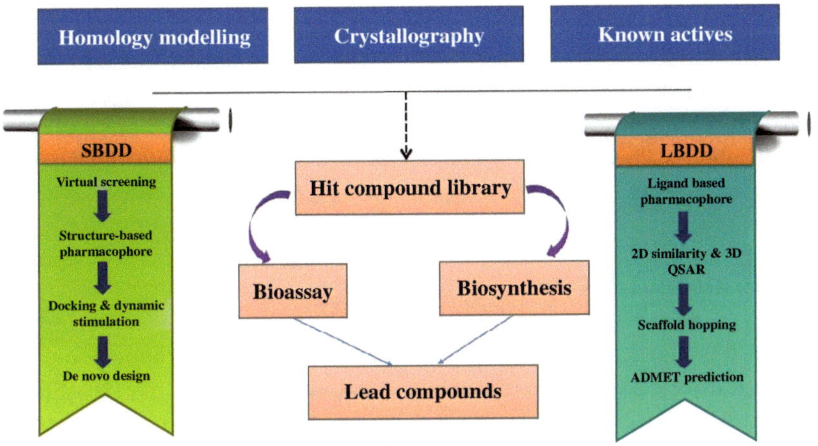

Figure 17.5 Workflow of pharmacophore modeling.

Validation and Optimization of Pharmacophore Models

Using dynamic simulation data, pharmacophore model optimization aims to fine-tune the spatial arrangement of chemical characteristics essential for biological activity. By capturing variable conformations and transient binding modes, MD simulations shed light on ligand–receptor interactions throughout time [38]. The pharmacophore model is updated by analyzing these dynamic snapshots to find recurring interaction patterns. Using receiver operating characteristic (ROC) curves and virtual screening, the model's capacity to distinguish between active and inactive substances is evaluated as part of the validation process. Common measures include area under the ROC area under the curve (AUC) and enrichment factors. A validated model improves drug discovery results by exhibiting robustness, specificity, and prediction accuracy [39].

Virtual Screening

MD uses virtual screening to improve hit detection by integrating dynamic structural information of target proteins. Intermittent binding locations and conformational flexibility not seen in static structures are revealed by MD simulations [40]. In order to better accommodate ligand variability, structure-based virtual screening employs ensembles of receptor conformations produced by MD. By docking ligand libraries into these many protein snapshots, the likelihood of finding genuine binders is increased. Top-ranked compounds are chosen for additional examination after scoring functions estimate binding affinities. By taking protein flexibility into consideration, quintet docking increases hit rates and pharmacological significance while also improving screening dependability [41].

Post-processing and Hit Analysis

In MD-guided drug discovery, hit analysis and post-processing entail assessing and improving possible lead compounds found by virtual screening. The stability, binding mode, and interaction profile of hits within the target binding site over time are all thoroughly revealed by MD simulations [42]. Hydrogen bonding patterns, binding free energy computations (e.g., MM/PBSA or MM/GBSA), root-mean-square deviation (RMSD), and root-mean-square fluctuation (RMSF) are important measurements. Compounds are ranked according to their dynamic stability and durability of interactions thanks to these investigations. Clustering of MD trajectories is another post-processing step used to find dominant binding conformations and possible allosteric effects. The goal is to validate and refine hit compounds for further experimental testing [43].

Key Computational Tools

Drug research has drastically changed as a result of the use of computer tools to identify and optimize potential medicinal molecules. AutoDock, Schrödinger Suite, Open Babel, and SwissADME are examples of applications that offer robust frameworks for cheminformatics, pharmacokinetic assumptions, and molecular docking.

The best possible techniques, procedures, and applications of these technologies are highlighted in this chapter's practical guide to their use [44].

AutoDock: A Simpler Approach to Molecular Docking

Scripps Research developed AutoDock, a well-known open-source molecular docking tool. It predicts how small substances, like substrates or potential drugs, will bind to a receptor with a specific 3D form. The two main components of the software package are AutoDock, which executes the docking simulations, and AutoGrid, which determines the docking search space by precalculating grid maps of target proteins. AutoDock can be installed on Windows, Linux, and macOS [45]. There are a few key steps in AutoDock's typical workflow. Initially the receptor and ligand are prepared. Water molecules are removed, polar hydrogens are added, and the receptor's 3D structure is obtained from the PDB utilizing AutoDock Tools. The ligand structure is created using software like ChemDraw, converted to three dimensions, and saved in PDB format. AutoDock tools (MGL Tools) are used to define rotatable bonds, and Gasteiger charges are used. Grid maps made using AutoGrid are then used to establish the docking search space, with the grid box dimensions modified to incorporate the receptor's active site. Following that, variables such as the quantity of runs and energy evaluations are chosen in order to set up docking simulations [46]. AutoDock runs the simulations and outputs the binding locations and pertinent energies. The docking results are then analyzed using visualization tools like PyMOL or Discovery Studio to assess binding interactions and affinities. AutoDock has contributed significantly to drug discovery, assisting in the development of therapies such as HIV-1 integrase inhibitors [47].

Schrödinger Suite: Superior Molecular Modeling

The Schrödinger Suite is a widely used molecular modeling software in computational drug discovery and materials research. It simplifies procedures like docking, pharmacophore modeling, and protein and ligand production by integrating several specialized tools, including Maestro, Glide, and LigPrep, into a single platform. Users can utilize Schrödinger's ML and predictive modeling procedures through Maestro. Scientists with varying degrees of experience may effectively view, manipulate, and analyze molecular structures due to its user-friendly design [48]. By adding hydrogens, assigning bond ordering, and optimizing hydrogen-bonding networks in raw PDB structures, the Protein Preparation Wizard simplifies the process of producing proteins. This ensures that protein models are accurate and ready to be used in later steps. LigPrep generates high-quality 3D structures of ligands by accounting for various ionization states, tautomers, and stereoisomers. Precise docking simulations require this preparation. High-throughput virtual screening (HTVS), standard precision (SP), and extra precision (XP) are some of the ligand docking options that Glide provides in order to balance precision and computational efficiency. In post-docking analysis, docking scores are analyzed and binding poses are visualized in order to identify significant interactions and optimize lead compounds [49]. This comprehensive procedure makes it easier to make well-informed decisions about drug design. The Schrödinger Suite can only be installed after downloading the software and required licenses from the official website. There are thorough instructions for establishing the system and setting up the license management to ensure seamless integration into various computing settings. Ultimately, the Schrödinger Suite offers a robust and all-inclusive foundation for molecular modeling, facilitating precise and fruitful computational studies in the biological sciences [50].

Open Babel: Cheminformatics Toolkit

Open Babel is a freely available cheminformatics toolkit designed to support various molecular modeling tasks and facilitate the conversion of various chemical file types. It offers a versatile platform for researchers and developers working in fields like organic chemistry, materials science, drug design, and computational chemistry [51]. By enabling seamless conversion between over 110 chemical file formats, including SMILES, SDF, and PDB, the toolkit simplifies data transfer between different programs. Along with format conversion, Open Babel offers tools for conducting substructure and similarity searches within chemical libraries, generating 3D molecular structures from 2D representations, and identifying molecular descriptors required for modeling the quantitative structure–activity relationship (QSAR) [52]. Numerous platforms for operating systems, including Windows, Linux, and macOS, can install Open Babel. Taking everything into account, Open Babel is an important resource for the scientific community because it provides a wide range of tools that enhance the efficacy and accuracy of cheminformatics research [53, 54].

SwissADME: Predicting Drug-likeness

The open-source online program SwissADME was developed by the Swiss Institute of Bioinformatics to evaluate the pharmacokinetics, drug-likeness, and medicinal chemistry friendliness of tiny molecules [55]. It provides rapid and accurate predictions of physicochemical characteristics, pharmacokinetics, drug-likeness, and medicinal chemistry friendliness, which is helpful in the initial phases of drug discovery when there is a lack of experimental data. SwissADME allows users to utilize and input molecular structures utilizing the integrated molecular sketcher or SMILES notation [56]. The tool is used to calculate parameters such as molecular weight, lipophilicity (log P), topological polar surface area (TPSA), and the number of hydrogen bond sources and acceptors. Pharmacokinetic predictions take into account gastrointestinal absorption, blood–brain barrier permeability, and P-glycoprotein substrate identification [57]. Medicinal chemistry evaluations include alarms for pan-assay interference compounds (PAINS) and synthetic accessibility, while filters such as Lipinski's Rule of Five are used to assess drug-likeness. SwissADME's comprehensive predictive capabilities and user-friendly interface make it a valuable tool for cheminformatics and drug discovery specialists [58].

GitHub Repository: Datasets and Scripts

A dedicated GitHub repository has been established to support experiential learning in computational drug discovery [59]. This repository includes example receptor and ligand structure datasets, Open Babel chemical format conversion scripts, AutoDock parameter files and docking scripts, Schrödinger workflow files, and sample SMILES inputs for SwissADME analysis. The repository makes it simpler to understand and apply a range of computational drug discovery methods and strategies by providing these resources in one accessible location [39].

Conclusion and Future Directions

The subject of computational drug discovery is quickly evolving, thanks to advancements in AI, big data analytics, and high-performance computing. The way that researchers think about drug discovery is changing due to a number of new recent developments. These include AI and ML, integration of multi-omics data, cloud-based platforms, quantum computing etc. Even with great advancements, computational drug discovery still has a number of issues and room for development. These are data quality and standardization, bridging the *in silico–in vivo* gap, scalability of workflows, ethical and regulatory considerations, and open science and accessibility. The field of computational drug discovery has evolved from a specialized domain of study to a vital part of contemporary pharmaceutical development. As the landscape of drug development evolves, computational techniques and approaches will play an increasingly significant role in everything from early target identification to clinical choice of candidates. The combination of AI, structural bioinformatics, chemoinformatics, and systems biology is opening up hitherto unimaginable prospects for drug creation. With the use of integrated platforms, robust databases, and innovative algorithms, researchers may investigate vast chemical regions, uncover biological insights, and accelerate the translation of scientific discoveries into therapeutic therapies. However, the future requires constant innovation, moral vision, and cooperation. Addressing the drawbacks of current models, enhancing data quality, and ensuring interdisciplinary training are all necessary to fully realize drug design promise. Furthermore, a balance between automation and human supervision must be maintained when we integrate these technologies into real drug development pipelines to safeguard patient safety and scientific integrity. In conclusion, despite not being a panacea, computational drug development is a vital source of innovation. Data, technological developments, and the shared goal of improving global public health will drive its bright future.

List of Abbreviations

ADME	Absorption, distribution, metabolism, and excretion
AI	Artificial intelligence
AMBER	Assisted Model Building with Energy Refinement

AUC	Area under the curve
BBB	Blood–brain barrier
CADD	Computer-aided drug design
CHARMM	Chemistry at HARvard Macromolecular Mechanics
ChEMBL	Chemical Database of Bioactive Molecules
CPU	Central processing unit
FDA	Food and Drug Administration
GBSA	Generalized Born surface area
GOLD	Genetic optimization for ligand docking
HTVS	High-throughput virtual screening
IC_{50}	Half-maximal inhibitory concentration
K_i	Inhibition constant
LBDD	Ligand-based drug design
log P	Logarithm of the partition coefficient
MD	Molecular dynamics
ML	Machine learning
MM/GBSA	Molecular mechanics/generalized Born surface area
MM/PBSA	Molecular mechanics/Poisson–Boltzmann surface area
NAMD	Nanoscale molecular dynamics
NPT	Constant number of particles, pressure, and temperature
NVT	Constant number of particles, volume, and temperature
OPLS	Optimized potentials for liquid simulations
PAINS	Pan-assay interference compounds
PDB	Protein Data Bank
QSAR	Quantitative structure–activity relationship
RMSD	Root-mean-square deviation
RMSF	Root-mean-square fluctuation
ROC	Receiver operating characteristic
SAR	Structure–activity relationship
SBDD	Structure-based drug design
SMILES	Simplified Molecular Input Line Entry System
SP	Standard precision
TPSA	Topological polar surface area
VMD	Visual molecular dynamics
XP	Extra precision

References

1 Gupta PK, Pal Y, Kumar P, et al. A critical review on computational techniques through in silico assisted drug design. *Int. J. Pharm. Investig.* 2024;14(4):1035–41.

2 Macalino SJ, Gosu V, Hong S, Choi S. Role of computer-aided drug design in modern drug discovery. *Arch. Pharm. Res.* 2015;38:1686–701.

3 Sabe VT, Ntombela T, Jhamba LA, et al. Current trends in computer aided drug design and a highlight of drugs discovered via computational techniques: a review. *Eur. J. Med. Chem.* 2021;224:113705.

4 Talevi A. Computer-aided drug design: an overview. In: *Computational Drug Discovery and Design*. Humana Press; 2018. pp. 1–9.
5 Shah M, Patel M, Prajapati M. Computational transformation in drug discovery: a comprehensive study on molecular docking and quantitative structure activity relationship (QSAR). *Intell. Pharm.* 2024;2:589–95.
6 Lombardino JG, Lowe JA III. The role of the medicinal chemist in drug discovery – then and now. *Nat. Rev. Drug Discov.* 2004;3(10):853–62.
7 Taft CA, Da Silva VB. Current topics in computer-aided drug design. *J. Pharm. Sci.* 2008;97(3):1089–98.
8 Talele TT, Khedkar SA, Rigby AC. Successful applications of computer aided drug discovery: moving drugs from concept to the clinic. *Curr. Top. Med. Chem.* 2010;10(1):127–41.
9 Leelananda SP, Lindert S. Computational methods in drug discovery. *Beilstein. J. Org. Chem.* 2016;12:2694–718.
10 Kitchen DB, Decornez H, Furr JR, Bajorath J. Docking and scoring in virtual screening for drug discovery: methods and applications. *Nat. Rev. Drug Discov.* 2004;3(11):935–49.
11 Clark DE. What has computer-aided molecular design ever done for drug discovery? *Expert Opin. Drug Discov.* 2006;1(2):103–10.
12 Song CM, Lim SJ, Tong JC. Recent advances in computer-aided drug design. *Brief. Bioinform.* 2009;10(5):579–91.
13 Ain QU, Batool M, Choi S. TLR4-targeting therapeutics: structural basis and computer-aided drug discovery approaches. *Molecules* 2020;25(3):627.
14 Kapetanovic I. Computer-aided drug discovery and development (CADDD): in silico-chemico-biological approach. *Chem. Biol. Interact.* 2008;171(2):165–76.
15 Kuntz ID, Blaney JM, Oatley SJ, et al. A geometric approach to macromolecule–ligand interactions. *J. Mol. Biol.* 1982;161(2):269–88.
16 Jain AN, Nicholls A. Recommendations for evaluation of computational methods. *J. Comput. Aided Mol. Des.* 2008;22:133–9.
17 Mursal M, Ahmad M, Hussain S, Khan MF. Navigating the computational seas: a comprehensive overview of molecular docking software in drug discovery. In: *Unravelling Molecular Docking – From Theory to Practice*. IntechOpen; 2014.
18 Madhavi Sastry G, Adzhigirey M, Day T, et al. Protein and ligand preparation: parameters, protocols, and influence on virtual screening enrichments. *J. Comput. Aided Mol. Des.* 2013;27:221–34.
19 Aja PM, Agu PC, Ezeh EM, et al. Prospect into therapeutic potentials of *Moringa oleifera* phytocompounds against cancer upsurge: molecular docking and ADMET studies. *Bull. Natl. Res. Cent.* 2021;45(1):99.
20 Morris GM, Huey R, Lindstrom W, et al. AutoDock4 and AutoDockTools4: automated docking with selective receptor flexibility. *J. Comput. Chem.* 2009;30(16):2785–91.
21 Friesner RA, Banks JL, Murphy RB, et al. Glide: a new approach for rapid, accurate docking and scoring. *J. Med. Chem.* 2004;47(7):1739–49.
22 Elokely KM, Doerksen RJ. Docking challenge: protein sampling and molecular docking performance. *J. Chem. Inf. Model.* 2013;53(8):1934–45.

23 Tiwari S, Prakash K. Molecular docking: a comprehensive review on current scenario and future initiatives. *Int. J. Chem. Res.* 2024;8:1–9.
24 Sakano T, Mahamood MI, Yamashita T, Fujitani H. Molecular dynamics analysis to evaluate docking pose prediction. *Biophys. Physicobiol.* 2016;13:181–94.
25 Agu PC, Afiukwa CA, Orji OU, et al. Molecular docking as a tool for the discovery of molecular targets of nutraceuticals in disease management. *Sci. Rep.* 2023;13(1):13398.
26 Klepeis JL, Lindorff-Larsen K, Dror RO, Shaw DE. Long-timescale molecular dynamics simulations of protein structure and function. *Curr. Opin. Struct. Biol.* 2009;19(2):120–7.
27 Abraham MJ, Murtola T, Schulz R, et al. GROMACS: high performance molecular simulations through multi-level parallelism. *SoftwareX* 2015;1:19–25.
28 Mikhailovskii O, Izmailov SA, Xue Y, et al. X-ray crystallography module in AMBER 2023. *J. Chem. Inf. Model.* 2023;64(1):18–25.
29 Cornell WD, Cieplak P, Bayly CI, et al. A second generation force field for proteins and nucleic acids. *J. Am. Chem. Soc.* 1995;117(19):5179–97.
30 Van Der Spoel D, Lindahl E, Hess B, et al. GROMACS: fast, flexible, and free. *J. Comput. Chem.* 2005;26(16):1701–18.
31 Phillips JC, Braun R, Wang W, et al. Scalable molecular dynamics with NAMD. *J. Comput. Chem.* 2005;26(16):1781–802.
32 Jo S, Kim T, Im W. Automated builder and database of protein/membrane complexes for molecular dynamics simulations. *PLoS One* 2007;2(9):e880.
33 Humphrey W, Dalke A, Schulten K. VMD: visual molecular dynamics. *J. Mol. Graph.* 1996;14(1):33–8.
34 Khedkar SA, Malde AK, Coutinho EC, Srivastava S. Pharmacophore modeling in drug discovery and development. *Med. Chem.* 2007;3(2):187–97.
35 Salo-Ahen OM, Alanko I, Bhadane R, et al. Molecular dynamics simulations in drug discovery and pharmaceutical development. *Processes* 2021;9(1):71.
36 Hollingsworth SA, Dror RO. Molecular dynamics simulation for all. *Neuron* 2018;99(6):1129–43.
37 Tuckerman ME, Martyna GJ. Understanding modern molecular dynamics. *J. Phys. Chem. B* 2000;104(2):159–78.
38 Gupta S, Varadwaj PK. Molecular dynamics simulation of biomolecular systems. *Int. J. Pharm. Sci. Res.* 2018;9(4):1333–50.
39 Allen MP. Introduction to molecular dynamics simulation. In: *Computational Soft Matter*. vol. 23. John von Neumann Institute for Computing; 2004. pp. 1–28.
40 Duay SS, Yap RC, Gaitano AL, et al. Virtual screening and molecular dynamics in antimalarial drug discovery. *Int. J. Mol. Sci.* 2023;24(11):9289.
41 Gan C, Jia X, Fan S, et al. Virtual screening and molecular dynamics to identify SARS-CoV-2 inhibitors. *Acta Virol.* 2023;67:12464.
42 Tovey S, Zills F, Torres-Herrador F, et al. MDSuite: post-processing tool for particle simulations. *J. Cheminform.* 2023;15(1):19.
43 Bissaro M, Sturlese M, Moro S. Molecular simulations in fragment-based drug design. *Drug Discov. Today* 2020;25(9):1693–701.

44 Vega A, Planas A, Biarnés X. Computational tools for engineering biocatalytic properties. *Int. J. Mol. Sci.* 2025;26(3):980.

45 Rizvi SM, Shakil S, Haneef M. AutoDock 4.2 docking protocol for non-bioinformaticians. *EXCLI J.* 2013;12:831.

46 Fan J, Fu A, Zhang L. Progress in molecular docking. *Quant. Biol.* 2019;7:83–9.

47 Sarkar A, Concilio S, Sessa L, et al. Modified AutoDock Vina algorithms. *Results Chem.* 2024;7:101319.

48 Ivanova L, Tammiku-Taul J, García-Sosa AT, et al. Molecular dynamics simulations of GFRα1–ligand interactions. *ACS Omega* 2018;3(9):11407–14.

49 Polo-Cuadrado E, Acosta-Quiroga K, Rojas-Pena C, et al. Molecular modeling of COX-2 inhibitors. *Arab. J. Chem.* 2022;15(2):103540.

50 O'Boyle NM, Morley C, Hutchison GR. Pybel: a Python wrapper for Open Babel. *Chem. Cent. J.* 2008;2:1–7.

51 Renganathan RRA, Vittal RR. Schrödinger Suite based molecular docking studies of metabolites isolated from endophytic fungi for multi-target therapy. In: *Proceedings of International Conference on Drug Discovery (ICDD).* 2020.

52 Ambure P, Aher RB, Roy K. Open-access cheminformatics tools and databases. In: *Computer-Aided Drug Discovery*. Springer Nature; 2016. pp. 257–96.

53 Yoshikawa N, Hutchison GR. Fragment-based coordinate generation for Open Babel. *J. Cheminform.* 2019;11(1):49.

54 Sardar H. Drug-like potential of daidzein using SwissADME. *Phytonutrients* 2023;02:02–8.

55 Daina A, Michielin O, Zoete V. SwissADME: pharmacokinetics and drug-likeness evaluation. *Sci. Rep.* 2017;7:42717.

56 Pattanayak P. SwissADME predictions of 5-nitroimidazole derivatives. *Russ. J. Bioorg. Chem.* 2022;48(5):949–57.

57 Jia CY, Li JY, Hao GF, Yang GF. Drug-likeness toolbox for ADMET study. *Drug Discov. Today* 2020;25(1):248–58.

58 Cosentino V, Luis J, Cabot J. Findings from GitHub: methods and datasets. In: *MSR '16: Proceedings of the 13th International Conference on Mining Software Repositories.* Association for Computing Machinery; 2016. pp. 137–41.

59 Spinellis D, Kotti Z, Mockus A. Dataset for GitHub repository deduplication. In: *MSR '20: Proceedings of the 17th International Conference on Mining Software Repositories.* Association for Computing Machinery; 2020. pp. 523–7.

18

Future Innovations in CADD: AI, Quantum Computing, and Beyond

*Nasir Vadia[1], Ramesh Parmar[1], Ashishkumar Kyada[1], and Vijaykumar Sutariya[2]**

[1]Department of Pharmaceutical Sciences, Faculty of Health Sciences, Marwadi University, Rajkot, Gujarat, India
[2]Department of Pharmaceutical Sciences, USF Health Taneja College of Pharmacy, University of South Florida, Tampa, FL, USA

*Corresponding author: nhvsudps@gmail.com

Introduction

Overview of Computer-aided Drug Discovery

Given the significant progress in biotechnology, pharmaceuticals, and related health sciences, acquiring a comprehensive understanding of disease biology and developing novel, effective small-molecule drugs remain a challenging, extensive, and expensive endeavor. This procedure involves collaboration among professionals from several multidisciplinary domains, including medicinal chemistry, computational chemistry, biology, drug metabolism, and clinical research. Moreover, research indicates that the effective discovery and development of a novel drug takes 1,215 years and significant financial resources [1]. The history of drug discovery goes back to the inception of human civilization. In antiquity, medicines were often discovered unexpectedly or through natural observation, typically using components sourced from plants or animals. They were utilized for both physical treatment and spiritual purposes. The contemporary epoch of pharmaceutical research commenced in the early 1900s. The development of a novel drug typically begins with foundational research, often conducted in academic environments, which identifies a macromolecule (such as a gene or protein), a dysfunctional signaling pathway, or a molecular mechanism seemingly linked to a disease (prediscovery stage) [2]. At this stage, research teams aim to identify therapeutic targets, typically proteins, associated with the illness state [3]. To determine a therapeutic target, researchers must also find therapeutic drugs capable of modifying the function of the impaired target to restore health or alleviate symptoms. Identifying the appropriate target, nonetheless, is exceedingly challenging. Moreover, drugs are efficacious in humans not only because of their unique activities at designated therapeutic targets but also because of interactions with additional, undesired, and frequently unidentified targets [4]. The process evolves with the identification of therapeutic molecules, followed by a preclinical phase in which potential drugs are evaluated in diverse animal models to establish safety and select drug candidates. Human clinical trials can then begin to assess the safety and efficacy of the drugs in patients, aiming to achieve an optimal benefit-to-risk ratio [5]. The studies are then submitted to regulatory agencies, which review the documents and determine whether to approve the market. If the review is favorable, the drug can be released to the market and administered to patients. Following drug approval, further studies are conducted to assess any long-term adverse reactions associated with the newly approved therapy. This concluding phase is referred to as pharmacovigilance studies, sometimes designated as "phase 4" clinical trials. The drug discovery and development process is expensive and encompasses multiple disciplines, requiring years of effort. It requires the generation and use of extensive data, typically acquired via diverse high-throughput methods [6]. A multitude of these evaluations and the subsequent analysis of results can be automated by computer-assisted techniques that accelerate some stages of the process, augment understanding, and minimize errors [7]. In the complex, tedious, and resource-demanding domain of drug development, a diverse array of computational methodologies is extensively utilized. The late 20th century was a crucial period for this field of study, characterized by the emergence of computer-aided drug discovery (CADD). The traditional approach of drug discovery is presented in Figure 18.1. This approach integrates complex characteristics of biological systems with the ability to predict using computational algorithms and the development of databases, including chemical and biological data. The core principle of CADD is the utilization of computer algorithms on chemical and biological data to simulate and predict the interaction between a therapeutic molecule and its target, typically a protein or DNA sequence within a biological system. This technique analyzes the molecular configurations of the drug or its

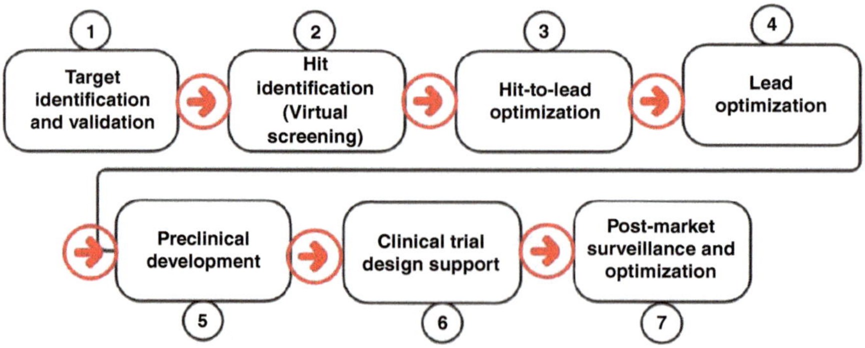

Figure 18.1 Traditional approach of drug discovery.

target, predicting binding interactions, pharmacological effects, and undesirable effects [8]. The emergence of CADD was driven by two significant advancements: the development of structural biology, which revealed the three-dimensional (3D) structures of biomolecules, and the rapid increase in computational power, enabling complex simulations to be conducted in shorter timeframes [9]. The use of CADD techniques by prominent pharmaceutical companies and research organizations has become essential in the early phase of drug discovery, speeding up the drug development process in a more economical manner and reducing failures in the final stage. Recent advancements in rational drug design have been reported in multiple sources [10]. Although the crucial role of high-throughput screening (HTS) in pharmaceutical drug discovery stems from its higher success rate, an insufficient understanding of the molecular mechanisms underlying the activity of discovered hits may hinder the selection of viable candidates [11, 12]. The drug discovery strategy using CAD is a knowledge-driven scientific approach that provides substantial insights into protein–ligand interactions and binding affinities. Furthermore, the availability of supercomputers, parallel computing, and advanced software has markedly improved lead identification in pharmaceutical research. The emergence of CADD caused a paradigm shift in drug discovery and development, transforming the approach from primarily empirical to more systematic and targeted [13]. Yet it has some limitations. When predicting the behavior of biological systems solely through computer simulations, it is crucial to acknowledge the inherent limitations. For instance, consider a scenario in which a data simulation accurately replicates molecular interactions between a receptor and its target [14]. However, predictions may be biased by accurate results if the simulation omits external environmental variables or lacks critical scientific information on unexpected biological responses. Despite their complexity, these models often require data validation to support their predictions. CADD signifies the collaborative integration of biology and technology that accelerates drug discovery [7]. Although significant progress has been made in the domain, the full promise remains incomplete as new calculation techniques and a deeper understanding of biological systems emerge.

Evolution of CADD Technology

Rule-based Systems

In the 1980s, CADD began with fundamental rule-based systems in chemistry and molecular modeling. These systems depended on predetermined chemical rules and empirical scoring functions to predict molecular interactions [15]. During those days, initial software such as SYBYL and DOCK used molecular docking to estimate how small molecules fit in protein binding sites [16]. Quantitative structure–activity relationship (QSAR) models correlate molecular features with biological activity [17]. The fundamental limitations were that these systems were more expensive, highly dependent on manual inputs, and limited by simplified assumptions about molecular behavior. They struggled with a complex biomolecular system and lacked flexibility [18].

Molecular Dynamics and Simulation-based Approaches

Following the 1980s, the use of computers in studying molecular dynamics (MD) simulations increased. During that period, extensive simulations of small proteins and nucleic acids were conducted using either CHARMM or AMBER force fields, comparing the experimentally observed conformations to those predicted by the computer [19]. Computer-aided tools, such as AutoDock Vina and Schrödinger, substantially improve the level of accuracy of the method being used [7, 20]. QSAR was further enhanced with 3D-QSAR, facilitating more accurate scientific investigations of molecular dynamics [21]. Virtual screening facilitates the identification of possible drug candidates. The combined approach of X-ray crystallography and NMR has led to high-resolution protein structures, enhancing docking accuracy [22]. Some limitations of this technology include the fact that simplified scoring functions often cannot predict binding affinities with precision. Using these static models leads to inflexibility and a lack of conformational results in the study. The study's unavailability/erroneous data significantly affects its accuracy [23].

Machine Learning and Artificial Intelligence

The emergence of machine learning (ML) and artificial intelligence (AI) has transformed CADD by facilitating predictive models that learn from extensive datasets, thereby overcoming the constraints of rule-based and docking methodologies [24]. Algorithms such as random forests and support vector machines predict molecular properties using training data [25]. This technology enables convolutional neural networks (CNNs) and graph neural networks (GNNs) to simulate detailed molecular architectures and interactions that traditional methods cannot achieve [26]. With variational autoencoders (VAEs) and generative adversarial networks (GANs), it is possible to create innovative compounds with specified attributes [27]. Public databases, such as ChEMBL and PubChem, as well as proprietary datasets, supply millions of molecular and biological data points [28]. AlphaFold is an AI system developed by Google DeepMind that predicts a protein's 3D structure from its amino acid sequence [29]. This technique solved the protein structure prediction problem, delivering precise 3D models of pharmaceutical targets. AI models efficiently manage extensive datasets and intricate systems. Deep learning (DL) elucidates nonlinear interactions, enhancing the accuracy of binding affinity and toxicity forecasts [30]. Generative AI formulates innovative compounds, broadening the investigation of chemical space. This technology is constrained by the need for high-quality, diverse datasets; data biases can lead to erroneous predictions. Training DL models necessitates considerable resources. AI has accelerated drug discovery timelines, encompassing applications in lead optimization, toxicity prediction, and personalized treatment [31].

Quantum Computing

Quantum computing has gained prominence since 2010, entering the field of data-aided drug design (DAD), and offers promise in solving quantum-mechanical problems beyond the capabilities of classical computers [32]. A highly promising application is quantum chemistry simulation, where quantum computers can accurately represent electron-level interactions, resulting in enhanced predictions of molecule properties [33]. Moreover, hybrid algorithms, such as Variational Quantum Eigensolver (VQE) and quantum learning, have been devised to combine the advantages of classical and quantum systems, thereby enhancing the efficacy of drug discovery and development processes [34]. Leading technology firms, like Google and IBM, together with pioneering startups, such as Zapata Computing, are significantly investing in the development of quantum algorithms that investigate the application of cloud computing in drug discovery [35]. In this context, the primary benefits of quantum calculation include the ability to simulate large biomolecular systems, providing rapid solutions to complex problems, enhanced accuracy, and facilitating molecular adaptation. Existing quantum computers are characterized by noise and a limited number of qubits, which constrain their practical applications [36]. Quantum algorithms for pharmaceutical discovery remain in the experimental phase. The prohibitive expenses and requisite expertise hinder adoption.

AI in Drug Discovery

AI changes traditional drug and development methods by effectively combining data, computer dynamics, and algorithms [37]. This collaboration improves drug research efficiency, accuracy, and success rate, accelerates development time, and reduces expenses. AI, as a combination of ML and DL, has demonstrated significant progress across multiple domains, including drug discovery and validation, small-molecule design, and clinical research [38]. AI can generate novel drug candidates and predict their functions and activities, while virtual screening enhances drug selection. Additionally, the AI enhances the effectiveness of clinical studies by predicting outcomes, collecting and structuring data, and interpreting the results. Pharmaceutical companies using AI should integrate biological science and algorithms to effectively merge wet- and dry-laboratory experiments. As AI technology advances and these obstacles are overcome, AI-operated medications will be distributed to support a more comprehensive and vital future in pharmaceuticals [39, 40]. Commonly used AI tools are given in Table 18.1. This table presents details about the technologies used, their areas of application, and the names of their developers.

ML and DL Approaches

Machine intelligence achieved through AI performs human cognitive abilities through specialized training using a specifically designed algorithm. These algorithms use a larger data pool for innovative drug discovery [41]. This integrates pattern recognition, probability theory, statistics, ML, and computational intelligence methodologies, considering various models and neural networks for generating, processing, and executing data commonly employed in clinical and nonclinical settings [42]. ML, a fundamental aspect of AI, excels by adjusting to new data. A model is initially established with stated parameters, allowing

Table 18.1 AI tools used in drug discovery.

AI tool name	Basic details	Application area	Developer
AlphaFold	A DL algorithm that forecasts 3D protein structures from amino acid sequences with near-experimental accuracy	Protein folding, structural biology, and target identification	DeepMind (Google)
IBM Watson for Drug Discovery	Uses NLP, ML, and semantic analysis to extract and connect insights from scientific literature and databases	Literature mining, target identification, and biomarker discovery	IBM
AtomNet	A deep CNN trained on protein–ligand structures to predict binding affinities	Structure-based virtual screening, hit identification	Atomwise
BenevolentAI Platform	Integrates biomedical literature, genomic, and patient data using AI to identify drug targets and repurpose compounds	Target identification, drug repurposing, and hypothesis generation	BenevolentAI
Chemputer/ ChemOS	An AI-guided chemical synthesis automation platform that interprets and executes synthetic chemical procedures	Automated chemical synthesis, compound preparation	University of Glasgow/ Cronin Lab
Exscientia AI Design Platform	Combines DL, active learning, and reinforcement learning (RL) for rapid design and optimization of drug molecules	De novo design, lead optimization, clinical candidate selection	Exscientia
DeepChem	An open-source toolkit offering DL libraries tailored to chemistry and life sciences	Bioactivity prediction, ADMET modeling, molecular property prediction	Stanford University (initially)
Insilico Medicine Platform	Integrates generative AI with multi-omics and biological data for end-to-end drug discovery and biomarker identification	Target identification, hit generation, and preclinical development	Insilico Medicine
MELLODDY Project	A federated learning platform allowing multiple pharma companies to train AI models on private data without sharing it	Predictive modeling (QSAR, ADMET), lead optimization	Consortium (e.g., Janssen, Novartis, Servier)
REINVENT	An RL-based platform for de novo molecular generation with specific desired properties.	Scaffold hopping, molecular optimization, and property-based design	AstraZeneca/ Molecular AI (Sweden)

AI tool name	Basic details	Application area	Developer
BioXcel BDAI	An AI-powered platform to identify novel applications for existing drugs through big data analytics and pattern recognition	Drug repurposing, behavioral health, immune-oncology	BioXcel Therapeutics
Cyclica's Ligand Express	Uses AI to predict polypharmacology and off-target interactions through proteome-wide screening	Target deconvolution, drug safety profiling, lead prioritization	Cyclica
Peptone AI	Uses protein biophysics and AI to design better biologics by understanding protein dynamics and misfolding	Biologic drug design, protein optimization	Peptone

machines to enhance their capabilities through training data [43]. This procedure enables the model to predict future trends and uncover significant insights. The efficacy of ML is evident in its expanding application across various sectors, as major pharmaceutical firms increasingly leverage their capabilities to enhance drug research and development. Integrating ML into future developments is vital due to its transformative potential in drug discovery. ML utilizes high-throughput screening technology to optimize drug development, minimizing risks and expenses [44]. Ultimately, it presents the possibility of diminishing or eliminating the need for live animal experimentation. These accomplishments underscore the pivotal role of ML in revolutionizing the field of drug discovery. In drug discovery and development, ML models and techniques are utilized to accelerate and enhance various stages, from target identification to clinical trials. Presented here is a comprehensive classification of frequently used ML models and techniques, along with their standard applications. Leading DL and ML technologies used in drug discovery are presented in Figure 18.2. DL has revolutionized the process of new drug discovery. It offers ample data analysis, pattern recognition, and predictive modeling capabilities that are not commonly achieved with conventional computational techniques [45]. DL utilizes neural networks to automatically identify complex relationships within large volumes of structured and unstructured data, unlike traditional methods that rely primarily on rule-based algorithms, manual feature extraction, and small datasets [46]. This enables more accurate predictions for target identification, molecular design, binding affinity estimation, and toxicity assessment [47]. A prominent application of DL is virtual screening, where CNNs and GNNs evaluate molecular structures to predict bioactivity, thereby significantly reducing the cost and time required for HTS [48]. Compared to conventional QSAR models, DL models, such as DeepChem and Chemprop, can categorize millions of compounds in silico and identify promising candidates with greater accuracy [49]. De novo drug design generates novel druglike compounds with enhanced characteristics utilizing generative DL models, including GANs

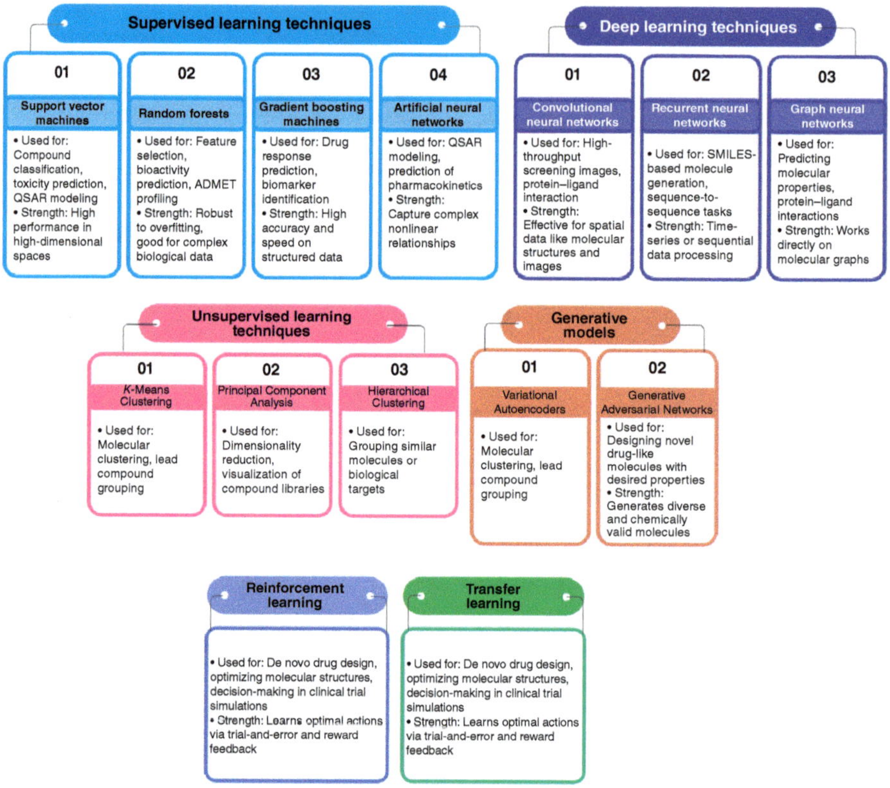

Figure 18.2 ML and DL technologies in drug discovery.

and VAEs [50]. These methodologies have been efficiently used to accelerate the development of preclinical drug candidates. Another significant development is the prediction of protein–ligand interactions, where DL models outperform traditional molecular docking tools such as AutoDock and Glide [51]. Even for proteins without known crystal structures, methods such as 3D-CNN can accurately predict binding affinity. Researchers can employ unexplored targets to develop a structure-based drug design that was previously impossible. The results suggested that AlphaFold and RoseTTAFold were successful in predicting protein 3D structures [52]. Traditional approaches fundamentally depend on heuristic rules and animal testing, which are time-consuming and often ineffective for humans. A DL model trained on bioassay, pharmacokinetic, and toxicogenomic datasets can forecast the drug safety profile, thereby minimizing the costs associated with late-phase errors [53]. For example, DeepTox and ADMET Predictor utilize DL to predict and evaluate blood–brain barrier penetration, hepatotoxicity, and cardiotoxicity [54, 55]. Drug repurposing study is also possible with DL. Using DL tools, it is possible to analyze omics data, scientific literature, and electronic health records (EHRs). DL will predict the potential new drug applications for already approved drugs [45]. For example, the repurposing of baricitinib for COVID-19 was established using BenevolentAI, a DL-based tool [56].

Applications in Target Identification and Validation

The most crucial step in drug discovery is target identification, which enables the identification of key biological molecules or cell pathways through which drugs can exert therapeutic action. Despite advancements in experimental techniques and omics-based technologies over the past few decades, therapeutically meaningful target identification remains a significant challenge. Recently, scientists have begun to leverage the potential of multi-omics datasets in conjunction with AI algorithms, providing new directions for identifying potential targets [57]. This is typically carried out using three main strategies: experimental, multi-omics, and computational approaches [58]. Combined, these approaches enhance the validity of exploratory target identification and open the door to generating new therapeutic hypotheses, thereby further enhancing our understanding of complex disease mechanisms. Different approaches used for target identification are illustrated in Figure 18.3. Since the 1960s, experimental methodologies for target identification, including affinity-based biochemical assays, comparative profiling, and chemical and genetic screening, have proven vital. These techniques, encompassing small-molecule affinity probes, stable isotope labeling by amino acids in cell culture (SILAC), and chemical and genetic screening, have been employed in different cancer types [59]. CRISPR-Cas9 gene editing has enhanced our comprehension of human diseases, with research elucidating its involvement in disease pathogenesis [60]. Multi-omics data, which encompasses static genetic information and dynamic expression

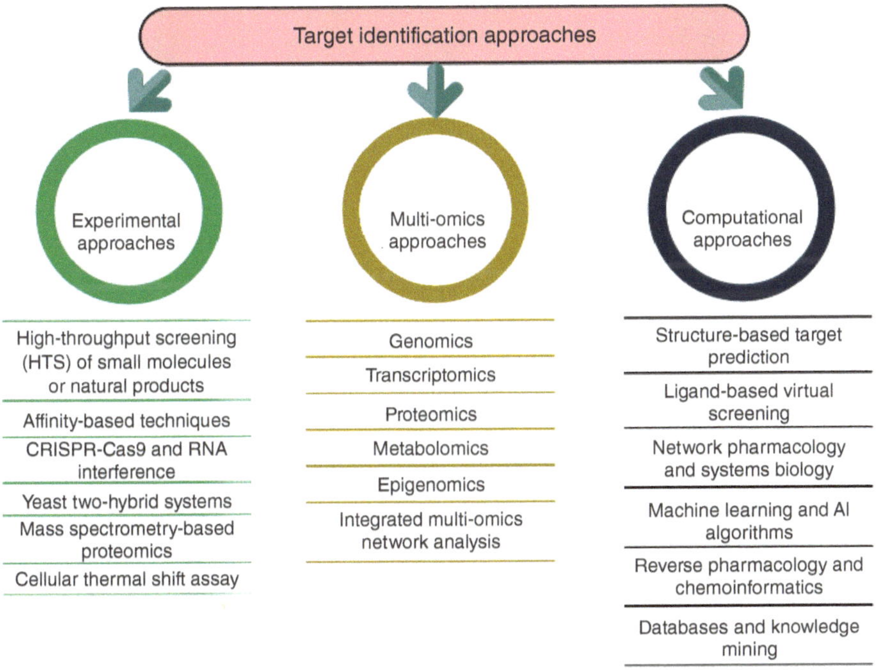

Figure 18.3 Strategies for target identification.

and metabolic profiles, provides researchers with combined molecular insights. Genomics examines genetic variations in DNA sequences, leading to innovative therapeutics such as cystic fibrosis modulators and new treatments for inflammatory bowel disease. Integrating various omics data can yield a more holistic understanding of disease mechanisms, facilitating the identification of biomarkers and therapeutic targets, as well as predicting treatment responses and patient outcomes [61]. Computerized techniques have emerged as effective alternatives for efficient target screening, including pharmacophore screening, reverse docking, and structural similarity evaluation. AI is an expanding field within computational science, particularly in target discovery, where ML plays a crucial role. Supervised learning employs labelled datasets for data classification and accurate outcome prediction, whereas unsupervised learning investigates the latent structure of unlabeled data without human intervention.

ML, DL, and AI-driven Virtual Screening and Lead Optimization

Virtual screening (VS) is a valuable computational technique for drug discovery, enabling the prediction of interactions between bioactive compounds and biological targets. It includes ligand-based and structure-based approaches for finding potential drug candidates. Nonetheless, difficulties persist in accurately predicting interactions, managing enormous chemical libraries, and enhancing hit identification accuracy. Structure-based virtual screening is essential for drug discovery, but only a few successful campaigns have been reported. Techniques include scalable platforms, DL, hierarchical structure-based screening, and GPU-accelerated ligand docking. The effectiveness of VS campaigns using the approaches above depends fundamentally on the accuracy of the ligand docking programs used to predict protein–ligand complex structures and distinguish true binders from nonbinders. Notable physics-based ligand docking tools, such as Schrödinger Glide and CCDC GOLD, and their virtual screening platforms for comprehensive library assessment are not accessible to researchers without incurring financial costs [62]. AutoDock Vina, a widely used free software, demonstrates slightly lower virtual screening accuracy than Glide [63]. Moreover, there is a lack of an open-source, scalable virtual screening platform that utilizes active learning for comprehensive virtual screening of chemical libraries. The emergence of DL technology has led to the development of numerous models designed to predict protein–ligand complex structures in a significantly reduced timescale. These approaches, however, are more suitable for blind docking, when the binding site of the small molecule is unknown. In scenarios where the binding site is discerned, as is often the case in virtual screening, physics-based ligand docking methodologies routinely outperform DL models. Moreover, DL approaches demonstrate reduced generalizability to new challenges. Figure 18.4 illustrates various traditional virtual screening techniques and advanced methods based on ML, DL, and AI. It also highlights key applications and commonly used software tools for virtual screening, as well as the evolution of virtual screening technologies. AI, ML, and DL are technologies designed to play a crucial

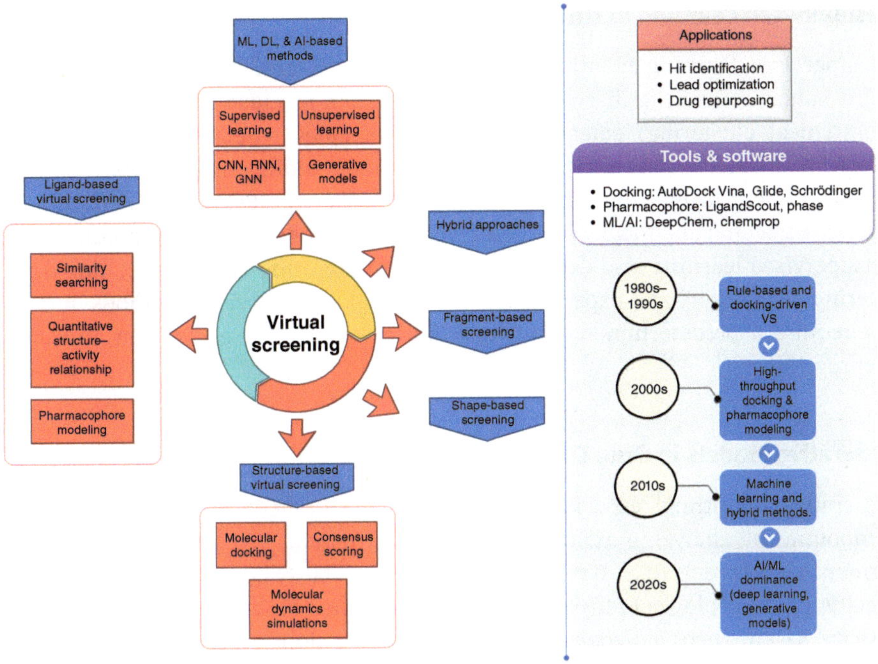

Figure 18.4 Virtual screening methodologies.

role in discovery. As mentioned earlier, AI encompasses broader areas such as rule-based systems, adaptation techniques, and ML and is used in literature mining to identify drug targets from educational publications. ML emphasizes statistical modeling approaches, such as random forests in AI, frequently employed in QSAR studies. This encompasses both supervised and unsupervised learning techniques. DL is a distinct subset of ML that utilizes deep neural networks, including CNNs and GNNs, to address complex problems such as molecule synthesis and protein folding, as exemplified by AlphaFold. Meanwhile, DL extraction functions from raw data require an extensive, well-annotated dataset.

Supervised Learning in Drug Discovery

Supervised learning techniques are widely used to predict molecular properties or biological activities of chemical compounds of interest in various databases. For example, QSAR utilizes functions such as molecular description or fingerprints to predict binding affinity, solubility, or toxicity. Algorithms, such as random forests, support vector machines (SVMs), and gradient boosted trees (e.g., XGBoost), are often used [64]. These models help prioritize promising drug candidates by estimating their efficiency and safety before incurring the expense of experimental verification. Applications include virtual screening, addressing ADME prediction, and target identification.

Unsupervised Learning in Drug Discovery

Unsupervised learning identifies hidden patterns in raw, unlabeled data, making it valuable for drug discovery. Techniques such as clustering (e.g., *k*-means, hierarchical clustering) categorize similar molecules based on their physicochemical properties and structural configurations, aiding in lead optimization and scaffolding. Dimensionality reduction methods (e.g., PCA, tSNE) visualize a high-dimensional chemical space and reveal connections among compounds [65]. Unsupervised learning also detects deviations in high-throughput screening data, filtering out outliers and experimental errors. Unlike supervised methods, it does not require a predetermined label, making it valuable for generation and data preprocessing.

Generative Models in Drug Discovery

Generative algorithms, especially those based on DL, create innovative drug-like compounds by analyzing existing chemical information. VAEs, GANs, and transformer-based models (e.g., GPT for molecules) can clarify compounds that possess specific pharmacological attributes, including elevated binding affinity or reduced toxicity. RL enhances molecules for particular applications or objectives [66]. These methods expedite de novo drug development by identifying extensive chemical spaces beyond human perception, thereby facilitating the discovery of structurally novel molecules devoid of medicinal prerequisites.

Recurrent Neural Networks in Drug Discovery

Recurrent neural networks (RNNs) process sequential data, which is well suited for tasks involving molecular sequences (e.g., SMILES strings for small molecules or amino acid sequences for proteins). Long short-term memory (LSTM) and gated recurrent Unit (GRU) networks forecast molecular characteristics, thereby advancing the development of potential drug candidates. RNN is utilized to forecast pharmacokinetic parameters and examine time series data from biological analyses [67]. However, their sequential nature limits parallelization, leading to slower training than other architectures.

CNNs in Drug Discovery

CNNs are distinguished by the type of input data, such as molecular illustration images (e.g., 2D chemical structures) or 3D voxelized protein–ligand complexes, which are depicted as a grid during processing. They automatically remove hierarchical features (e.g., function groups, pharmacophores) for tasks such as binding affinity prediction, protein–ligand interaction prediction, and toxicity classification. CNN is also used in the cryo-EM image analysis for structural determination [68]. Although it traditionally applies to image data, adaptation via graph convolution networks (GCNs) distinguishes it from non-Euclidean molecular representations.

GNNs in Drug Discovery

GNNs work directly with molecular graphs, where molecular nodes and bonds retain structural information. It predicts molecular properties (e.g., solubility and bioactivity), optimizes lead connections, and models protein–protein interactions. Message-passing neural networks (MPNNs) and graph attention networks (GATs) effectively capture local and global dependencies, surpassing conventional approaches in tasks such as virtual screening and reaction prediction. GNN structure-based drug design is compelling, as it naturally handles the relationships between chemistry and biology [69].

Quantum Computing in Drug Discovery

Principles of Quantum Computing Relevant to Drug Discovery

The principles of quantum mechanics enable the formulation of quantum algorithms for drug discovery and development. These algorithms facilitate rapid examination of large datasets, thereby reducing overall research time, costs, and manpower requirements. Traditional chemical modeling techniques are often rejected due to their limited ability to accurately represent molecular interactions. The quantum approach employed in drug discovery improves researchers' capacity to forecast molecular interactions, therefore enhancing the identification of potential drug candidates [70]. Quantum computing is built on principles of quantum mechanics, distinguishing itself from classical computing in fundamental ways. Classical computers utilize bits, which represent information as either 0 or 1. In contrast, quantum computers employ qubits, which can exist in a superposition of states, enabling them to define and process many combinations of 0 and 1 simultaneously. This attribute allows quantum computers to explore multiple potential solutions to a problem concurrently, potentially offering exponential speedup for specific tasks. Quantum gates are fundamental components of quantum circuits, altering qubits to perform operations. Quantum gates can generate entanglement among qubits, a phenomenon wherein the state of one qubit is contingent upon the state of another, regardless of their spatial separation. This attribute is essential for quantum parallelism and interference, enabling quantum computers to efficiently process complex datasets [71]. This property is essential for quantum parallelism and interference, enabling the quantum computer to effectively analyze complex datasets. Figure 18.5 represents the principles of quantum mechanics and their applications in drug discovery.

Quantum Algorithms for Molecular Simulation

Challenges in molecular simulation pose substantial barriers to the precise prediction of molecular reactions, particularly in quantum chemical simulations. A significant difficulty is the scalability of existing approaches, which inadequately model huge systems due to the exponential increase in necessary processing

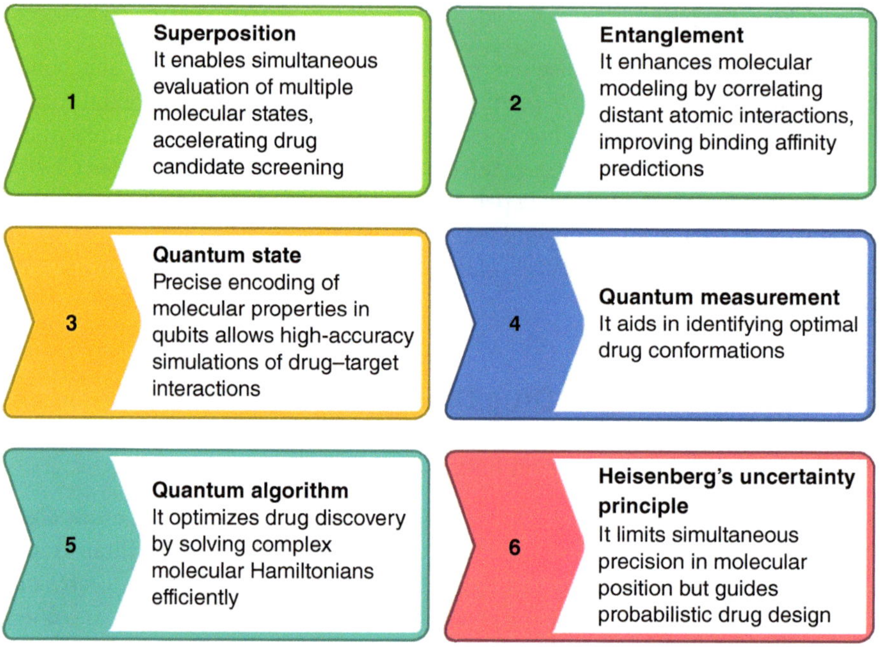

Figure 18.5 Principles of quantum mechanics.

resources. This constraint necessitates the development of innovative algorithms and methodologies that can efficiently manage complex molecular systems. Another significant challenge is the precise representation of molecular interactions, which is essential for accurately replicating chemical reactions. The management of electron correlation presents a considerable difficulty in molecular simulations. Electron correlation effects are crucial for determining the electronic structure of molecules; however, precisely capturing them with existing approaches can be computationally prohibitive. Consequently, researchers often rely on approximations or simplifications that can compromise the accuracy of simulations [36]. Quantum computing offers potential solutions to several of these challenges. Quantum algorithms, including the quantum phase estimation technique, may resolve specific problems more efficiently than classical computers. Quantum simulation can accurately model molecular interactions, providing valuable insights into drug interactions, bonding mechanisms, and other aspects of molecular dynamics. Quantum algorithms analyze large datasets to identify patterns and predict molecular properties, aiding the search for new medications and the understanding of disease pathophysiology. Quantum computing is transforming the domain of MD simulations. These tools are not merely augmenting current approaches but radically revolutionizing our approach to computational biology [32]. AI and ML enhance the efficiency of processing massive datasets from MD simulations by simplifying complex data analysis and optimizing simulation settings, thereby expediting hypothesis testing and adaptation. This integration enhances the accuracy and speed of simulations, thereby enriching our understanding of complex biological systems. Quantum algorithms for chemical applications have the potential to transform quantum chemistry by facilitating the modeling of molecular reactions with remarkable accuracy. The quantum phase estimation (QPE) method effectively computes the eigenvalues of a Hamiltonian operator. This method is particularly beneficial for chemical applications, as it enables the effective simulation of molecular systems. The quantum approximate optimization algorithm (QAOA) has also addressed problems in chemistry. QAOA is a hybrid quantum-classical method that utilizes both classical and quantum computational resources to obtain approximate solutions to optimization challenges. This method has successfully resolved optimization challenges in chemistry, including the computational identification of molecular ground states. The VQE is a significant quantum algorithm utilized in chemical applications. It is a hybrid quantum-classical method that uses a classical optimizer to determine the optimal parameters for a quantum circuit, which is then used to estimate the eigenvalues of a Hamiltonian operator [72]. This method has been effective in simulating molecule configurations and elucidating their electronic structure. Quantum algorithms have been employed to simulate chemical reactions, encompassing the calculation of reaction rates and mechanisms. The quantum circuit learning (QCL) algorithm leverages a combination of classical and quantum computing resources to determine the optimal parameters for a quantum circuit that simulates a chemical reaction [73]. The application of quantum algorithms in chemistry has led to the development of innovative, quantum-inspired classical algorithms. The density matrix embedding theory (DMET) is a method that utilizes both classical and quantum computational resources to simulate molecular systems. Quantum algorithms for chemical applications have the potential to revolutionize quantum chemistry by enabling the precise modeling of molecular reactions with remarkable accuracy.

Emerging Quantum Hardware and Platforms

Quantum computing presents a remarkable opportunity to transform drug discovery by accelerating and enhancing multiple phases of the drug development process, including identifying potential drug candidates and optimizing their efficacy and safety. This approach can overcome the constraints of conventional drug discovery techniques by facilitating more precise and effective simulations of molecular interactions and intricate biological systems. Notable applications include accelerated molecular modeling and simulation, optimization of drug design and development, enhancement of clinical trial design and analysis, personalized medicine and disease modeling, and streamlining the drug discovery process. Figure 18.6 illustrates prominent global companies engaged in quantum computing, AI, and ML technologies. These companies offer hardware, software, and associated services, actively participating in numerous aspects of drug development, encompassing illness selection, target validation, hit identification, lead identification, and lead optimization [74]. Roots Analysis, a business research and consulting group, has comprehensively analyzed the quantum computing market, highlighting the present market environment and prospective opportunities for service providers delivering

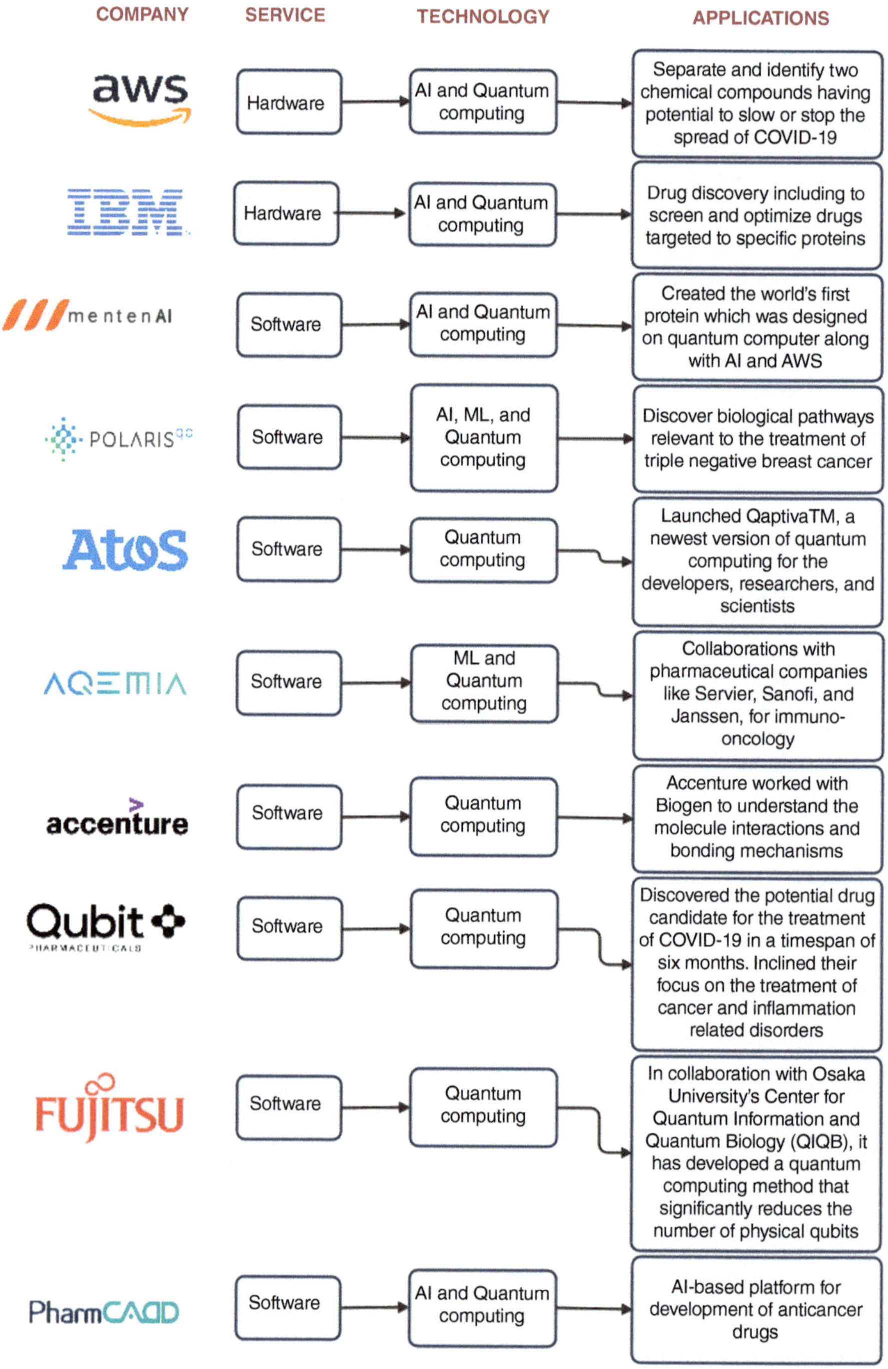

Figure 18.6 Leading global companies involved in the development and application of quantum computing, AI, and ML technologies and their applications.

software and hardware solutions for drug development [75], given the significant potential of quantum computing to minimize costly expenditures by decreasing preclinical failures and efficiently analyzing giant, complex molecules in a reduced time frame. Consequently, numerous industry players anticipate incorporating this methodology into their drug discovery and development processes. The growing application of quantum computing in drug discovery is expected to drive substantial demand for quantum computers in the near future.

Integrative Approaches: AI Meets Quantum Computing

AI methods, including ML and natural language processing, can intensify this process by facilitating more efficient and accurate analysis of extensive drug discovery datasets. AI-operated approaches have successfully predicted the toxicity of drug candidates. Novel bioactive compounds pose challenges for AI applications due to their vast range, rapid expansion, structural variability, and inherent unpredictability. Pharmaceutical companies have datasets for drug research that may include millions of molecules, and traditional ML methods for processing such data can be inadequate. Nevertheless, these models fail to predict complex biological properties, including the effectiveness of medicines and their harmful consequences. In addition, QSAR-based models face challenges, including limited, inaccurate, or inadequate training datasets and inadequate experimental data. To address these challenges, integrating AI with other advanced digital technologies, such as quantum computing, DL, blockchain, and digital twins, can enhance safety and efficiency assessments of potential drug compounds. These methods leverage extensive data modeling and analysis, as demonstrated in recent studies. AI is a crucial driver of quantum computation, helping alleviate some of the field's most significant challenges. In particular, AI contributes to the following vital areas: hardware customizing AI automates the design and calibration of quantum processors, enabling precise control over qubit systems. This optimization enhances the simulation of complex molecular interactions, a cornerstone of drug discovery, while minimizing quantum noise that could distort results. Error detection and correction, AI models trained to identify quantum errors, improve the reliability of molecular simulations. For instance, Google's AlphaQubit demonstrates how AI can mitigate errors in quantum calculations, which is critical for accurately predicting drug–target binding or protein folding dynamics [76]. ML, particularly RL, tailor quantum algorithms to simulate pharmacological systems more efficiently. Hybrid quantum-classical algorithms, streamlined by AI, are already being used to model drug–receptor interactions and optimize lead compounds, thereby reducing reliance on costly trial-and-error experimentation [31].

Hybrid Quantum-classical Models

A paradigm shift has been observed in the drug discovery process nowadays. Previously, traditional approaches have been used for drug discovery and development. However, over the last few years, hybrid quantum-classical models have been explored in drug discovery, integrating principles of quantum and classical computing to overcome challenges posed by traditional methods [70]. Quantum computing enhances the accuracy of molecular properties, such as binding energy, while classical methods handle data processing and wide-ranging workflows. These hybrid approaches enhance tasks such as virtual screening, molecular design, and reaction optimization, providing faster and more accurate solutions than traditional methods [77]. Although the current quantity is limited by hardware noise and qubit barriers, hybrid models serve as a practical bridge, enabling practical applications in drug discovery and paving the way for future successes with advanced quantum operations. Some leading applications of this hybrid approach are briefly discussed below. Transforming molecular optimization challenges, drug discovery includes essential functions, such as molecular docking and novo drug design, which evaluate countless possible compound configurations to identify optimal candidates. Quantum computing's ability to detect multiple molecular states simultaneously enables researchers to identify promising drug candidates and address challenging problems in the rapidly evolving fields of molecular dynamics and protein–ligand interactions. Securing sensitive pharmaceutical data, quantum-enhanced AI provides robust solutions for protecting valuable drug discovery data through advanced encryption. Quantum key distribution (QKD) ensures secure transmission of proprietary compound data. In contrast, blind quantum computing enables collaborative research by allowing quantum servers to process confidential molecular datasets without exposing sensitive structural information or research parameters. Revolutionizing drug discovery through accelerated computational workflows: traditional AI technology requires extensive computational resources to study, especially for molecular modeling and structural relationships. Quantum computing addresses these limitations by enabling parallel processing of molecular data, dramatically reducing the time required for virtual screening and enabling the analysis of extensive, complex chemical libraries [78].

Other Emerging Technologies

Digital Twins in Drug Discovery

The integration of diverse omics data and single-cell sequencing information with ML, DL, and AI enables the creation of digital twins of human cells for drug discovery and investigation. DeepLife is a startup that provides a cellular digital twin platform utilizing single-cell omics data to enhance and intensify target search. The cloud-based system can quickly assess the reactions of 10 million cells to billions of drugs, as well as their combinations with different compounds. DeepLife is improving in vitro testing for drug discovery to maximize target identification. DeepLife has developed a collaborative, data-driven drug development ecosystem to accomplish that goal. The all-in-one solution uses data from DeepLife, its clients, and public sources, and applies an ML-based data curation process. Digital twins can aid in forecasting drug distribution, absorption, metabolism, and excretion. They may utilize patient-centric data that encompasses physiology, genetics, lifestyle factors, and other relevant factors. Digital twins may also help researchers identify potential future problems and optimize their formulas accordingly. They may also help in simulating drug delivery systems, such as nanocarriers or implantable devices, facilitating the optimization of medication doses and release rates accordingly [79]. Some current examples of this technology in the real world include AstraZeneca, which has partnered with Insilco Medicine to utilize digital twin technology in its drug candidate identification and development processes [80]. Pfizer has established a comparable collaboration with BioLingus to enhance the formulation and distribution of biologics through digital twins [81]. Roche has invested in Physiomics to leverage digital twins, while Novartis is employing this technology in collaboration with Lattice AI [82]. Sanofi has invested in Owkin, a company that utilizes digital twin technology to accelerate drug research and discovery [83].

Blockchain for Data Integrity and Sharing in Research

Blockchain technology is a decentralized, irreversible digital ledger system. It comprises several interlinked blocks, each having numerous transactions. Since data cannot be modified once inscribed in a block, both transparency and security are guaranteed. Due to its decentralized structure, blockchain is maintained by a network of computers rather than a single authority [84]. Clinical trials are a crucial component of drug research and development, providing the evidence needed to establish a new drug's safety and efficacy. Nonetheless, transparency and data integrity in clinical trials have been key considerations. Blockchain technology has the potential to profoundly transform the critical stage of drug development. Blockchain technology enables secure and transparent documentation of each phase of clinical research. The blockchain's immutability ensures that trial data remains unchanged and can be independently verified, thereby enhancing trust in the integrity of the results. Furthermore, blockchain facilitates secure data sharing and collaboration among trial stakeholders, including researchers, ethics committees, and regulatory authorities [85]. This accelerates the trial process and ensures the highest ethical and scientific standards are upheld. Figure 18.7 presents several elements useful in drug discovery.

Augmented and Virtual Reality in Molecular Visualization

The conventional drug development method is protracted, costly, and resource-demanding. VR/AR technologies offer a viable alternative by enabling researchers to observe and interact with intricate 3D representations of cell, tissue, or organ architecture, cellular structures, and molecular interactions. Virtual reality can replicate disease models, such as psoriasis or eczema, enabling researchers to test pharmacological responses in a controlled virtual environment. This capability diminishes dependence on animal models and conventional wet lab experiments, enhancing preclinical research [86]. A practical example is the application of

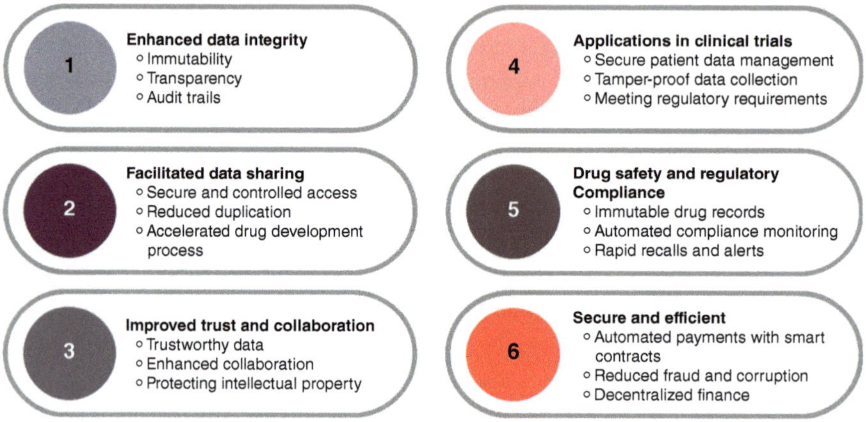

Figure 18.7 Elements of blockchain in drug discovery.

augmented reality (AR) to view molecular docking in pharmaceutical research. Augmented reality applications, exemplified by Molecular Rift, enable scientists to interact with 3D molecular models, thereby enhancing their comprehension of drug–target interactions. Utilizing augmented reality to model drug penetration and cellular response allows researchers to refine compound compositions before undertaking expensive clinical studies [87]. In the pharmaceutical sector, virtual reality (VR) can enhance CADD by making it more efficient, data-driven, and interactive, thereby addressing issues encountered in structure-based drug design (SBDD) methodologies. The prospects for medicinal chemistry and drug design have been examined from multiple perspectives, including biomolecular structure teaching, virtual drug design practice, and immersive interactivity. Numerous separate software packages tailored for specialized VR applications are recorded in the literature, including ChimeraX, Nanome, UnityMol, DockIT, VRPharmer, and VRChem. Web-based solutions include ProteinVR, VRmol, and VISPER. VR's key advantage is its ability to analyze and comprehend intricate 3D ligand–protein structures [88]. Some important software based on this technology is listed in Table 18.2. This often involves identifying precise spatial relationships, interactions, and structural changes that are difficult to measure, even with conventional automated computing methods.

Challenges and Future Directions

The future of drug discovery is undergoing a significant transformation due to the implementation of advanced digital technologies, including AI, ML, DL, AR, VR, blockchain, digital twins, and quantum computing. Despite the current impossibility of full implementation and ongoing challenges faced by researchers and professionals, foundational efforts must be undertaken to translate potential into reality. The application of AI and ML expedites drug development, but success

Table 18.2 Different VR software used in drug discovery.

Software platform	Company	Application in Drug discovery
NanoVerse	Nanome	Molecular modeling, protein–ligand docking, and collaborative drug design in VR
BiologicsVR	Nanome	Visualization and manipulation of large biomolecules (antibodies, proteins) in VR
ProteinVR	University of Illinois (NIH-funded)	Interactive exploration of protein structures in VR for drug target analysis
SAMSON VR	SAMSON	Atomistic simulations, MD, and drug design in VR
Arivis VR	Arivis AG	3D visualization of microscopy data (e.g., cellular interactions) for drug research
Virtonomics	Virtonomics (SimBioSys)	Virtual lab simulations for drug development and biochemical pathway analysis
InsightXR	Chemovision	AR/VR-based drug molecule visualization and collaborative research
Cresset's Flare™ VR	Cresset	SBDD with electrostatic potential mapping in VR
Unreal Engine for Pharma	Epic Games (Custom Solutions)	Virtual lab environments for drug testing and training
Genedata VR	Genedata	HTS data visualization in VR for lead optimization
BioVR	Labster (education-focused)	Virtual lab training for drug discovery techniques (e.g., ELISA, PCR)
Arquitecto Molecular	IndieBio (SynBio)	Assisted in synthetic biology and drug molecule assembly
Molecular Rift	Oculus (research use)	Early-stage VR tool for molecular dynamics and drug docking studies

depends on the availability of high-quality data. Numerous models continue to function as "black boxes," complicating researchers' and regulators' ability to rely on their predictions. Likewise, deep learning can provide numerous possible medication candidates, but laboratory validation is costly and time-intensive. Digital twins can revolutionize personal medicine, but accurately reflecting human physiology is remarkably complex. These models require substantial amounts of real-time clinical data, and it's often challenging to align the predictions with actual patient outcomes [89]. Blockchain can bring transparency to drug development, tracking everything from clinical studies to the supply chain. However, the pharmaceutical industry has decreased its use of it, due to concerns about balanced data privacy with the need for partially open cooperation [90]. AR/VR tools are transforming how scientists visualize molecular interactions, but their widespread adoption depends on more affordable and user-friendly hardware [91]. Quantum calculation has excellent potential for simulating complex molecules, but today's noise and incorrectly exposed hardware limit its practical use [92]. The use of advanced digital technologies, such as AI, quantum computing, and digital twins, can significantly drive drug discovery. However, many obstacles must be addressed for their efficient implementation. Since the majority of these advanced digital technologies rely heavily on data, the availability of authentic data, data quality, and data security are critical parameters. It is commonly observed that researchers often struggle with unauthenticated, incomplete, and biased datasets. Even the most sophisticated algorithms cannot be performed as per the study objectives due to a lack of high-quality data [93]. It was observed that a significant gap exists between academics and industry. Innovative research from educational institutions accelerates innovation, but converting these findings into practical applications requires collaboration with the pharmaceutical industry. Often, these collaborative approaches are absent, or if they are present, their progress is relatively slow due to differences in priorities, intellectual property, and financial constraints. The success rate can be enhanced by strengthening fruitful, active collaboration, leveraging dedicated funding sources, and implementing more transparent commercialization pathways. Interdisciplinary collaboration and skills development are other areas that can effectively accelerate drug discovery and development [94]. Currently, drug discovery is not limited to traditional principles of biological and chemical sciences. The conventional approach of drug discovery must integrate expertise with AI, computational modeling, quantum physics, and allied technologies. The current environment lacks interdisciplinary research and training, hindering innovation. Hence, the investments in educational programs that integrate life sciences with quantitative skills in a collaborative work environment are the need of the hour for drug discovery and development. This will also nurture and cultivate the next generation of researchers and professionals [95]. The process of drug discovery is now integrated with quantum computing, AI, ML, DL, and other advanced technological innovations. However, it also raises significant concerns, such as data privacy, security, and intellectual property protection. Regulatory authorities worldwide are now more aware and vigilant in addressing issues related to sensitive biomedical data to prevent breaches and abuse. Hence, various regulatory bodies have been established and aligned with standards, ensuring safety and effectiveness while promoting innovation. Strong cybersecurity is another central avenue that needs to be protected from theft or unwanted attacks. A significant issue concerning IP rights is that, to address the complexities of technological innovation in IP protection, traditional laws need to be revised and updated to reflect current realities. For permanent progress in the discovery of drugs, it is vital to balance rapid adoption with moral use, compliance, and security [96].

Conclusion

This chapter provides insight into the potential transformations of conventional drug discovery, advancing AI-enabled drug discovery. Technologies such as AI, ML, and DL facilitate traditional drug discovery by enabling rapid virtual screening, predictive modeling, and de novo drug design. Quantum computing enables faster solution of complex molecular simulations than classical computers, potentially unlocking new drug candidates. Blockchain enhances data safety and integrity, ensuring the development of secure and transparent partnerships for sensitive biomedical data. Digital twins enable real-time, patient-centric drug response. Meanwhile, AR/VR technologies offer a more effective platform for studying potential drug compounds at the molecular level, with the added benefit of 3D visualization and simulation. Collectively, all technological advancements improve researchers' understanding and decision-making. Despite these advancements, some associated issues include data privacy and security, regulatory compliance, ethical use of data, and computational scalability. This technological transformation is only efficiently implemented if issues such as cybersecurity, intellectual property, and regulatory compliance are addressed effectively. The collaboration of researchers, decision makers, and industry leaders will cultivate a future in which AI-driven, quantum-enhanced, and digitally integrated medicines are realized. The future of CADD lies at the intersection of AI, quantum computing, blockchain, digital twins, and AR/VR, offering exceptional speed, accuracy, and innovation in drug discovery and development. In contrast, this technological transformation is necessary to address moral, legal, and technical challenges. By promoting an interdisciplinary work environment and adaptive regulations, the pharmaceutical industry can use these cutting-edge tools to accelerate drug discovery, reduce costs, and distribute safe and effective treatments. The next era of CADD will be defined not only by computational power but also by responsible and fair technological progress, paving the way for a new paradigm in medicine.

List of Abbreviations

ADME	Absorption, distribution, metabolism, and excretion
AI	Artificial intelligence
AR	Augmented reality
CADD	Computer-aided drug discovery
CNN	Convolutional neural network
DL	Deep learning
DMET	Density matrix embedding theory
GAN	Generative adversarial network
GAT	Graph attention network
GCN	Graph convolutional network
GNN	Graph neural network
GRU	Gated recurrent unit
HTS	High-throughput screening
LSTM	Long short-term memory
MD	Molecular dynamics
ML	Machine learning
MPNN	Message-passing neural network
NMR	Nuclear magnetic resonance

QAOA	Quantum approximate optimization algorithm
QCL	Quantum circuit learning
QKD	Quantum key distribution
QPE	Quantum phase estimation
QSAR	Quantitative structure–activity relationship
RL	Reinforcement learning
RNN	Recurrent neural network
SBDD	Structure-based drug design
SILAC	Stable Isotope Labeling by Amino acids in Cell culture
SVMs	Support vector machines
VAE	Variational autoencoder
VQE	Variational Quantum Eigensolver
VR	Virtual reality
VS	Virtual screening

References

1 Singh N, Vayer P, Tanwar S, et al. Drug discovery and development: Introduction to the general public and patient groups. *Front. Drug. Discov. Front.* 2023;3:1201419.

2 Mohs RC, Greig NH. Drug discovery and development: role of basic biological research. *Alzheimers Dement. (NY)* 2017;3:651–7.

3 Jia ZC, Yang X, Wu YK, et al. The art of finding the right drug target: emerging methods and strategies. *Pharmacol. Rev.* 2024;76:896–914.

4 Berger SI, Iyengar R. Role of systems pharmacology in understanding drug adverse events. *Wiley Interdiscip. Rev. Syst. Biol. Med.* 2011;3:129–35.

5 Selker HP, Gorman S, Kaitin KI. Efficacy to effectiveness of clinical trials. *Trans. Am. Clin. Climatol. Assoc.* 2018;129:279–300.

6 Dhudum R, Ganeshpurkar A, Pawar A. Revolutionizing drug discovery: A comprehensive review of AI applications. *Drug Dev. Res.* 2024;3:148–71.

7 Niazi SK, Mariam Z. Computer-aided drug design and drug discovery: a prospective analysis. *Pharmaceuticals (Basel)* 2023;17:22.

8 Bassani D, Moro S. Past, present, and future perspectives on computer-aided drug design methodologies. *Molecules* 2023;28:3906.

9 Sliwoski G, Kothiwale S, Meiler J, Lowe EW. Computational methods in drug discovery. *Pharmacol. Rev.* 2014;66:334–95.

10 Shah M, Patel M, Shah M, et al. Computational transformation in drug discovery: A comprehensive study on molecular docking and quantitative structure activity relationship (QSAR). *Intell. Pharm.* 2024;2:589–95.

11 Szymański P, Markowicz M, Mikiciuk-Olasik E. Adaptation of high-throughput screening in drug discovery – Toxicological screening tests. *Int. J. Mol. Sci.* 2011;13:427–52.

12 Ayon NJ. High-throughput screening of natural product and synthetic molecule libraries for antibacterial drug discovery. *Metabolites* 2023;13:625.

13 Oselusi SO, Dube P, Odugbemi AI, et al. The role and potential of computer-aided drug discovery strategies in the discovery of novel antimicrobials. *Comput. Biol. Med.* 2025;169:107927.

14 Wooley JC, Lin HS. Computational modeling and simulation as enablers for biological discovery. In: *Catalyzing Inquiry at the Interface of Computing and Biology*. National Academies Press (US); 2005.

15 Rocha VN, Sant'Anna CMR. From origin to current methods: an overview of molecular modeling applied to medicinal chemistry in the last 30 years. *J. Braz. Chem. Soc.* 2024; 35:e-20240103.

16 Pagadala NS, Syed K, Tuszynski J. Software for molecular docking: a review. *Biophys. Rev.* 2017;9:91–102.

17 Bastikar V, Bastikar A, Gupta P. Quantitative structure–activity relationship-based computational approaches. In: *Computational Approaches for Novel Therapeutic and Diagnostic Designing to Mitigate SARS-CoV-2 Infection*. Academic Press; 2022. pp. 191–205.

18 Sur S, Nimesh H. Challenges and limitations of computer-aided drug design. *Adv. Pharmacol.* 2025;103:415–28.

19 Yu W, MacKerell AD. Computer-aided drug design methods. *Methods Mol. Biol.* 2017;1520:85–106.

20 Bhatkhande K, Vysyaraju NN, Alavala RR, Gokhale KM. Recent advances in drug-likeness screening by using the software and online tools. In: *Applications of Computational Tools in Drug Design and Development*, SN Koteswara Rao G and RR Alavala (eds.). Springer Nature; 2025. pp. 683–717.

21 Perkins R, Fang H, Tong W, Welsh WJ. Quantitative structure-activity relationship methods: Perspectives on drug discovery and toxicology. *Environ. Toxicol. Chem.* 2003;22:1666–79.

22 Mozaffari S, Moen A, Ng CY, et al. Structural bioinformatics for rational drug design. *Res. Pract. Thromb. Haemost.* 2025;9:102691.

23 Gimeno A, Ojeda Montes MJ, Tomás Hernández S, et al. The light and dark sides of virtual screening: what is there to know? *Int. J. Mol. Sci.* 2019;20:1375.

24 Bekbolatova M, Mayer J, Ong CW, Toma M. Transformative potential of AI in healthcare: Definitions, applications, and navigating the ethical landscape and public perspectives. *Healthcare (Basel)* 2024;12:125.

25 Rodríguez-Pérez R, Bajorath J. Evolution of support vector machine and regression modeling in chemoinformatics and drug discovery. *J. Comput. Aided Mol. Des.* 2022;36:355–62.

26 Khemani B, Patil S, Kotecha K, Tanwar S. A review of graph neural networks: concepts, architectures, techniques, challenges, datasets, applications, and future directions. *J. Big Data* 2024;11:18.

27 Sun H, Zhu M, Wang X. Combining variational autoencoders with generative adversarial networks to adaptively adjust the electromagnetic compatibility of biomechanical data analysis platforms. *Mol. Cell. Biomech.* 2025;22:1332.

28 Guha R, Nguyen DT, Southall N, Jadhav A. Dealing with the data deluge: Handling the multitude of chemical biology data sources. *Curr. Protoc. Chem. Biol.* 2012;4:193–209.

29 Varadi M, Bertoni D, Magana P, et al. AlphaFold Protein Structure Database in 2024: providing structure coverage for over 214 million protein sequences. *Nucleic Acids Res.* 2024;52:D368–75.

30 Ocana A, Pandiella A, Privat C, et al. Integrating artificial intelligence in drug discovery and early drug development: a transformative approach. *Biomark. Res.* 2025;13:45.

31 Chen Fu, Chen, Q. The future of pharmaceuticals: artificial intelligence in drug discovery and development. *J. Pharm. Anal.* 2025;15:101248.

32 Kumar G, Yadav S, Mukherjee A, et al. Recent advances in quantum computing for drug discovery and development. *IEEE Access* 2024;12:64491–509.

33 Weidman JD, Sajjan M, Mikolas C, et al. Quantum computing and chemistry. *Cell Rep. Phys. Sci.* 2024;5:102105.

34 Sivakumar A, Nair HK, Joshi A, et al. A computational study and analysis of Variational Quantum Eigensolver over multiple parameters for molecules and ions. *EPJ Quantum Technol.* 2024;11:1–21.

35 Bova F, Goldfarb A, Melko RG. Commercial applications of quantum computing. *EPJ Quantum Technol.* 2021;8:1–13.

36 Abu-Ghanem M. IBM quantum computers: evolution, performance, and future directions. *J. Supercomput.* 2025;81:687.

37 Paul D, Sanap G, Shenoy S, et al. Artificial intelligence in drug discovery and development. *Drug Discov. Today* 2021;26:80–93.

38 Patel V, Shah M. Artificial intelligence and machine learning in drug discovery and development. *Intell. Med.* 2022;2(3):134–140.

39 Vora LK, Gholap AD, Jetha K, et al. Artificial intelligence in pharmaceutical technology and drug delivery design. *Pharmaceutics* 2023;15: 1916.

40 Suriyaamporn P, Pamornpathomkul B, Patrojanasophon P, et al. The artificial intelligence-powered new era in pharmaceutical research and development: a review. *AAPS PharmSciTech* 2024;25:188.

41 Soori M, Arezoo B, Dastres R. Artificial intelligence, machine learning and deep learning in advanced robotics: a review. *Cogn. Robot.* 2023;3:54–70.

42 Alizadehsani R, Roshanzamir M, Hussain S, et al. Handling of uncertainty in medical data using machine learning and probability theory techniques: a review of 30 years (1991–2020). *Ann. Oper. Res.* 2024;339:1077–118.

43 Haleem A, Javaid M, Qadri MA, et al. Artificial intelligence (AI) applications for marketing: a literature-based study. *Int. J. Intell. Networks* 2022;3:119–32.

44 Patel V, Shah M. Artificial intelligence and machine learning in drug discovery and development. *Intell. Med.* 2022;2:134–40.

45 Askr H, Elgeldawi E, Aboul Ella H, et al. Deep learning in drug discovery: an integrative review and future challenges. *Artif. Intell. Rev.* 2023;56:5975–6037.

46 Sarker IH. Deep learning: A comprehensive overview on techniques, taxonomy, applications and research directions. *SN Comput. Sci.* 2021;2:420.

47 Ali N, Hanif N, Khan HA, et al. Deep learning and artificial intelligence for drug discovery, application, challenge, and future perspectives. *Discov. Appl. Sci.* 2025;7:533.

48 Noor F, Junaid M, Almalki AH, et al. Deep learning pipeline for accelerating virtual screening in drug discovery. *Sci. Rep.* 2024;14:28321.

49 Djoumbou-Feunang Y, Wilmot J, Kinney J, et al. Cheminformatics and artificial intelligence for accelerating agrochemical discovery. *Front. Chem.* 2023;11:1292027.

50 Meyers J, Fabian B, Brown N. De novo molecular design and generative models. *Drug Discov. Today* 2021;26:2707–15.

51 Cai H, Shen C, Jian T, et al. CarsiDock: a deep learning paradigm for accurate protein–ligand docking and screening based on largescale pretraining. *Chem. Sci.* 2024;15:1449–71.

52 Kumar H, Kim P. Artificial intelligence in fusion protein three-dimensional structure prediction: Review and perspective. *Clin. Transl. Med.* 2024;14:e1789.

53 Amorim AMB, Piochi LF, Gaspar AT, et al. Advancing drug safety in drug development: bridging computational predictions for enhanced toxicity prediction. *Chem. Res. Toxicol.* 2024;37:827–49.

54 Mayr A, Klambauer G, Unterthiner T, Hochreiter S. DeepTox: toxicity prediction using deep learning. *Front. Environ. Sci.* 2016;3:80.

55 Schapin N, Majewski M, VarelaRial A, et al. Machine learning small molecule properties in drug discovery. *AIChem* 2023;1:100020.

56 Jarallah S, Almughem F, Alhumaid N, et al. Artificial intelligence revolution in drug discovery: A paradigm shift in pharmaceutical innovation. *Int. J. Pharm.* 2025;680:125789.

57 Yetgin A. Revolutionizing multi-omics analysis with artificial intelligence and data processing. *Quant Biol.* 2025;13:e70002.

58 Mansoor S, Hamid S, Tuan T, et al. Advance computational tools for multi-omics data learning. *Biotechnol. Adv.* 2024;77:108447.

59 Sun B, Yu He Q. Hunting molecular targets for anticancer reagents by chemical proteomics. In: *Novel Approaches and Strategies for Biologics, Vaccines and Cancer Therapies*. Academic Press; 2015. pp. 347–63.

60 Jinka C, Sainath C, Babu S, et al. CRISPR-Cas9 gene editing and human diseases. *Bioinformation* 2022;18:1081–6.

61 Chen C, Wang J, Pan D, et al. Applications of multi-omics analysis in human diseases. *MedComm* 2023;4:e315.

62 Zhou G, Rusnac DV, Park H, et al. An artificial intelligence accelerated virtual screening platform for drug discovery. *Nat. Commun.* 2024;15:7761.

63 Ruiz-Carmona S, Alvarez-Garcia D, Foloppe N, et al. rDock: a fast, versatile and open source program for docking ligands to proteins and nucleic acids. *PLoS Comput. Biol.* 2014;10:e1003571.

64 Omar ED, Mat H, Abd Karim AZ, et al. Comparative analysis of logistic regression, gradient boosted trees, SVM, and random forest algorithms for prediction of acute kidney injury requiring dialysis after cardiac surgery. *Int. J. Nephrol. Renovasc. Dis.* 2024;17:197–204.

65 Orlov AA, Akhmetshin TN, Horvath D, et al. From high dimensions to human insight: exploring dimensionality reduction for chemical space visualization. *Mol. Inform.* 2025;44:e202400265.

66 Mao J, Wang J, Zeb A, et al. Transformer-based molecular generative model for antiviral drug design. *J. Chem. Inf. Model.* 2024;64:2733–45.

67 Grisoni F, Moret M, Lingwood R, Schneider G. Bidirectional molecule generation with recurrent neural networks. *J. Chem. Inf. Model.* 2020;60:1175–83.

68 Zhang S, Tong H, Xu J, Maciejewski R. Graph convolutional networks: a comprehensive review. *Comput. Soc. Netw.* 2019;6:11.

69 Reiser P, Neubert M, Eberhard A, et al. Graph neural networks for materials science and chemistry. *Commun. Mater.* 2022;3:1–18.

70 Chow JCL. Quantum computing in medicine. *Med. Sci. (Basel)* 2024;12:67.

71 Rietsche R, Dremel C, Bosch S, et al. Quantum computing. *Electron Markets* 2022;32:2525–36.

72 Hidary JD. Quantum computing methods. In: *Quantum Computing: An Applied Approach*, JD Hidary (ed.). Springer International Publishing; 2019. pp. 131–88.

73 Hatakeyama-Sato K, Igarashi Y, Kashikawa T, et al. Quantum circuit learning as a potential algorithm to predict experimental chemical properties. *Digit. Discov.* 2023;2:165–76.

74 Visan AI, Negut I. Integrating artificial intelligence for drug discovery in the context of revolutionizing drug delivery. *Life (Basel)* 2024;14:233.

75 RootsAnalysis. https://www.rootsanalysis.com/

76 Wheeler K. How Google is tackling quantum computing challenges. 2024. Available from: https://technologymagazine.com/articles/how-google-is-tackling-quantum-computing-challenges

77 Gentile F, Yaacoub JC, Gleave J, et al. Artificial intelligence–enabled virtual screening of ultra-large chemical libraries with deep docking. *Nat. Protoc.* 2022;17:672–97.

78 Li W, Yin Z, Li X, et al. A hybrid quantum computing pipeline for real world drug discovery. *Sci. Rep.* 2024;14:16942.

79 Bordukova M, Makarov N, Rodríguez-Esteban R, et al. Generative artificial intelligence empowers digital twins in drug discovery and clinical trials. *Expert Opin. Drug Discov.* 2024;19(1):33–42.

80 Mirakhori F, Niazi SK. Harnessing the AI/ML in drug and biological products discovery and development: the regulatory perspective. *Pharmaceuticals (Basel)* 2025;18:47.

81 M-Star. How Pfizer leveraged digital twins to create a process scale-up roadmap. 2024. Available from: https://mstarcfd.com/resources/case-study/how-pfizer-leveraged-digital-twins-to-create-a-process-scale-up-roadmap/

82 INT Marketing. Digital twins and virtual simulations in drug development. 2024. Available from: https://intglobal.com/blogs/digital-twins-and-virtual-simulations-in-drug-development/

83 Zhavoronkov A. Sanofi goes all-in on AI. *Forbes*. Jun. 21, 2023. Available from: https://www.forbes.com/sites/alexzhavoronkov/2023/06/21/sanofi-goes-all-in-on-ai/

84 Punia A, Gulia P, Gill NS, et al. A systematic review on blockchain-based access control systems in cloud environment. *J. Cloud Comput.* 2024;13:146.

85 Hang L, Chen C, Zhang L, Yang J. Blockchain for applications of clinical trials: Taxonomy, challenges, and future directions. *IET Commun.* 2022;16:2371–93.

86 Gasmi A, Benlamri R. Augmented reality, virtual reality and new age technologies demand escalates amid COVID-19. In: *Novel AI and Data Science Advancements for Sustainability in the Era of COVID-19*. Academic Press; 2022. pp. 89–111.

87 Ventola CL. Virtual reality in pharmacy: opportunities for clinical, research, and educational applications. *P T* 2019;44:267–76.

88 Baaden M, Glowacki DR. Virtual reality in drug design: Benefits, applications and industrial perspectives. *Curr. Opin. Struct. Biol.* 2025;92:103044.

89 Meijer C, Uh HW, El Bouhaddani S. Digital twins in healthcare: Methodological challenges and opportunities. *J. Pers. Med.* 2023;13:1522.

90 Andrew J, Isravel D, Sagayam K, et al. Blockchain for healthcare systems: architecture, security challenges, trends and future directions. *J. Netw. Comput. Appl.* 2023;215:103633.

91 Eswaran M, Bahubalendruni M. Challenges and opportunities on AR/VR technologies for manufacturing systems in the context of Industry 4.0: A state of the art review. *J. Manuf. Syst.* 2022;65:260–78.

92 Blunt NS, Camps J, Crawford O, et al. Perspective on the current state-of-the-art of quantum computing for drug discovery applications. *J. Chem. Theory. Comput.* 2022;18(12):7001–23.

93 Norori N, Hu Q, Aellen FM, et al. Addressing bias in big data and AI for health care: A call for open science. *Patterns (N Y)* 2021;2:100347.

94 Singh A, Singh S. The connection between academia and industry. *Mens. Sana. Monogr.* 2005;3:5–35.

95 O'Dwyer M, Filieri R, O'Malley L. Establishing successful university–industry collaborations: barriers and enablers deconstructed. *J. Technol. Transf.* 2022;48:900–31.

96 Filkins BL, Kim JY, Roberts B, et al. Privacy and security in the era of digital health: what should translational researchers know and do about it? *Am. J. Transl. Res.* 2016;8:1560–80.

Index

f

g

h

m

q

t

Printed in the USA/Agawam, MA
October 8, 2026

911622.019